The Mountain Pass Theorem

Variational methods are very powerful techniques in nonlinear analysis and are extensively used in many disciplines of pure and applied mathematics (including ordinary and partial differential equations, mathematical physics, gauge theory, and geometrical analysis).

This book presents min-max methods through a comprehensive study of the different faces of the celebrated Mountain Pass Theorem (MPT) of Ambrosetti and Rabinowitz. The reader is gently led from the most accessible results to the forefront of the theory, and at each step in this walk between the hills, the author presents the extensions and variants of the MPT in a complete and unified way. Coverage includes standard topics: the classical and dual MPT; second-order information from (PS) sequences; symmetry and topological index theory; perturbations from symmetry; convexity; and more. But it also covers other topics covered nowhere else in book form: the nonsmooth MPT; the geometrically constrained MPT; numerical approaches to the MPT; and even more exotic variants. Each chapter has a section with supplementary comments and bibliographical notes, and there is a rich bibliography and a detailed index to aid the reader. The book is suitable for researchers and graduate students. Nevertheless, the style and the choice of the material make it accessible to all newcomers to the field.

ENCYCLOPEDIA OF MATHEMATICS AND ITS APPLICATIONS

The Mountain Pass Theorem

Variants, Generalizations and Some Applications

YOUSSEF JABRI

University of Oujda, Morocco

CAMBRIDGE
UNIVERSITY PRESS

PUBLISHED BY THE PRESS SYNDICATE OF THE UNIVERSITY OF CAMBRIDGE
The Pitt Building, Trumpington Street, Cambridge, United Kingdom

CAMBRIDGE UNIVERSITY PRESS
The Edinburgh Building, Cambridge CB2 2RU, UK
40 West 20th Street, New York, NY 10011-4211, USA
477 Williamstown Road, Port Melbourne, VIC 3207, Australia
Ruiz de Alarcón 13, 28014 Madrid, Spain
Dock House, The Waterfront, Cape Town 8001, South Africa

http://www.cambridge.org

First published 2003

Printed in the United States of America

Typeface Times New Roman PS 10/12.5 pt.   *System* LaTeX 2_ε [TB]

A catalog record for this book is available from the British Library.

Library of Congress Cataloging in Publication Data
Jabri, Youssef, 1970
The mountain pass theorem : variants, generalizations and some applications / Youssef Jabri.
p.   cm. – (Encyclopedia of mathematics and its applications)
Includes bibliographical references and index.
ISBN 0-521-82721-3

1. Mountain pass theorem.   2. Critical point theory (Mathematical analysis)
3. Hamiltonian systems.   4. Variational principles.   5. Variational inequalities.
6. Maxima and minima.   7. Nonsmooth optimization.   I. Title.   II. Series.
QA614.7.J33   2003
515′.64–dc21   2003043597

ISBN 0 521 82721 3 hardback

To the memory of my mother

Contents

Introduction

The methods we will take up here are all variations on a basic result known to everyone who has done any walk in the hills: *the mountain pass lemma*.

L. Nirenberg, Variational and topological methods
in nonlinear problems, *Bull. Am. Math. Soc.*, **4**, 1981

Why a Book on the Mountain Pass Theorem?

The *mountain pass theorem* (henceforth abbreviated as MPT) is a "phenomenal result" that marks the beginning of a new approach to critical point theory. It constitutes a particularly interesting model for the abstract minimax principle known since the pioneering work of Ljusternik and Schnirelman in the 1940s. It is also the grandfather and the prototype of all the "postmodern" critical point results from the *linking* family. As early as it appeared, it attracted attention by raising up a lot of theoretical development and serving to solve a very large number of problems in many areas of nonlinear analysis.

The MPT has been intensively investigated. Indeed, there is actually a huge amount of references specifically devoted to its study or presenting one of its variants, generalizations, or applications. Its influence can be measured by the fact that you will rarely find a recent paper or book dealing with variational methods that do not cite it. Our aim here is to provide an expanded publication fully devoted to present some of its various forms and shed light on its numerous faces, comparing and classifying them when possible.

Who Should Read It?

The monograph may be used as a complementary textbook in a course on variational methods in nonlinear analysis. The reader is only supposed to be familiar with some elementary notions of topology and analysis. The first part, especially destined for beginners, aims to be a connection with the MPT starting from very simple notions of analysis. It consists of a very accessible expository of the basic background and main principles of critical point theory. We continue then gradually with more advanced

1

topics including many very recent references, connecting the reader this way with up-to-date material not available anywhere in book literature.

More advanced readers working on critical point theory should also find it useful to have the large amount of information on the MPT that is scattered in the literature collected together and classified. They should appreciate as well the extended description in the final notes and comments of the items appearing in the very large bibliography on the MPT.

The Style and Format

In general, chapters begin with short abstracts, followed if necessary by the background material needed within, with pointers to the main references. Then the main results and their most important consequences and applications are given. We emphasize the basic ideas and principles.

> We premeditated to focus our attention on the abstract results rather than applications for three reasons. First, applications are very well documented in many recent books. (See a list following this discussion.) Second, the amount of details and technicalities they involve may sometimes hide the simple abstract ideas on which they rely. And third, this would have enlarged excessively the size of this book.

When judged to be too complicated or when requiring technical material not directly connected to the subject, the proofs are omitted and the corresponding results are only given in outline form for completeness. This is the case for the material involving algebraic topology notions or Morse or Morse-Conley theories, for example.

Chapters end in a systematic way with many complementary remarks and additional bibliographical references divided in blocks. There should be specific pointers (when appropriate) to

- alternative ways that could be used to present the material discussed in each chapter. This is an invitation to investigate the very rich existing literature.
- important "historical" contributions, so that the reader can trace the origin of these notions.
- the most recent developments, to follow the directions they are actually taking without having to go to other references.

This has been done with the intention to provide the reader with a comprehensive and as complete a reference as possible.

The Approach

The diversity and very large quantity of references dealing with the MPT makes it hard to find a pertinent and satisfactory classification that can serve as a plan. In particular, from a pure pedagogical point of view, chronological order does not seem to be a good choice because we noticed that ideas became easier to see with time and also that results are less burdened nowadays by technical details that may be omitted in a first contact.

We remarked that the MPT was, for some reason, a fantastic testing tool used by anyone who has a new idea about a possible development in critical point theory, and we could not resist the temptation to try to present yet another look at minimax theorems in critical point theory through a special study of the MPT.

Our approach there is the following. In the first chapters, we will get a close look at the different ingredients involved in the elaboration of a critical point theorem, whereas in the subsequent chapters, we will discover how they are actually pushed to the limits, using each time some particular form of the MPT. This fantastic tool has grown so much so that this is indeed possible and this program really works! We could even treat some subjects that are not generally found in books on critical point theory.

The so-called "minimax methods" characterize a critical value c of a functional Φ by

$$c = \inf_{K \in \mathcal{K}} \sup_{u \in K} \Phi(u).$$

The choice of the sets K must reflect some change in the topology of the (sub-)level sets $\Phi^\alpha = \{u; \; \Phi(u) < \alpha\}$ for the values α near to c, as we will see in Chapter 4.

If we carefully analyze minimax theorems, we will notice that independently of the level of smoothness (C^1, Lipschitz continuity, etc.) of the functional they follow always the same scheme:

1. We require some geometric conditions where a relation appears between the values (levels) of the functional over sets that *link*.
2. Then, using either a (quantitative) deformation lemma or Ekeland's variational principle, we show that for some value c characterized by a minimax argument, there exists a Palais-Smale sequence of level c, that is, a sequence such that $\Phi(u_n) \to c$ and $\Phi'(u_n) \to 0$.
3. Last, meaning some compactness condition of Palais-Smale type, we bring the amount of compactness required to conclude that c is a critical value.

The following figure describes this process. You must think of it as a graph in three dimensions, the third dimension (not represented) being the smoothness of Φ. This is very important, because the form of these principles and the techniques available vary dramatically according to the smoothness of Φ.

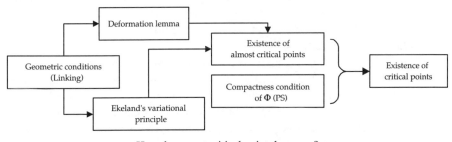

How do we get critical point theorems?

As a constant in this book, we try always to go from the more elementary to the more sophisticated, gradually adding new elements, our aim being to exhibit clearly the exact role of each assumption and how it intervenes in the proof. In general, consecutive results will seem to be natural and facile generalizations of each other. This will indeed be visible in the first part. Chapters constitute different faces of the MPT. Although there is some logical ordering, they are independent and must not necessarily be read linearly. Nevertheless, Chapters 2, 3, and 4, relating respectively to (a first contact with) the Palais-Smale condition and both Ekeland's variational principle and the deformation lemma, are of critical importance. They present basic features of critical point theory and should be well known before any further reading.

> Beginners *should not* linger, in a first reading, on the final comments and bibliographical remarks of the first chapters because they are linked with more advanced topics discussed in the chapters that follow. They should come back after mastering the new material.

How Is the Book Organized?

As we said before, the first part is initiative; it is intended to present the basics of critical point theory. After Chapter 1, where we briefly expose a historical description of the subject, we get a first and brief contact in Chapter 2, with a compactness condition on functionals that plays a central role in critical point theory, known as the Palais-Smale (PS) condition. It was introduced by R. Palais and S. Smale in the 1960s to allow the calculus of variations in the large to deal with mappings on general Banach manifolds. Chapters 3 and 4 recall two fundamental results in critical point theory. Chapter 3 is on Ekeland's variational principle while Chapter 4 is on the deformation lemma. They are behind the scenes in the statements and proofs of all the abstract results that will be covered in this monograph.

The second part begins with some "elementary" versions of the MPT appropriate enough to introduce its different aspects. Chapter 5 describes a finite dimensional ancestor of the MPT due to Courant (1950), very similar to the version of Ambrosetti-Rabinowitz both in its statement and its proof. Then Chapter 6 presents, in a topological adaptation of the concepts of Chapter 5, a purely topological version of the MPT. We continue then in Chapter 7 with the Ambrosetti-Rabinowitz MPT, the result properly known as *the* MPT in the literature. We also present two models of *standard* applications of the MPT to variational problems for illustration. The final chapter in this part, Chapter 8, contains some of the earliest variants of the MPT, including a dual form. It also presents some details of one of the first extensions of the MPT to higher dimensions, destined essentially to provide a more appropriate tool to treat some particular kinds of semilinear elliptic equations.

In the third part, we should relate more deeply the *topological* consideration involved in the MPT. Chapter 9 gives a detailed account of the results that accumulated gradually during years concerning what should be the right geometry to answer in the affirmative the question of the "limiting case" in the MPT. We will have the opportunity to see non-linear analysts at work on an exciting example. Chapter 10 is a continuation of Chapter 2. It focuses on the asymptotic behavior of functionals satisfying (PS) when some control is imposed on the level sets and presents some second-order information on functionals that satisfy the geometric conditions of the MPT. Chapter 11 discusses in detail the symmetric MPT, a multiplicity result in which the functional is supposed to be invariant under the action of a group of symmetries. It also discusses some extensions: the fountain theorem and its dual form and a procedure that inductively uses the MPT to obtain multiplicity results without passing by any *Index theory*. In Chapter 12, we describe the structure of the critical set in the situation of the MPT without requiring any nondegeneracy condition. In Chapter 13, we first present a minimax theorem that uses a "weighted" form of the (PS) condition. Then, we recall a very interesting procedure, attributed to Corvellec, for deducing new critical point theorems with weighted (PS) condition from older ones with the standard (PS) condition just by performing a change of metric.

The fourth part is devoted to some versions of the MPT that can be described as *nonsmooth* in many senses. They are motivated by applications to variational problems for functionals lacking regularity. Chapter 14 is consecrated to a situation of functionals $\Phi + \Psi$ considered as *semismooth*, where Φ is a C^1-functional and Ψ is proper, lower semicontinuous, and convex. In Chapter 15, we present a version of the MPT for locally Lipschitz functionals on Banach spaces. Chapter 16 goes further in nonsmoothness. We consider *continuous functionals* defined on *metric spaces*.

The fifth part is devoted to some speculations about the mountain pass geometry. In Chapter 17, a special extension of critical point theory to smooth functionals defined on *convex sets* is recalled briefly, and a corresponding version of the MPT with two different proofs is then given. The first proof is based on an appropriate form of the deformation lemma, while the second uses Ekeland's variational principle. While in Chapter 18, some variational methods in ordered Banach spaces are investigated. In particular, a variant of the MPT in order intervals in the spirit of some pioneering work by Hofer that exploited the natural *ordering*, intrinsic to semilinear elliptic problems, is given. In Chapter 19, we review the notion of linking that proved to be very important in critical point theory. This is a unified formulation of the geometric conditions that appear, among other results, in the MPT. We will see various definitions of this notion that led to many new results extending the MPT in different contexts: homotopical, homological, local, isotopic, and so forth. Chapter 20 is devoted to the "intrinsic MPT" and to one of its metric extensions. And in Chapter 21, we present some bounded variants of the MPT where the minimaxing paths are confined in a bounded region.

In the sixth and last part (Technical Climbs), we take the risk to go a little farther from the main road to discover the neighboring landscape. We investigate some topics that require the user to have a more advanced level and broader interests. In Chapter 22, we present three numerical implementations of the MPT. We present first a "mountain pass algorithm" that begins to be widely used. Then, we describe a partially interactive algorithm for computing unstable solutions of differential equations and a third algorithm used in quantum chemistry. In Chapter 23, we expose two approaches relying on the MPT to investigate the stability of the multiplicity results obtained by the symmetric MPT when the symmetry is broken. While in Chapter 24, we indicate how the MPT was used by Rabinowitz to treat some bifurcation problems. The last chapter in this part, Chapter 25, contains a series of short descriptions of many interesting variants of the MPT and some atypical or ingenious applications that did not find a place in the text in its actual form. This may be done in forthcoming editions.

For the convenience of the reader, we include at the end an appendix where we recall some definitions and basic properties of Sobolev spaces. We also investigate Nemystskii operators. Mastering these two topics is essential to treat nonlinear differential problems by critical point theorems in general and by the MPT in particular. Finally, we give a *large* bibliography on the subject, whose size might be explained by the growing interest stirred up by this specific area of analysis. The subject is so healthy that it is impossible to be exhaustive and any attempt in this direction gives only a momentary snapshot that may become obsolete in little time. An index is also given to help in navigating the book.

We would like to mention a certain number of very interesting and useful books on variational methods and critical point theory focusing on some particular aspects that appeared these past years, enhancing the existing bibliography and confirming that the theory is passing through a new age. We cite, among others, the following excellent references [43, 74, 197, 205, 315, 360, 411, 425, 517, 534, 623, 628, 654, 700, 748, 816, 882, 957, 982]. Of course standard books consecrated to nonlinear functional analysis, to cite only a few of them [49, 121, 131, 145, 285, 430, 520, 641, 825, 983, 984, 986], are also an indispensable part of the bibliography to be consulted by any serious "critical point theorist" and should certainly not be neglected.

Conventions and Notations

The typographical conventions used are standard. Theorems, lemmata, corollaries, and propositions are numbered consecutively and the counter they use is reset each chapter. For example, Theorem 5.3 refers to the third theorem in the fifth chapter. Equations are also numbered according to their appearance in chapters; for example, (6.4) refers to the fourth equation in the sixth chapter. The Notes at the end of the different chapters are also numbered according to their order of appearance within each chapter with the symbol $\diamond$ before their number; for example $\diamond$ 4.3. The sans serif font is used to report quotation from the existing literature.

By the end of this mountaineering expedition, I hope sincerely that you will feel all the beauty and elegance of the subject and all the pleasure experienced by the mathematicians who discovered the different versions of the MPT during their climbs.

Acknowledgments

I would like to express my gratitude and recognition to the kind people who transmitted their enthusiasm to me and expressed their interest in the project of writing this monograph. I. Ekeland, P.H. Rabinowitz, and M. Willem deserve a special mention. I am indebted also to all those who kindly sent me materials in connection with the subject. Of course, none else is to blame for the misprints and errors which are to be found.

Should you have any remark, suggestion, or correction, please do not hesitate to contact me.

Oujda, April 2001

1

Retrospective

> ... it was Riemann who aroused great interest in them [problems of the calculus of variations] by proving many interesting results in function theory by assuming Dirichlet's principle ...
>
> C.B. Morrey Jr., *Multiple integrals in the calculus of variations,* Springer-Verlag, 1966.

Variational and topological methods have proved to be powerful tools in the resolution of concrete nonlinear boundary value problems appearing in many disciplines where classical methods may fail. This is the case in particular for critical point theory, which became very successful these past years. Its success is due, in addition to its theoretical interest, to the large number of problems it handles.

To understand how the interest arose in this discipline, let us recall some of the main evolutions of its underlying principles in a series of historical events.

An Algorithm for Finding Extrema by Fermat

In a pure chronological order, the first *variational* treatments may be traced to the Greeks, who were interested in isoperimetric problems. Hero of Alexandria discovered in 125 B.C. that the light reflected by a mirror follows the shortest possible path. Fermat proved in 1650 that the light follows the path that takes the *least* time to go from one point to an other.

A little time before, in 1637, he published without proof in a small treatise entitled *Methodus ad Disquirendam Maximam et Minimam* an algorithm for finding the extrema of algebraic functions. It may be described as follows:

> We want to find a maximum or a minimum of a function f whose variable is A. We replace A with $A + E$ in the expression of f (E plays the role of a little Δx) and suppose that $f(A + E) \approx f(A)$. Then, we divide each term by E and eliminate all terms where E appears (i.e., we take $E = 0$). The values for which the result vanishes correspond then to a minimum or a maximum.
>
> This algorithm will certainly be more clear with an example. Let us consider a rectangle $\mathcal{R}$ with sides A and B, and perimeter $P = 2(A + B)$. The area of $\mathcal{R}$ is

$AB = A(P/2 - A)$. We want to find the lengths of A and B for which $\mathcal{R}$ has a maximal area, for a fixed parameter P.

Set

$$f(A) = A\left(\frac{P}{2} - A\right).$$

Then, $f(A + E) = (AP + EP)/2 - E^2 - X^2 - 2XE$. When taking

$$f(A) = f(A + E),$$

we get

$$0 = \frac{EP}{2} - E^2 - 2XE.$$

Dividing then by E, we get

$$0 = \frac{P}{2} - E - 2X.$$

Take $E = 0$. Then, $X = P/4$, i.e.,

$$X = Y.$$

So, $\mathcal{R}$ is the *square* of side $P/4$.

The procedure of Fermat turns out to be just evaluating

$$\lim_{E \to 0} \frac{f(A + E) - f(A)}{E}$$

and looking for the extrema of f at the points where the derivative of f vanishes. Notice that at that time, nobody knew what a *limit* or a *derivative* was.

Appearance of Calculus

Calculus and derivatives were first discovered in connection with the study of the variation of functions (a concept which was also not yet well comprehended), simultaneously and independently by two exceptional mathematicians: Newton and Leibnitz (see, for example, [124, 467]).

The approach of Newton, in 1672, relied on kinematics. He imagined an *auxiliary* moving point M following the curve describing the real function to study, like a car moving on a road. He supposed that the "speed" of the projection of M on the X-axis moves *uniformly*, and he noticed that, as a consequence, M should move forward slowly when the curve is flat and quickly elsewhere. And instead of following the point M in its trajectory on the curve, Newton discovered that he would learn as much while following its projection on the Y-axis. The advantage was that he would have to work on a line, which was the only type of curve that one could really treat in those times. The speed with which one explores the X-coordinates does not have an absolute sense and is used only to give some mental support.

The approach of Leibnitz does not rely on kinematics and is more abstract than that of Newton. It is essentially the one we use nowadays. In 1684 he had a publication that appeared in *Acta eriditorum*, entitled *Nova methodus proximamis et minimis, itemque tangentibus, qua nec irrationales quantiates moratur* (A new method for maxima and minima, and also tangents, which can be used with fractional and irrational quantities too), which gave some general rules of calculus for *differentials* using the symbol d. He presented among other things, the formulas he had already obtained in 1677:

$$d(xy) = x\,dy + y\,dx, \qquad d(x/y) = (y\,dx - x\,dy)/y^2, \qquad dx^n = nx^{n-1}.$$

As geometrical applications, he studied tangents, minima, and maxima. In particular, he gave the conditions $dv = 0$ for a minimum or a maximum.

Nevertheless, these two founders of modern analysis did not convince the whole mathematical community. The reason was that they did not get control of a concept at the heart of this process: the *limit*. To be accepted by all, calculus had to wait until 1820, when Cauchy gave the final and unassailable definition of this notion.

Meanwhile, in 1743, Euler submitted "*A method for finding curves possessing certain properties of maximum or minimum* [376]". And, in 1744, he published the first book on the calculus of variations, in which he expressed his conviction that **the** nature acts everywhere following some rule of maximum or minimum[1]:

> ... je suis convincu que partout **la** nature agit selon quelque principe d'un maximum ou minimum ...

This book was a source of inspiration for the mathematicians who came later (according to [882]).

Dirichlet Principle at the Roots of Modern Critical Point Theory

Critical point theory is concerned with *variational problems*. These are problems $(\mathcal{P})$ such that there exists a smooth functional Φ whose critical points are solutions of $(\mathcal{P})$.

The *abstract process* followed in modern critical point theorems has its roots in the Dirichlet principle. Dirichlet postulated at Göttingen that, given an open bounded set Ω in the plane and a continuous function $h : \partial\Omega \to \mathbb{R}$, the boundary value problem

$$\begin{cases} -\Delta u = 0 \text{ in } \Omega \\ \quad u = h \text{ on } \partial\Omega \end{cases} \tag{1.1}$$

admits a *smooth* solution u that minimizes the functional[2]

$$\Phi(u) = \int_\Omega \sum_{i=1}^{2} (D_i h(x))^2 \, dx \tag{1.2}$$

in the set of smooth functions defined on Ω that are equal to h on $\partial\Omega$. This principle was called the Dirichlet principle by Riemann in his thesis in 1851. He used it as a basis for his theory of analytic functions of a complex variable. "He studied the properties

---

[1] Note also a similar quotation by Maupertuis in Chapter 25.

[2] By *functional* we mean a function defined on a space whose elements are functions, and by *smooth* that it is continuous on $\overline{\Omega}$ and that its Laplacian exists in the usual sense using Fréchet derivatives, so that $u \in \mathcal{C}^2(\Omega; \mathbb{R}) \cap \mathcal{C}(\overline{\Omega}; \mathbb{R})$. This particular one is known as the *Dirichlet integral*.

of analytic functions by investigating harmonic functions in the plane," to quote
Brézis and Browder [150].

The *Euler equation* corresponding to (1.2) is the equation (1.1). This appellation
is due to the fact that Euler discovered the first general necessary condition $f'(u) = 0$
which must be satisfied by a smooth functional f at an extremum u. The condition was
known to hold for polynomials since Fermat.

And any *smooth minimizer* of (1.2), such that $u = h$ on $\partial\Omega$, is a solution of (1.1).
This very important principle was already observed for the Laplace operator, some
time before Dirichlet did, by Green in 1833. The idea was defended by Gauss in 1839
(in his study of magnetism) and (the future Lord Kelvin) W. Thomson in 1847. (The
reference [642] is entirely dedicated to the history of Dirichlet principle.)

Weierstrass pointed out in 1870, that the existence of the minimum is not assured in
spite of the fact that the functional Φ may be bounded from below. The subtle difference
between minimum and *infimum*, not yet perceived in these early times, was made. He
proved that the functional

$$\Psi(u) = \int_{-1}^{1} (x.u'(x))^2 \, dx$$

possesses an infimum but does not admit any minimum in the set

$$\mathfrak{C} = \left\{ u \in C^1[-1, 1]; \, u(-1) = 0, u(1) = 1 \right\}.$$

Indeed, if we consider the sequence

$$u_n = \frac{1}{2} + \frac{\arctan(x/n)}{2\arctan(1/n)}, \qquad n = 1, 2, \ldots,$$

then, $u_n \in \mathfrak{C}$ and $\Psi(u_n) \to 0$. If some u was a minimum, then $xu'(x) = 0$ on $[-1, 1]$.
Therefore, $u = $ constant, in contradiction with $u(-1) = 0$ and $u(1) = 1$.

Another nice counterexample to the Dirichlet principle, attributed to Courant [279], is
the following. Consider the (one-dimensional) integral

$$\Phi(u) = \int_0^1 \left(1 + (u'(x))^2\right) \, dx,$$

for $\Omega =]0, 1[$, where the admissible functions u are those in $C^1([0, 1]; \mathbb{R})$ with $u(0) = 0$
and $u(1) = 1$.

Stating correctly and justifying the Dirichlet principle became a challenge for the mathematicians in the second half of the 19th century. After many partially successful tentative
attempts to solve the problem by many mathematicians, Arzela used his famous compactness theorem in 1897 to treat the problem, under some conditions, and was not far
from succeeding. Only few times after, following Arzela's ideas, the Dirichlet principle
was established rigorously in certain important cases by Hilbert [470], Lebesgue [556],
and others in what is considered the beginning of the *direct methods of the calculus of
variations*.

Tonelli is the author of the three volumes "Fondamenti di calcolo delle variazioni"
[923] in 1921–23, one of the main references used by the mathematicians of the 1930s.

He used *uniform convergence* in the *interior of domains* and considered *absolutely continuous functions* satisfying the given boundary conditions as *admissible functions* (for one-dimensional problems). Then Morrey, who was interested in multidimensional problems and also in regularity, tried to develop the theory to allow more general functions than Tonelli's as admissible functions and to allow a more general type of convergence. He used Sobolev spaces and was able to obtain very general existence theorems. However, the solutions obtained were known at that time to be only continuous. But in fact, these were of class C^2 when some growth condition was satisfied. The interested reader may consult Morrey's paper [646] for a simplified presentation of this work.

Modern Critical Point Theory

Major contributions to critical point theory were also made by pioneers like Lagrange, Legendre, Jacobi, Hamilton, Poincaré, etc. Until the beginning of the 20th century, mathematicians were looking only for absolute minimizers of functionals bounded from below. The methods they found belong now to the heritage of the direct method in the calculus of variations. (Many specific works were developed to review this part of the theory; see, for example [284, 442, 922, 960].) In 1905, Poincaré, since his thesis where he developed some ideas of Hilbert on the Dirichlet principle, made a valuable contribution to the calculus of variations [627]. He treated a variational problem whose solution corresponded neither to a minimum nor to a maximum. This approach was revisited by Birkhoff in 1917 who succeeded to obtain a *minimax principle* where critical points u are such that $\Phi(u) = \inf_{A \in \mathcal{A}} \sup_{x \in A} \Phi(x)$ and $\mathcal{A}$ is a family of particular sets.

An important evolution that occurred in the beginning of the 1930s is the introduction of functional analysis by Volterra, which unified and clarified the intrinsic principles behind the different results. Functional analysis was in a very mature state by the late 1930s, thanks to the work of Banach and his school. A *theory of minimax* was elaborated in the late 1920s and early 1930s independently by Morse and by Ljusternik and Schnirelman. They extended minimax results to functionals that are not necessarily quadratic, they studied C^2-functionals on finite dimensional spaces. Inspired by Birkhoff's work, Morse used algebraic topology to study nondegenerate critical points. Ljusternik and Schnirelman developed a more general theory without the nondegeneracy of critical points, but lost some additional information and obtained less precise results. Their results all contain the basic ingredients of modern minimax theorems, which embody a notion of *compactness* on the functional, introduced during the 1960s by Palais, Smale, and Rothe to play the role of local compactness in infinite dimensional spaces. It permitted the extension of Ljusternik and Schnirelman and Morse methods to Banach and Hilbert spaces instead of working only in finite dimensional spaces or on bounded regions [694, 824]. These were too-severe restrictions that did not allow the treatment of boundary value problems like those considered using the direct method.

These contributions allowed contemporary mathematicians to obtain important results that form a substantial part of modern critical point theory. They also served to solve numerous nonlinear problems: elliptic problems, Hamiltonian systems, nonlinear wave equations, and so forth.

The Beginning of Postmodern Critical Point Theory

In this series of results, one became famous and influenced its "successors" so much that it may be considered the beginning of a postmodern era in the theory. This is the *mountain pass theorem* of Ambrosetti and Rabinowitz [50]. It is worth it, in these brief historical notes, to say that this result has a less known finite dimensional *ancestor*, as we will see later. The Ambrosetti-Rabinowitz' theorem was revealed to be very useful and was used as a model in other critical point theorems, due to Rabinowitz, in the late 1970s. They proved now, thanks to Benci and Rabinowitz [113], to be different faces of *one* same principle, known as the *linking principle*. In this monograph, you will get an idea of some very recent developments of critical point theory related to the MPT beginning from the finite dimensional MPT until the up-to-date forms of the MPT that appeared by the time of this writing.

So, what is this famous MPT and what are the ideas behind it? Stay tuned and continue reading if you want to know.

I

First Steps Toward the Mountains

2

Palais-Smale Condition:
Definitions and Examples

> (PS) is also crucial for the MPT. One can frequently, but not always, verify the condition (PS) for nonlinear partial differential equations.
>
> E. Zeidler, *Nonlinear functional analysis*, **III**,
> Springer-Verlag, (p. 163)

This chapter is a first attempt to introduce a compactness condition on functionals. It permits the extension of some interesting properties, proper to functionals, defined on finite dimensional spaces and infinite dimensional ones. It will play a central role in subsequent chapters. More elaborate aspects of this condition involving some material not yet presented are discussed in two later chapters (Chapters 10 and 13).

The Palais-Smale condition is a condition that appears in all the chapters, so it deserves this place at the beginning. The references $[628, 683, 748, 882, 956]$ can be consulted for some material on the Palais-Smale condition.

A detailed chapter on the subject could seem rather technical to people new to critical point theory, so we decided to split it into three parts. This first one is very elementary, the second one (Chapter 10) assumes the reader has some background in the theory, while the third one (Chapter 13) is destined for more advanced readers.

2.1 Definitions

We begin by defining what is generally meant by the Palais-Smale condition. The original condition that appears in the works of Palais and Smale $[693, 694, 699, 848]$ and that known in the literature, for historical reasons, as the condition (C), is the following.

Definition 2.1. Let X be a Banach space and $\Phi \colon X \to \mathbb{R}$ be a C^1-functional. Then Φ is said to satisfy the condition (C), if for any subset $S \subset X$ such that the restriction $\Phi|_S$ of Φ to S is bounded but the restriction of $\|\Phi'\|$ to S is not bounded away from 0, Φ admits a critical point on the closure of S.

What actually passes for the (classical) Palais-Smale condition and is denoted by (PS) is the following condition.

Definition 2.2. Let X be a Banach space and $\Phi \colon X \to \mathbb{R}$ a C^1-functional. We say that Φ satisfies the *Palais-Smale condition*, denoted (PS), if any sequence $(u_n)_n$ in X such that

$$(\Phi(u_n))_n \text{ is bounded and } \Phi'(u_n) \to 0, \tag{2.1}$$

admits a convergent subsequence.

Any sequence satisfying (2.1) is called a *Palais-Smale sequence*.

When dealing with abstract critical point theorems, we need in general a *weaker* condition, introduced by Brézis, Coron and Nirenberg in [155].

Definition 2.3. Let X and Φ be as in the former definition, and $c \in \mathbb{R}$. The functional Φ is said to satisfy the *(local)* Palais-Smale condition at the level c, denoted by $(PS)_c$, if any sequence $(u_n)_n$ in X such that

$$\Phi(u_n) \to c \text{ and } \Phi'(u_n) \to 0 \tag{2.2}$$

admits a convergent subsequence.

Remark 2.1. The condition (PS) is stronger than (C), which we can check immediately. But for the converse, it suffices to consider $\Phi \equiv 0$ to see that Φ satisfies (C) but not (PS).

Denoting by $\mathbb{K}$ the set of all critical points of Φ in X, we have that

the condition (PS) implies that any set of critical points $B \subset \mathbb{K}$ such that $\Phi|_B$ is uniformly bounded is relatively compact.

And this property, if required together with the condition (C), will give us a new condition that is equivalent to (PS) [882, p. 78].

Remark 2.2. The following two properties are easy consequences of the definitions.

- When (PS) is satisfied, we can check immediately that $(PS)_c$ holds for all $c \in \mathbb{R}$, while the converse is not true in general.
- If a functional Φ satisfies $(PS)_c$ for all c, then this does not imply that the critical set $\mathbb{K}$ of Φ is bounded.

Nowadays, more and more authors take the following as an alternative definition of (PS):

The functional Φ is said to satisfy (PS) if and only if it satisfies $(PS)_c$ for all $c \in \mathbb{R}$.

Remark 2.3. The condition $(PS)_c$ is a *compactness condition on the functional* Φ, in the sense that the set $\mathbb{K}_c$ of critical points of Φ at the level c,

$$\mathbb{K}_c = \{u \in X; \ \Phi(u) = c \text{ and } \Phi'(u) = 0\},$$

is compact.

2.2 Examples

To develop some intuition about this compactness notion, we recall some illustrative examples from [956] stressing the fact that the (PS) condition does not have any influence on the size of the critical set of a given functional that can be either empty, finite, or infinite.

1. The identity functional on $X = \mathbb{R}$ satisfies (PS), while the critical set $\mathbb{K} = \varnothing$.
2. The functional $\Phi \equiv 0$ on $X = \mathbb{R}$ satisfies neither $(PS)_0$ nor (PS), and the set $\mathbb{K}$ is the whole space.
3. The functional $\Phi(u) = \sin(u)$ on $X = \mathbb{R}$ satisfies $(PS)_c$ for all $c \in \mathbb{R} - \{-1, 1\}$ and $\mathbb{K}$ is an infinite unbounded set.

As a more interesting and less obvious example, consider the functional $L: \mathcal{D}(L) \subset L^2(\Omega) \to L^2(\Omega)$ where Ω is a bounded open domain of $\mathbb{R}^N$ and $Lu = -\Delta u$ for $u \in \mathcal{D}(L)$ where

$$\mathcal{D}(L) = \left\{ u \in H_0^1(\Omega); \ \Delta u \in L^2(\Omega) \right\}.$$

Denote by

$$0 < \lambda_1 < \lambda_2 \leq \lambda_3 \leq \cdots$$

the sequence of eigenvalues of L that will be denoted by $\sigma(L)$.

Identifying $L^2(\Omega)$ with its dual, we have $H_0^1(\Omega) \hookrightarrow L^2(\Omega) \hookrightarrow H^{-1}(\Omega)$. Fix a number $\lambda \in \mathbb{R}$ and a functional $f \in H^{-1}(\Omega)$, and set

$$\Phi_\lambda(u) = \frac{1}{2} \int_\Omega \left[|\nabla u(x)|^2 - \lambda u^2(x) \right] dx - \langle f, u \rangle.$$

Then, the values c where $(PS)_c$ holds for Φ_λ are intimately related to the spectrum of L.

1. If $\lambda \notin \sigma(L)$, that is, $\lambda \neq \lambda_k$ for all $k \in \mathbb{N}$, then Φ_λ satisfies (PS) in $H_0^1(\Omega)$.
2. If λ is in the spectrum of L, that is, $\lambda = \lambda_k$ for some $k \in \mathbb{N}$ and $f = 0$, then Φ_λ does not satisfy (PS).

For the first point, let us denote by $\tilde{L}$ the extension of L to the whole $H_0^1(\Omega)$ such that $\tilde{L}(u) = -\Delta u$, whose existence follows easily by the Lax-Milgram theorem (cf. Brézis book [145], for example).

We have that $\Phi_\lambda'(u) = \tilde{L} - \lambda u - f$ while $\tilde{L} - \lambda I_d$ is a homeomorphism of $H_0^1(\Omega)$ onto $H^{-1}(\Omega)$. Hence if $(u_n)_n \subset H_0^1(\Omega)$ is a Palais-Smale sequence for the level c, that is,

$$\begin{cases} \Phi_\lambda(u_n) \to c, \text{ and} \\ \Phi_\lambda'(u_n) = \tilde{L} u_n - \lambda u_n - f \to 0 \text{ in } H^{-1}(\Omega), \end{cases}$$

then

$$u_n = \left(\tilde{L} - \lambda I_d \right)^{-1} \left[f + \left(\tilde{L} u_n - \lambda u_n - f \right) \right] \to u = \left(\tilde{L} - \lambda I_d \right)^{-1} f \in H_0^1(\Omega).$$

And for the second point, consider an eigenfunction $\varphi_k \neq 0$ associated to the eigenvalue λ_k. Then, the sequence $(u_n)_n = (n\varphi_k)_n$ does not contain any converging subsequence, while

$$\begin{cases} \Phi_\lambda(n\varphi_k) = 0, \text{ and} \\ \Phi'_\lambda(n\varphi_k) = 0. \end{cases}$$

□

The different nonlinear differential problems treated in the sequel constitute other interesting examples.

When the existence of *bounded* Palais-Smale sequences is guaranteed, we can consider a weaker condition that has been introduced in [628].

Definition 2.4. Let X, Φ, and c be as above. The functional Φ is said to satisfy the *weak Palais-Smale condition* (WPS) if any *bounded* sequence $(u_n)_n$ in X such that

$$\left(\Phi(u_n)\right)_n \text{ is bounded and } \Phi'(u_n) \to 0 \tag{2.3}$$

admits a convergent subsequence.

The condition (WPS)$_c$ is defined by analogy to (PS)$_c$.

Remark 2.4. Pay attention that in the definition of the (local) "weak" Palais-Smale condition (WPS)$_c$, the sequences $(\Phi(u_n))_n$ and $(\Phi'(u_n))_n$ are supposed to converge for the topology induced by the norm (strongly, not weakly!).

In all these cases, the limit of the convergent subsequence is a critical point of Φ, because Φ is supposed to be of class C^1.

When using a critical point theorem involving the local Palais-Smale condition (PS)$_c$ for some value c, in general it is not known explicitly. Thus we have to verify (PS)$_c$ for all (possible) values c or more generally we have to check that (PS) holds. To check the former condition, some practical criteria have been developed as we will see in the next section.

To describe the relations between different types of conditions, we can use some of the examples seen earlier. In particular, it is obvious that (PS) implies (PS)$_c$, (WPS), and (WPS)$_c$, while (WPS) implies (WPS)$_c$, and (PS)$_c$ implies (WPS)$_c$. By the third example shown ($\Phi(u) = \sin(u)$ on $\mathbb{R}$), we see that a functional may satisfy (WPS) and (WPS)$_c$ for all values c and fail to satisfy (PS)$_c$ for some values c (there, it fails to satisfy (PS)$_c$ for $c = \pm 1$) and, hence, (PS).

In general, it is not very difficult to check if a functional satisfies (PS), but in some cases this is not so easy, and we have to do a lot of calculations. But, as remarked by Mawhin and Willem [628, Remark 6.3], "it is in general easier to verify (PS) than to find a priori bounds for all possible solutions of $\Phi'(u)$ since in (PS) the sequence $(\Phi(u_n))_n$ has to be bounded."

2.3 Some Criteria for Checking (PS)

Checking (PS) using its definition is not always the best way to proceed. We present some prototypes of functionals that satisfy (PS).

When the dimension of the space X is finite, say, $X = \mathbb{R}^N$, one has the following result.

Proposition 2.1. *Let* $\Phi \in C^1(\mathbb{R}^N; \mathbb{R})$ *where* X *is a Banach space. If the function*

$$|\Phi| + \|\Phi'\| : \mathbb{R}^N \to \mathbb{R}$$

is coercive, that is, it tends to $+\infty$ *as* $\|x\|$ *goes to* $+\infty$, *then* Φ *satisfies* (PS).

Proof. Since X is finite dimensional, it is locally compact. So, if we suppose that $|\Phi| + \|\Phi'\|$ is coercive, then any Palais-Smale sequence is bounded and hence contains a convergent subsequence. $\square$

In particular if Φ is coercive, it satisfies (PS). This fact has a connection with the statement of the finite dimensional MPT (Theorem 5.2) as we will see in Chapter 5.

When X is a general Banach space, the former criterion does not apply. Nevertheless, in such cases, a result that has proved to be very useful in applications is the following.

Proposition 2.2. *Let* $\Phi \in C^1(X; \mathbb{R})$. *Suppose that*

$$\Phi'(u) = Lu + K(u),$$

where L *is an invertible linear operator and* K *is* compact. *And suppose that any Palais-Smale sequence for* Φ *in* X *is bounded. Then,* Φ *satisfies the* (PS) *condition.*

Remark 2.5. Recall that a compact functional from X to Y is a continuous functional that maps bounded sets of X into relatively compact sets of Y.

Proof. Consider a Palais-Smale sequence $(u_n)_n \subset X$, that is,

$$\begin{cases} (\Phi(u_n))_n \text{ is bounded, and} \\ \Phi'(u_n) = Lu_n + K(u_n) \to 0 \text{ as } n \to \infty. \end{cases}$$

Then, $u_n + L^{-1}K(u_n)$ tends to 0. But by assumption, $(u_n)_n$ is bounded. Hence, since K is compact, the sequence $(L^{-1}K(u_n))_n$ is relatively compact, that is, it admits a convergent subsequence and then $(u_n)_n$ also. $\square$

Remark 2.6. If X is a Hilbert space, it can be identified with its dual X^*, so we can *choose* L to be the identity I_d. This gives to Φ', the derivative of Φ, the form of *compact perturbation of the identity*

$$\Phi' = I_d + K,$$

a well-known form that allows the use of the Leray-Schauder *topological degree theory*.

Remark 2.7. In Chapter 10, a complement is given to what was presented here. It assumes the reader is acquainted with a part of the material of forthcoming chapters. We will discuss in particular the relation between the (PS) condition for a functional, its level sets and coercivity, and see a critical point theorem of functionals Φ defined on a Hilbert space H such that $I_d + \Phi'$ maps a closed convex subset C of H into itself. We will discuss also the role that is played by (PS) in the MPT, the duality between "geometry and compactness," and give a qualitative description of (PS) as a natural condition rather than being a purely mathematical artifice.

Comments and Additional Notes

◇ 2.I Qualitative Meaning of (PS)

We close this chapter by reporting the following remark of Struwe concerning a qualitative meaning of the (PS) condition [882, p. 169]:

> Condition (PS) may seem rather restrictive. Actually, as Hildebrandt [471, p. 324] records, for quite a while many mathematicians felt convinced that in spite of its success in dealing with one-dimensional variational problems like geodesics (see Birkhoff's Theorem I.4.4, for example, or Palais' [695] work on closed geodesics), the Palais-Smale condition could never play a role in the solution of "interesting" variational problems in higher dimensions.
>
> Recent advances in the Calculus of Variations have changed this view and it has become apparent that the methods of Palais and Smale apply to many problems of physical and/or geometric interest and – in particular – that the Palais-Smale condition will in general hold true for such problems in a broad range of energies. Moreover, the failure of (PS) at certain levels reflects highly interesting phenomena related to internal symmetries of the systems under study, which geometrically can be described as "separation of spheres", or mathematically as "singularities", respectively as "change in topology". Again speaking in physical terms, we might observe "phase transitions" or "particle creation" at the energy levels where (PS) fails.

◇ 2.II The Morse and Sard Theorems

We saw that although (PS) implies the compactness of the set of critical points $\mathbb{K}_c$ for some level c, it has no influence on the *size* of critical points. Nevertheless, we should point out the following facts about the size of critical values for smooth mappings.

a. In finite dimensional spaces

Theorem (Morse Theorem). *If $\Phi: U \to \mathbb{R}^N$ is of class C^N on the open set U of $\mathbb{R}^N$, then the set of critical values of Φ has measure zero.*

Theorem (Sard Theorem). *If $\Phi: U \subset \mathbb{R}^N \to \mathbb{R}^M$ is of class C^r on the open set U of $\mathbb{R}^N$, then the set of critical values of Φ has measure zero provided $r \geq N - M + 1$.*

b. In infinite dimensional spaces

Theorem. *If $\Phi: X \to \mathbb{R}$ is analytic on the Banach space X and Φ' is Fredholm, then the set of critical values of Φ is at most countable.*

The interested reader may also consult the following stronger form of Sard theorem (e.g., cf. Schwartz [825, p. 55])

Theorem. *Let $\Phi: U \subset \mathbb{R}^N \to \mathbb{R}^N$ be of class $\mathcal{C}^1$ and let $J(x)$ be the Jacobian determinant of Φ at x. Then, for any measurable set $D \subset U$, the set $\Phi(D)$ is measurable and*

$$\operatorname{meas}(\Phi(D)) \leq \int_D |J(x)|\, dx.$$

This theorem is useful in particular when defining the topological degree, in the sense of Brouwer, of continuous mappings (e.g., cf. [412, 825])

For more notes and comments on the (PS) condition, look at the section with the same name in Chapters 10 and 13.

3

Obtaining "Almost Critical Points" – Variational Principle

Measuring instruments always have limited resolution. For this reason, in real world applications of mathematics, approximate solutions are often as good as, and frequently indistinguishable from, true solutions. Since mathematical laws governing real world situations are often in variational form, it is natural to develop a theory of "almost critical points", and the paper under review can be regarded as an interesting and successful step in that direction.

> From a Review of [I. Ekeland, On the variational principle.
> *J. Math. Anal. Appl.*, **47**, 324–353 (1974)] by R.S. Palais.

In a substantial part of modern analysis characterized by the tendency to avoid differentiability assumptions, this principle will likely play at least the same role as, say, the contraction mapping principle plays in "smooth" analysis. The elegance of the proofs and the natural way the principle appears in them lend much support to this belief.

> From a Review of [I. Ekeland, Nonconvex minimization problems.
> *Bull. Am. Math. Soc.*, **1**, 443–474 (1979)] by A.D. Ioffe.

This principle discovered in 1972 has found a multitude of applications in different fields of Analysis. It has also served to provide simple and elegant proofs of known results. And as we see, it is a tool that unifies many results where the underlying idea is some sort of approximation.

> D.G. de Figueiredo, *Lectures on the Ekeland variational principle
> with applications and detours*, Springer-Verlag, 1981.

Critical point theorems that suppose the *compactness condition* (PS) introduced in the previous chapter generally follow the scheme that consists of finding "almost critical points" via Ekeland's variational principle or via a deformation lemma. And then, by (PS), one concludes that the functional under study indeed possesses a critical point. In this chapter, we will see in detail the first one these two indispensable approximation tools.

All minimax results belonging to the family of linking theorems, including the MPT, suppose some *geometric* conditions that for some "inf max" value $c \in \mathbb{R}$, there exists a $(PS)_c$ sequence, that is, a sequence of almost critical points converging to c.

The program from the geometric conditions to the almost critical point theorems is realized using either Ekeland's variational principle, known also as the ε-principle (or Phelps-Bishop-Ekeland principle in some old references), or some quantitative deformation lemma.

3.1 Ekeland's Variational Principle

The best references for Ekeland's variational principle seem to be the papers [353,355] by Ekeland himself and his book with Aubin [74]. The reader may also consult the references [296,425,623,628].

The variational principle was established in 1972 by Ekeland [353]. It is an extraordinary result, like the MPT, that has proved to be a powerful tool in many areas of analysis. It was used, among other things, to give simpler and elegant proofs[1] to some known results.

When a functional Φ defined on a Banach space X is bounded from below, it is interesting to get sufficient conditions for its infimum to be attained. Recall that the minimum of a regular functional is a critical point. But without supplementary conditions on Φ or on X, a priori, nothing can be said.

In the direct method of the calculus of variations that looks for absolute minima of a functional Φ bounded from below through some manipulation of minimizing sequences, the functional to minimize Φ is in general supposed to be *weakly lower semicontinuous* (w.l.s.c.) and defined on a *reflexive* Banach space. When these conditions are met, some standard and well-known results of minimization (see [284], for example) can be used.

Nevertheless, this procedure can no longer be applied if we are working in a nonreflexive Banach space or if the functional is not w.l.s.c. Such situations occur often when dealing with applications.

Ekeland's result tells us that when Φ is only l.s.c. and bounded from below in a complete metric space, it possesses a minimizing sequence satisfying an interesting property that reduces, when the space admits a norm and Φ is of class $\mathcal{C}^1$, to the fact that this minimizing sequence is also "almost critical." In other words, it is a Palais-Smale sequence for the level "infimum of Φ." The exact statement is the following.

Theorem 3.1 (Ekeland's variational principle). *Let X be a complete metric space and $\Phi\colon X \to \mathbb{R} \cup \{+\infty\}$ a l.s.c. functional, bounded from below, and not identically equal to $+\infty$ ($\Phi \not\equiv +\infty$).*

Let $\varepsilon > 0$ and $x \in X$ such that

$$\Phi(x) \leq \inf_{u \in X} \Phi(u) + \varepsilon.$$

Then, for all $\delta > 0$ there exists $y = y(\varepsilon) \in X$ such that

a. $\Phi(y) \leq \Phi(x)$,

---

[1] To quote de Figueiredo [296].

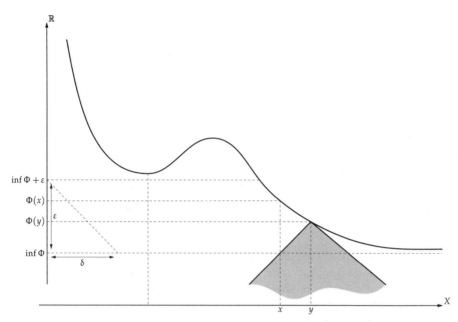

Figure 3.1. Ekeland's variational principle. The cone delimited by $\tilde{\Phi}(y) = \Phi(y) - \frac{\varepsilon}{\delta}$ dist (y, z) touches the graph of Φ from below at y.

   b.   dist $(x, y) \leq \delta$, and

   c. $\Phi(y) < \Phi(u) + \dfrac{\varepsilon}{\delta}$ dist (u, y) for all u in X such that $u \neq y$.

So, we can find a better approximation of the infimum of Φ as near as we want to our initial point with the supplementary property of minimizing the perturbation that appears in c. (See Figure 3.1.)

Proof. Consider the relation defined in X by

$$u \prec v \iff \Phi(u) \leq \Phi(v) - \frac{\varepsilon}{\delta} \text{ dist } (u, v),$$

where $\prec$ defines a partial ordering in X depending on δ. The reflexivity and antisymmetry are obvious. For the transitivity, suppose that both $u \prec v$ and $v \prec u$ hold. Since dist $(u, u) \leq$ dist $(u, v) +$ dist (v, u), we conclude immediately that $u \prec u$.

We will define by induction a decreasing sequence of closed sets $(S_n)_n$ in X using the ordering $\prec$, such that the intersection

$$\bigcap_n S_n = \{y\} = \{y(\delta)\}.$$

The sets defined by

$$I_v = \{u \in X; \ u \prec v\}$$

are nonempty $(v \in I_v)$ and closed because of the l.s.c. of Φ. Indeed, let $(u_n)_n \subset I_v$ such that $u_n \to u$ as $n \to \infty$; then

$$\Phi(u_k) + \frac{\varepsilon}{\delta} \, \text{dist}\,(u_k, v) \leq \Phi(v) \qquad \text{for all } k \in \mathbb{N}.$$

Therefore,

$$\Phi(v) \geq \liminf_{k \to \infty} \left[\Phi(u_k) + \frac{\varepsilon}{\delta} \, \text{dist}\,(u_k, v) \right]$$

$$\geq \liminf_{k \to \infty} \Phi(u_k) + \frac{\varepsilon}{\delta} \, \text{dist}\,(u, v)$$

$$\geq \Phi(u) + \frac{\varepsilon}{\delta} \, \text{dist}\,(u, v).$$

Start with $z_1 = x$ (the x appearing in the statement of the principle) and set

$$S_1 = \left\{ u \in X; \; u \prec z_1 \right\}.$$

Then, take $z_2 \in S_1$ such that

$$\Phi(z_2) \leq \inf_{S_1} \Phi + \frac{\varepsilon}{2}.$$

By induction, define

$$S_n = \left\{ u \in X; \; u \prec z_n \right\}$$

and pick a point $z_{n+1} \in S_n$ satisfying

$$\Phi(z_{n+1}) \leq \inf_{S_n} \Phi + \frac{\varepsilon}{n+1}.$$

Since $z_{n+1} \in S_n$, that is, $z_{n+1} \prec z_n$, for all n, we have

$$S_1 \supset S_2 \supset \cdots \supset S_n \supset S_{n+1} \supset \cdots .$$

We claim that diam $S_n \to 0$.

Indeed, for $u \in S_{n+1}$, we have $u \prec z_{n+1}$ and $u \in S_n$. So,

$$\Phi(u) + \frac{\varepsilon}{\delta} \, \text{dist}\,(u, z_{n+1}) \leq \Phi(z_{n+1}) \leq \Phi(u) + \frac{1}{n+1}.$$

Therefore,

$$\text{dist}\,(u, z_{n+1}) \leq \frac{\delta}{\varepsilon \cdot (n+1)},$$

which means that

$$\text{diam } S_{n+1} \leq \frac{2\delta}{\varepsilon \cdot (n+1)}.$$

Hence, diam $S_n \to 0$ as $n \to \infty$.

Since the metric space X is complete and $(S_n)_n$ is a decreasing sequence of closed sets such that diam $S_n \to 0$, we conclude that their intersection reduces to a single point:

$$\bigcap_n S_n = \{y\}.$$

We will verify that the point y satisfies the properties a, b, and c.

- Since $y \in S_1$, we have that $y \prec x$. Then property a holds.
- For b, we have

$$\frac{\varepsilon}{\delta} \, \text{dist}\,(y, x) \le \Phi(x) - \Phi(y),$$
$$\le \inf_{u \in X} \Phi(u) + \varepsilon - \Phi(y),$$
$$\le \varepsilon.$$

- For the property c, if for some $u \in X$, we had $u \prec y$, then we would have $u \prec z_n$ for all n. And then $u \in \bigcap_n S_n$, that is, $u = y$.

$\square$

Remark 3.1. Notice that the closer the point y to x the larger its derivative may be. And conversely, the smaller is its derivative, more imprecise is its position since it has to be in a larger ball of center x. Naturally, we cannot get the two together without requiring additional assumptions – *yet another uncertainty principle.*

When X is normed, the notion of derivative makes sense. The last relation c in Ekeland's principle implies that

$$\|\Phi'(y)\|_{X^*} \le \frac{\varepsilon}{\delta}.$$

More precisely, we get, as a consequence, the useful result.

Theorem 3.2. *Let X be a Banach space and $\Phi \colon X \to \mathbb{R}$ a l.s.c. functional, bounded from below and Gâteaux differentiable. Then, for all $\varepsilon > 0$ and any $u \in X$ such that*

$$\Phi(u) \le \inf_{x \in X} \Phi(x) + \varepsilon,$$

there exists a point v in X satisfying

a'. $\Phi(v) \le \Phi(u),$
b'. $\|v - u\| \le \sqrt{\varepsilon}$ *and*
c'. $\|\Phi'(v)\| \le \sqrt{\varepsilon}.$

Proof. The relations a' and b' are obviously the a and b in Ekeland's principle for $\delta = \sqrt{\varepsilon}$, while for c', from the relation c one has that, for all $w \in X$ and any $t > 0$,

$$\frac{\Phi(v) - \Phi(v + tw)}{t} \le \sqrt{\varepsilon} \, \|w\|.$$

Passing to the limit as t tends to 0, we get that for each w in X

$$-\langle \Phi'(v), w \rangle \le \sqrt{\varepsilon} \, \|w\|.$$

This holds also for $-w$, so

$$\left| \langle \Phi'(v), w \rangle \right| \le \sqrt{\varepsilon} \, \|w\|,$$

and then

$$\|\Phi'(v)\| \leq \sup_{\|w\|=1} |\langle \Phi'(v), w \rangle| \leq \sqrt{\varepsilon}.$$

$\square$

Remark 3.2. We would like to point out that in general the Gâteaux differentiability of a functional does not guarantee its l.s.c.:

The relations a′ and c′ mean that we obtained an "almost minimizer" of Φ, which is also an "almost critical point."

Corollary 3.3. *Let X be a Banach space and $\Phi: X \to \mathbb{R}$ a l.s.c. functional, bounded from below and Gâteaux differentiable. Then, there exists a minimizing sequence $(u_n)_n$ of almost critical points of Φ in the sense*

$$\Phi(u_n) \to \inf_{x \in X} \Phi(x) \quad and \quad \Phi'(u_n) \to 0. \tag{3.1}$$

When we add some compactness, even when the space is not reflexive and the functional is only l.s.c., the variational principle implies the following direct consequence.

Corollary 3.4. *Let X be a Banach space and $\Phi: X \to \mathbb{R}$ a C^1-functional that is bounded from below and satisfies* $(PS)_{\inf_X \Phi}$. *Then, there exists $u \in X$ such that*

$$\Phi(u) = \inf_{x \in X} \Phi(x) \quad and \quad \Phi'(u) = 0.$$

Proof. By Corollary 3.3, Φ possesses a minimizing sequence of almost critical points $(u_n)_n$, which by $(PS)_{\inf_X \Phi}$ admits a converging subsequence. $\square$

Comments and Additional Notes

◇ *3.1 The Original Proof of Ekeland's Principle*

A famous result of Bishop and Phelps [130] says that

the set of continuous linear functionals on a *Banach space E* which attain their maximum on a given closed convex bounded set $C \subset E$ is dense for the topology induced by the norm in E^*. Such spaces were known as "subreflexive" spaces.

The name is due to Phelps [720] who conjectured in this paper that every Banach space is subreflexive.

In the proof of this result appears a certain *convex cone* in E, associated with a partial ordering, to which a transfinite argument is applied (Zorn's lemma). This argument was adapted to l.s.c. functionals by Ekeland in the original proof of his variational principle. More precisely, for $s \in \mathbb{R}$, consider the closed convex cone with nonempty interior,

$$\mathcal{C}(s) = \{(u, a) \in X \times \mathbb{R}; \ a + s\|u\| \leq 0\}.$$

The associated ordering in $X \times \mathbb{R}$ is defined by

$$(u, a) \preceq (v, b) \iff (v - u, b - a) \in \mathcal{C}(s).$$

Using Zorn's lemma, we show that any closed subset S of $X \times \mathbb{R}$, such that

$$\inf\{s; \ (u, s) \in S\},$$

has a maximal element.

The variational principle follows by taking the closed set $S = \text{epi } \Phi$ for the ordering relation induced by ε/λ (see [354, 370] for details).

Later, as reported by Ekeland in [355], Lasry pointed out that the transfinite induction is in fact not needed, as we could see earlier.

◇ 3.II Smooth Extensions of Ekeland's Principle

The perturbation that appears in Ekeland's variational principle is Lipschitz but is not differentiable. Borwein and Preiss stated in [136] a *smooth* version where the perturbation is *quadratic*, and when the space X is a Banach space that admits a differentiable norm (away from the origin of course) it is differentiable everywhere. This principle has been used to study differentiability properties on Banach spaces. There are still other generalizations in this direction, including [325] and [571]. The version in [571] has been published recently.

Theorem 3.5 (Variational Principle, Li and Shi). *Let (X, dist) be a complete metric space and $\Phi \colon X \to \mathbb{R} \cup \{+\infty\}$ be a l.s.c. function bounded from below. Suppose that $\rho \colon X \times X \to \mathbb{R}_+ \cup \{+\infty\}$ is a function satisfying*

i. $\forall x \in X$ \qquad\qquad\quad $\rho(x, x) = 0$

ii. $\forall \{x_n, y_n\} \in X \times X$ \quad $\rho(y_n, z_n) \to 0 \Rightarrow \text{dist}(y_n, z_n) \to 0$ \qquad (3.2)

iii. $\forall z \in X$ \qquad\qquad\quad $y \mapsto \rho(y, z)$ *is l.s.c.*

and that $\delta_0 > 0$, $\delta_n \geq 0$, $n = 1, 2, \dots$ is a nonnegative number sequence. Then, for every $x_0 \in X$ and $\varepsilon > 0$ with

$$\Phi(x_0) \leq \inf_X \Phi + \varepsilon, \tag{3.3}$$

there exists a sequence $(x_n)_n \subset X$ which converges to some $x_\varepsilon \in X$ such that

$$\rho(x_\varepsilon, x_n) \leq \varepsilon/2^n \delta_0. \tag{3.4}$$

When for infinitely many n, $\delta_n > 0$,

$$\Phi(x_\varepsilon) + \sum_{n=0}^{\infty} \delta_n \leq \Phi(x_\varepsilon) \leq \inf_X \Phi + \varepsilon, \tag{3.5}$$

$$\forall x \neq x_\varepsilon, \ \ \Phi(x) + \sum_{n=0}^{\infty} \delta_n \rho(x, x_n) > \Phi(x_\varepsilon) + \sum_{n=0}^{\infty} \delta_n \rho(x_\varepsilon, x_n). \tag{$\star$}$$

And when $\delta_k > 0$ and for all $j > k \geq 0$, $\delta = 0$, $(\star)$ is replaced by

$$\forall x \neq x_\varepsilon, \exists m \geq k,$$

$$\Phi(x) + \sum_{i=0}^{k-1} \delta_i \rho(x, x_i) + \delta_k \rho(x, x_m) > \Phi(x_\varepsilon) + \sum_{i=0}^{\infty} \delta_{k-1} \delta_i \rho(x_\varepsilon, x_i) + \delta_k \rho(x_\varepsilon, x_m).$$

$$(3.6)$$

The proof is similar to that of Ekeland's principle and provides a new and simpler one for the Borwein-Preiss principle. It contains both the Ekeland and Borwein-Preiss variational principles. The function ρ that appears in (3.2) is *gauge type*. For example, it may be $f(\mathrm{dist}\,(x, y))$ where $f : \mathbb{R}^+ \to \mathbb{R}^+$ is strictly increasing, continuous, and $f(0) = 0$. For $\rho(x, y) = (\varepsilon/\lambda)\,\mathrm{dist}\,(x, y)$, $\delta_0 = 1$ and $\delta_n = 0$ for all $n > 0$, the theorem of Shi and Li (in this case (3.6) does not exist) recaptures Ekeland's principle with some improvement.

The book by Ghoussoub [425] contains various variational principles: Ekeland's principle, Borwein and Preiss's principle, and also the *mountain pass principle* (see Chapter 9), which is presented as a "multidimensional extension" of Ekeland's variational principle.

◇ *3.III A Nice Generalization of Ekeland's Principle*

A very simple and elegant generalization of Ekeland's principle to a general form on ordered sets was given by Brezis and Browder [149] that enables us to use other ordering relations instead of the one that appears in Ekeland's principle. It was applied by its authors to nonlinear semigroups and to derive diverse results from nonlinear analysis, including the variational principle and one of its equivalent forms, the Bishop-Phelps theorem. We recall here its first corollary [149, Corollary 1]; the interested reader is invited to look at the basic result in [149].

Theorem 3.6 (General Variational Principle on Ordered Sets, Brézis and Browder).
Suppose that X is an ordered set such that
(i) *any increasing sequence in X has an upper bound:*

$(u_n \preceq u_{n+1}$ *for any* $n \in \mathbb{N})$ *implies that* (*there exists* $v \in X$ *such that* $u_n \preceq v$,

for any $n \in \mathbb{N}$).

(ii) *The functional* $\Phi : X \to \mathbb{R} \cup \{-\infty\}$ *is bounded from above and increasing, that is,*

$$u \preceq v \text{ implies that } \Phi(u) \leq \Phi(v).$$

Then there exists u in X such that for any $v \in X$

$$u \preceq v \qquad implies \qquad \Phi(u) \leq \Phi(v).$$

It is easy to check that this result implies Ekeland's principle. It suffices to consider the order

$$u \preceq v \qquad \text{if and only if} \qquad \Phi(u) - \mathrm{dist}\,(u, v) = \Phi(v).$$

Proof. The proof is similar to that of Ekeland's principle. Indeed, choose an arbitrary element $u_0 \in X$ and inductively construct an increasing sequence $(u_n)_n$. Suppose u_n to be known and set $M_n = \{u \in X; \ u_n \preceq u\}$ and $c_n = \sup_{M_n} \Phi$. If the conclusion of the theorem holds for u_n, we are done. Otherwise, we have $c_n > \Phi(u_n)$ and we can choose $u_{n+1} \in X$ such that

$$c_n - \Phi(u_{n+1}) \leq \frac{c_n - \Phi(u_n)}{2}. \tag{3.7}$$

We get in this way an increasing sequence (u_n) that has by (i) an upper bound u, that is, $u_n \preceq u$ for all n. This element u is the desired solution. Otherwise, there would exist some $v \in X$ such that $u \preceq v$ and $\Phi(u) < \Phi(v)$. Since $(\Phi(u_n))_n$ is monotone increasing and bounded above, by (ii), it is convergent. Then, by the monotonicity of Φ, we have that $\lim_n \Phi(u_n) \leq \Phi(u)$. And since $v_n \in M_n$ for all n, we have by (3.7) that

$$\Phi(u_{n+1}) - \Phi(u_n) \geq c_n \geq \Phi(v) \qquad \text{for all } n.$$

As n tends to infinity, we obtain the contradiction $\Phi(v) \leq \Phi(u)$.   □

A Physical Interpretation – An Abstract Entropy Principle

A physical interpretation of this theorem with a connection with the second law of thermodynamics is given in Zeidler's monumental work [982]. "For each closed system, the entropy is a monotone increasing function of the time tending to a maximum."

> The values taken by Φ correspond to the values of the entropy at different states of the system. The relation $u \preceq v$ means that the system can pass from the state u to the state v at a later time. And the relation (ii) means that the entropy is a monotone increasing function of the time. And the states of maximal entropy correspond to stable equilibrium states of the system. So, the preceding theorem yields the existence of a stable equilibrium of the system at which the entropy can increase no more.

◇ 3.IV Other Forms of Ekeland's Principle

Many equivalent results to Ekeland's principle were discovered independently. The first of all these Ekeland-type theorems seems to have been formulated by Phelps in 1963 in Haussdorff locally convex vector spaces [721, Lemma 1].

Theorem 3.7 (Formulation of Phelps Lemma in Banach Spaces). *Let X be a Banach space. Let $A \subset X$ be a nonempty closed set and $B \subset X$ be a nonempty closed bounded and convex set such that $0 \notin B$. Let K be the cone $K = \mathbb{R}_+ B = \{\alpha \cdot b; \ \alpha \geq 0, \ b \in B\}$. Then, for each $x_0 \in A$ such that $A \cap (x_0 + K)$ is bounded and nonempty, there exists $x^* \in X$ such that*

$$x^* \in A \cap (x_0 + K) \qquad and \qquad \{x^*\} = A \cap (x^* + K).$$

Ekeland's principle is also equivalent to the maximal point version of Phelps' lemma ([722, Lemma 1.2]), to the Krasnosel'skii-Zabreiko theorem on normal solvability of operator equations [978], to the equilibrium point theorem of Oettli and Théra [690], and to the Takahashi minimization principle [896]:

Theorem 3.8 (Minimization Principle of Takahashi). *Let Φ be a l.s.c. function bounded from below on a complete metric space (X, dist). Given $x \in X$ with $\Phi(x) > \inf \Phi(X)$, let $\text{dist}(x, y) \le \Phi(x) - \Phi(y)$ for some $y \ne x$. Then, $\Phi(z) = \inf \Phi(X)$ for some $z \in X$.*

It is also equivalent to the drop theorem of Daneš related to the theory of normal solvability of nonlinear equations [293], to the flower petal theorem of Penot [712], and to the Caristi fixed point theorem, also called the Kirk-Caristi fixed point theorem [182].

Theorem 3.9 (The Drop Theorem, Daneš). *Given two closed nonempty sets A, B in a Banach space, with B bounded and convex and $\text{dist}(A, B) > 0$, there exists a point a in A such that there is no other point between a and B, that is, $D(a, B) \cap A = \{a\}$, where $D(x, B) = \text{clco}[\{x\} \cup B]$, and where clco refers to the closure of the convex hull; this set is called a "drop" because of its geometry.*

Theorem 3.10 (The Flower Petal Theorem of Penot). *If X is a complete subset of a metric space (E, dist), $x_0 \in X$, $b \notin X$, $r \le \text{dist}(b, X)$, $s = \text{dist}(b, x_0)$, then for each $\gamma > 0$ there exists $a \in X \cap P(x_0, b)$ such that $P(a, b) \cap X = \{a\}$ where $P(u, v) = \{x \in E; \ \gamma \, \text{dist}(x, u) + \text{dist}(x, v) \le \text{dist}(u, v)\}$.*

Theorem 3.11 (Caristi Fixed Point Theorem). *Let X be a complete metric space and $\Phi \colon X \to \{+\infty\}$ a l.s.c. functional that is bounded from below. Let $T \colon X \to 2^X$ be a multivalued mapping such that*

$$\Phi(y) \le \Phi(x) - \text{dist}(x, y), \qquad \forall x \in X, \ \forall y \in Tx. \tag{3.8}$$

Then, there exists $\overline{x} \in X$ such that $\overline{x} \in T\overline{x}$.

For illustration of the strength and elegance of these results, we give the short proofs of the equivalence of Caristi's fixed point theorem with Ekeland's principle.

Proof of Caristi's fixed point theorem using Ekeland's principle. For $\varepsilon = \delta$, we get that $\overline{x}$ is the minimal point that appears in c, that is

$$\Phi(\overline{x}) < \Phi(x) + \text{dist}(\overline{x}, x), \qquad \forall x \ne \overline{x}.$$

Otherwise, all $y \in T\overline{x}$ satisfy $y \ne \overline{x}$. So, by the preceding relation and by (3.8), we get the contradiction

$$\Phi(y) \le \Phi(\overline{x}) - \text{dist}(\overline{x}, y) \qquad \text{and} \qquad \Phi(\overline{x}) < \Phi(y) + \text{dist}(\overline{x}, y).$$

$\square$

Proof of Ekeland's principle using Caristi's fixed point theorem. Suppose by contradiction that no point in X satisfies c. Then, for each $x \in X$, the set $Tx = \{y \in X; \ \Phi(x) \ge \Phi(y) + (\varepsilon/\delta) \, \text{dist}(x, y), \ y \ne x\}$ is not empty. We have that $(\varepsilon/\delta) \, \text{dist}$ is an equivalent distance to dist, for which (3.8) is satisfied, so there must exist some point $\overline{x} \in X$

such that $\overline{x} \in T\overline{x}$: a contradiction with the definition of Tx which was not supposed to contain $\overline{x}$.   □

The latter fixed point theorem implies the existence part in the statement of the *contraction mapping principle* in complete metric spaces, which ensures the existence of fixed points for k-contraction mappings T (for $0 \le k < 1$). It suffices to consider $\Phi(x) = \text{dist}(x, Tx)/(1 - k)$.

◇ *3.V A Weighted Variational Principle*

We want to cite the paper [999], wherein the authors give a form of the variational principle closely related to some form of the deformation lemma we will see in the Notes of the next chapter. Its proof is similar to that of the classical principle.

Theorem 3.12 (Weighted Variational Principle, Zhong and Zhao). *If Φ is a Gâteaux-differentiable functional bounded from below, then, for every $\varepsilon > 0$, there exists some point z_ε such that*

$$\|\Phi'(z_\varepsilon)\| \le \frac{\varepsilon}{1 + h(\|z_\varepsilon\|)},$$

where $h: [0, \infty) \to [0, \infty)$ is a continuous function such that $\int_0^\infty \frac{1}{1+h(r)}\, dr = \infty$.

This may be used in critical point theorems with a weighted (PS) condition à la Cerami. In [999], Zhong and Zhao apply it to prove the existence of minimal points for some functionals that satisfy a weak (PS) condition and to some surjective mappings. Zhong used it also

- in [996, 998] to analyze the link between this compactness condition and the notion of coercivity (as in Chapter 10); and
- in [997] (on Ekeland's variational principle and a minimax theorem) to prove a minimax theorem with a (weighted) form of the (PS) condition (see Chapter 13).

4

Obtaining "Almost Critical Points"
– The Deformation Lemma

Locating critical values for a smooth functional I on a manifold X essentially reduces to capturing the changes in the topology of the sublevel sets $I^a = \{x \in X; \ I(x) < a\}$ as a varies in $\mathbb{R}$.

I. Ekeland and N. Ghoussoub, New aspects of the calculus of variations in the large. *Bull. Am. Math. Soc.*, **39**, no. 2, 207–265 (2001)

This chapter is a continuation of the previous one. The "deformation lemma" we will study is very important because, as we will see later, an important part of the inherent topological aspects of critical point theorems is always expressed in terms of deformations.

Another important tool, older than Ekeland's principle and widely used (in his quantitative form) to get "almost critical points," is the deformation lemma. Its form used nowadays seems to be due to Clark [242] following some ideas of Rabinowitz [734]. But the idea of deforming level sets near regular values to cross these values following the steepest descent direction of the function was already known before and is a basic tool in Morse theory. We will see it in action even in the original proof of the finite dimensional mountain pass theorem (MPT) of Courant that goes back to 1950. (See the notes of the next chapter.)

4.1 Regularity and Topology of Level Sets

Deformation lemmata may be found in many references relating minimax methods in critical point theory. You can consult in particular the paper by Ambrosetti-Rabinowitz [50], the monograph by Rabinowitz [748], and also the survey paper by Willem [956] where an interesting *quantitative* form that does not require the Palais-Smale condition is given. We adopted here the beautiful statement that appears in the latter reference.

All the different forms of deformation lemmata exploit and express the fact that *near a regular value $c \in \mathbb{R}$* and for $\varepsilon > 0$ sufficiently small, the level sets

$$\Phi^{c-\varepsilon} = \left\{ u \in X; \ \Phi(u) < c - \varepsilon \right\}$$

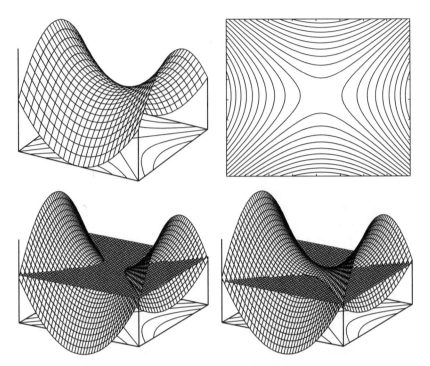

Figure 4.1. Crossing irregular values breaks level sets topology (I).

and

$$\Phi^{c+\varepsilon} = \big\{u \in X;\; \Phi(u) < c + \varepsilon\big\}$$

are *topologically the same*, in the sense that we can find a continuous deformation, a homeomorphism when Φ is smooth enough, that transforms $\Phi^{c+\varepsilon}$ into $\Phi^{c-\varepsilon}$ (or $\Phi^{c-\varepsilon}$ into $\Phi^{c+\varepsilon}$). While, when c is critical, this is no longer true, there must be some change in the topological nature of these level sets.

Example 4.1. Consider the function $\Phi(x, y) = x^2 - y^2$ on $\mathbb{R}^2$. Then, $\Phi'(x, y) = 0$ if and only if $(x, y) = (0, 0)$. For $\varepsilon > 0$ small enough, as we may see in Figure 4.1,

- the level set Φ^{ε} is connected
- while the set $\Phi^{-\varepsilon}$ has two components.

Example 4.2. Now consider the function $\Phi(x, y) = (x^2 + y^2)^2 - 2(x^2 + y^2)$ on $\mathbb{R}^2$. The only critical levels of Φ are -1 and 0 (see Figure 4.2).

- If $a < -1$, then Φ^a is empty.
- If $-1 < a < 0$, then Φ^a is a ring $r^2 < x^2 + y^2 < R^2$.
- If $0 < a$, then Φ^a is a ball $B(0, R)$.

In the first of these two illustrative examples borrowed from [517], we have a change in the number of components of the level sets Φ^a near the critical value, while in the

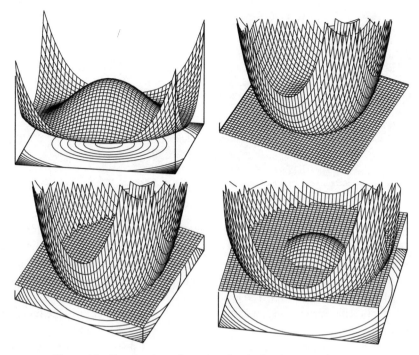

Figure 4.2. Crossing irregular values breaks level sets topology (II).

second one, it is more discrete; the number of components of Φ^a did not change, and it is always equal to 1. But in one situation the level set is simply connected whereas in the other it is not.

4.2 A Technical Tool

The deformation that appears in the deformation lemma is obtained as a solution of the differential equation

$$\dot{\eta}(t, u) = \frac{d}{dt}\eta(t, u) = -\alpha(u(t)).\nabla\Phi(\eta(u, t)) \qquad \text{where } (\alpha \geq 0),$$

for the case of a Hilbert space X when Φ is of class C^1 and $\nabla\Phi$ is locally Lipschitz continuous. This method is called the *steepest descent* method (la méthode de descente la plus rapide). It was introduced by Cauchy [188], as reported by Willem [956]. This is impossible when Φ is C^2, and we may take $\alpha(t)$ the constant mapping 1. In that case, the gradient vector field $-\nabla\Phi$ gives the steepest descent direction for Φ. Indeed,

$$\begin{aligned}
\frac{d}{dt}\Phi(\eta(t, u)) &= \langle\Phi'(\eta(t, u)), \dot{\eta}(t, u)\rangle \\
&= -(\nabla\Phi(\eta(t, u)), \nabla\Phi(\eta(t, u))) \\
&= -\|\nabla\Phi(\eta(t, u))\|^2.
\end{aligned}$$

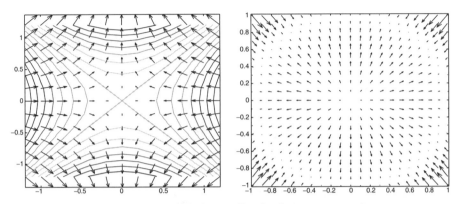

Figure 4.3. The steepest descent direction in our two examples.

Heuristically, you may imagine the graph of the functional to be materialized by some solid material, iron, for example. To find the steepest direction at some point, put a drop of some viscous liquid like honey on the corresponding point on the graph and follow its movement. For the examples of the two functions seen earlier, these directions look like the depictions in Figure 4.3. The length of the arrows represents the "speed" of the descent. Notice the very important fact that it is not possible to deform the functional on critical points.

When X is a Banach space, the notion of gradient does not exist. A notion of *pseudo-gradient vector field* was introduced by Palais [696] especially to handle this case. Moreover, it requires the functional Φ to be only of class $\mathcal{C}^1$.

Denote the set of regular points of Φ in X by

$$\tilde{X} = \big\{ u \in X; \; \Phi'(x) \neq 0 \big\},$$

where $\Phi \colon X \to \mathbb{R}$ is supposed to be a $\mathcal{C}^1$-functional.

Definition 4.1. A *pseudo-gradient vector* $v_0 \in X$ for Φ at $u \in \tilde{X}$ is a vector that satisfies

1. $\|v_0\| < 2\|\Phi'(u)\|$,
2. $\langle v_0, \Phi'(u) \rangle \geq \|\Phi'(u)\|^2$.

And a *pseudo-gradient vector field* for Φ is a locally Lipschitz continuous functional

$$v \colon \tilde{X} \to X$$

such that for all $u \in \tilde{X}$, $v(u)$ is a pseudo-gradient vector of Φ at u.

Remark 4.1. Notice that

1. Any *convex combination* of pseudo-gradient vectors (resp. of pseudo-gradient vector fields) is a pseudo-gradient vector (resp. a pseudo-gradient vector field). Hence, such a functional may exist but is not necessarily unique.
2. Since we require v to be locally Lipschitz continuous, even in a Hilbert space the gradient $\nabla \Phi$ of a functional Φ is not necessarily a pseudo-gradient vector field for Φ. But if Φ is $\mathcal{C}^2$, then $\nabla \Phi$ is a pseudo-gradient vector field for Φ.

This last point makes the following result due to Palais, which is true in general for Banach spaces, very interesting even in the case of a Hilbert space.

Lemma 4.1. *Let $\Phi \in \mathcal{C}^1(X; \mathbb{R})$. Then, Φ admits a pseudo-gradient vector field on $\tilde{X}$.*

Proof. Take $\tilde{u}$ in $\tilde{X}$, then there exists $w \in X$ such that

$$\|w\| = 1 \quad \text{and} \quad \langle \Phi'(\tilde{u}), w \rangle > \frac{2}{3} \|\Phi'(\tilde{u})\|.$$

Set $v = \dfrac{3}{2} \|\Phi'(\tilde{u})\| \cdot w$, then

$$\begin{cases} \|v\| < 2\|\Phi'(\tilde{u})\|, \\ \langle \Phi'(\tilde{u}), v \rangle > \|\Phi'(\tilde{u})\|^2. \end{cases}$$

By the continuity of Φ', there exists an open neighborhood $N(\tilde{u})$ of $\tilde{u}$ where the relations

$$\begin{cases} \|v\| < 2\|\Phi'(u)\| \\ \langle \Phi'(u), v \rangle > \|\Phi'(u)\|^2 \end{cases} \tag{4.1}$$

hold for all u in $N(\tilde{u})$. The family $\{N(u)\}_{u \in \tilde{X}}$ is obviously an open cover of $\tilde{X}$. Since $\tilde{X}$ is a metric space and hence paracompact, there exists an open cover $\{N_i\}_{i \in I}$ that is locally finite and is a refinement of $\{N(u)\}_{u \in \tilde{X}}$. For all i, there exists $\tilde{u} \in \tilde{X}$ such that $N_i \subset N(\tilde{u})$ and then (4.1) holds for some $u = u_i$ in each N_i.

Set for all u in $\tilde{X}$

$$\rho_i(u) = \mathrm{dist}\,(u, X \setminus N_i)$$

and

$$v(u) = \sum_{i \in I} \frac{\rho_i(u)}{\sum\limits_{j \in I} \rho_j(u)} u_i.$$

Since the cover $\{N_i\}_{i \in I}$ is locally finite, the sum makes sense.

The functions $(\rho_i)_i$ are Lipschitz continuous with support in N_i and are such that $0 \le \rho_i(u) \le 1$ and $\sum_{i \in I} \rho_i \equiv 1$ on $\tilde{X}$. Then, at each $\tilde{u} \in \tilde{X}$, the vector $v(\tilde{u})$ is a convex combination of pseudo-gradient vectors for Φ at u. Moreover, the functional v is locally Lipschitz continuous. $\square$

Remark 4.2. When Φ is *even*, that is, $\Phi(u) = \Phi(-u)$, it admits an *odd* pseudo-gradient vector field $\tilde{v}(-u) = -\tilde{v}(u)$. It suffices to set

$$\tilde{v}(u) = \frac{v(u) - v(-u)}{2},$$

where v is any pseudo-gradient vector field for Φ on $\tilde{X}$.

More generally, if there is an action of a compact topological group G on X, that is, there is a continuous function from $G \times X$ to X denoted by

$$(g, u) \mapsto g.u$$

and satisfying

$$g(h.u) = (g.h).u \qquad \text{for all } g, h \in G \text{ and all } u \in X,$$

and

$$e.x = x \qquad \text{for all } x \in X$$

where e is the unity of G. If Φ is G-invariant,

$$\Phi(g.u) = \Phi(u) \qquad \text{for all } (g, u) \in G \times \tilde{X},$$

then, the pseudo-gradient vector field $\tilde{v}$ may be constructed G-equivariant, that is,

$$\tilde{v}(g.u) = g.\tilde{v}(u) \qquad \text{for all } (g, u) \in G \times \tilde{X}.$$

Indeed, if dg is a normalized Haar's measure on G, it suffices to set

$$\tilde{v}(u) = \int_G g^{-1}.v(g.u)\, dg,$$

the average of any pseudo-gradient vector field v for Φ on $\tilde{X}$.

4.3 Deformation Lemma

Many versions of the deformation lemma exist in the literature. We will see a *quantitative* one proved by Willem in 1983 [952], whose advantage is that no a priori compactness condition, understand (PS), is needed.

Consider a set $S \subset X$ and $\delta > 0$, which we denote by $S_\delta = \{u \in X;\ \text{dist}\,(u, S) \leq \delta\}$ its δ-neighborhood, by $\overline{S}$ its closure, and by $\complement S = X \setminus S$ its complementary in X.

Theorem 4.2 (Quantitative Deformation Lemma, Willem). *Let X be a Banach space and $\Phi \colon X \to \mathbb{R}$ a C^1-functional. Let $S \subset X$, $c \in \mathbb{R}$, $\varepsilon > 0$, and $\delta > 0$ be such that for all $u \in \Phi^{-1}\left([c - 2\varepsilon, c + 2\varepsilon]\right) \cap S_{2\delta}$ we have*

$$\|\Phi'(u)\| \geq \frac{4\varepsilon}{\delta}. \tag{4.2}$$

Then, there exists a continuous deformation $\eta \in C([0, 1] \times X; X)$ such that

  i. $\eta(0, u) = u$ for all u in X,
  ii. $\eta(t, u) = u$ for all $u \notin \left\{ \Phi^{-1}\left([c - 2\varepsilon, c + 2\varepsilon]\right) \cap S_{2\delta} \right\}$ and all $t \in [0, 1]$,
  iii. $\eta(1, \Phi^{c+\varepsilon} \cap S) \subset \Phi^{c-\varepsilon} \cap S_\delta$, and
  iv. $\eta(t, .)$ is a homeomorphism for every $t \in [0, 1]$.

Proof. Using Lemma 4.1, the functional Φ admits a pseudo-gradient vector field v on $\tilde{X}$, while by (4.2), the set

$$A_2 = \Phi^{-1}\left([c - 2\varepsilon, c + 2\varepsilon]\right) \cap S_{2\delta}$$

is in $\tilde{X}$.

Consider a locally Lipschitz function $\psi : X \to \mathbb{R}$ such that $0 \leq \psi(u) \leq 1$ for all u in X and that satisfies

$$\psi(u) = \begin{cases} 1 & \text{on} & A_1 = \Phi^{-1}\left([c - 2\varepsilon, c + 2\varepsilon]\right) \cap S_\delta, \\ 0 & \text{on} & \overline{X \setminus A_2}. \end{cases}$$

Then, the vector field $f : X \to X$ defined by

$$\varphi(u) = \begin{cases} -\dfrac{\psi(u)}{\|v(u)\|} . v(u) & \text{on} & A_2 \\ 0 & \text{on} & \overline{X \setminus A_2} \end{cases}$$

is locally Lipschitz and bounded. Hence, for every $u \in X$, the Cauchy problem

$$\begin{cases} \dot{w}(t) = \varphi(w(t)) & t \geq 0 \\ w(0) = u \end{cases}$$

possesses a solution $w(., u)$ defined on $[0, \infty[$.

Define $\eta : [0, 1] \times X \to X$ by

$$\eta(t, u) = w(\delta t, u).$$

That way i and ii are trivially satisfied since $\varphi \equiv 0$ in $\overline{X \setminus A_2}$, and also iv holds true.

For iii,

$$w(\delta, \Phi^{c+\varepsilon} \cap S) \subset \Phi^{c-\varepsilon} \cap S_\delta.$$

When $t > 0$, we have

$$\|w(t, u) - u\| = \left\| \int_0^t \varphi(w(s, u))\, ds \right\|$$

$$\leq \int_0^t \|\varphi(w(s, u))\|\, ds$$

$$\leq t.$$

Then $w(t, S) \subset S_\delta$ for any $t \in [0, \delta]$.

Using the definition of φ, we get that

$$\frac{d}{dt}\Phi(w(t, u)) = \langle \Phi'(w(t, u)), \dot{w}(t, u) \rangle$$

$$= \langle \Phi'(w(t, u)), \varphi(w(t, u)) \rangle$$

$$\leq 0$$

Thus, $\Phi(w(., u))$ is decreasing.

Let $u \in \Phi^{c+\varepsilon} \cap S$. If there is some $t \in [0, \delta[$ such that $\Phi(w(t, u)) < c - \varepsilon$, then $\Phi(w(\delta, u)) < c - \varepsilon$ and $w(\delta, u) \in \Phi^{c-\varepsilon} \cap S_\delta$. Otherwise, for all $t \in [0, \delta]$,

$$c - \varepsilon \leq \Phi(w(t, u)) \leq \Phi(w(0, u)) = \Phi(u) \leq c + \varepsilon.$$

And then $w(t, u) \in A_1$ for all $t \in [0, \delta[$.

But then, by (4.2) and the definitions of φ and v, we deduce that

$$\Phi(w(\delta, u)) = \Phi(u) + \int_0^\delta \frac{d}{dt} \Phi(w(t, u)) \, dt$$

$$= \Phi(u) + \int_0^\delta \langle \Phi'(w(t, u)), \varphi(w(t, u)) \rangle \, dt$$

$$= \Phi(u) + \int_0^\delta \langle \Phi'(w(t, u)), v(w(t, u))/\|v(w(t, u))\| \rangle \, dt$$

$$\leq c + \varepsilon - \int_0^\delta \|\Phi'(w(t, u))\|^2/\|v(w(t, u))\| \, dt$$

$$\leq c + \varepsilon - \frac{1}{2} \int_0^\delta \|\Phi'(w(t, u))\| \, dt$$

$$\leq c + \varepsilon - \frac{\delta}{2} \frac{4\varepsilon}{\delta} = c - \varepsilon.$$

Thus, $w(\delta, u) \in \Phi^{c-\varepsilon} \cap S_\delta$. □

It is remarkable that we obtain as a corollary a form of Theorem 3.2 proved before using Ekeland's principle.

Corollary 4.3. *Let $\Phi \in C^1(X; \mathbb{R})$ be a functional bounded from below. Let $\varepsilon > 0$, and $u \in X$ such that*

$$\Phi(u) \leq \inf_{x \in X} \Phi(x) + \varepsilon.$$

Then, for every $\delta > 0$, there exists v in X such that

   i. $\Phi(v) \leq \inf_{x \in X} \Phi(x) + 2\varepsilon$,
   ii. $\|v - u\| \leq 2\delta$,
   iii. $\|\Phi'(v)\| \leq 4\varepsilon/\delta$.

Proof. It suffices to apply the quantitative deformation lemma (Theorem 4.2) with $S = \{u\}$ and $c = \inf_{x \in X} \Phi(x)$, if by contradiction, for all $u \in \Phi^{-1}([c, c + 2\varepsilon]) \cap S_{2\delta}$, we had

$$\|\Phi'(v)\| \geq \frac{4\varepsilon}{\delta}.$$

Then, $\eta(1, v)$ would be in $\Phi^{c-\varepsilon} \cap S_\delta$, which is impossible since $\Phi^{c-\varepsilon} = \varnothing$. □

Thus, also by using the deformation lemma we can conclude that a C^1-functional bounded from below admits a "minimizing sequence" of "almost critical points."

When combining the quantitative deformation lemma with the compactness condition $(PS)_c$, we obtain the useful result known widely in literature as *the* deformation lemma.

Theorem 4.4 (Standard Deformation Lemma). *Let $c \in \mathbb{R}$ and consider $\Phi: X \to \mathbb{R}$ a C^1-functional satisfying $(PS)_c$. If c is a regular value of Φ, then for every ε sufficiently small there exists $\eta \in C([0, 1] \times X; X)$ such that*

    i. $\eta(0, u) = u$ for all u in X,
    ii. $\eta(t, u) = u$ for all $u \notin \Phi^{-1}([c - 2\varepsilon, c + 2\varepsilon])$ and all $t \in [0, 1]$,
    iii. $\eta(1, \Phi^{c+\varepsilon}) \subset \Phi^{c-\varepsilon}$,
    iv. $\eta(t, .)$ is a homeomorphism for every $t \in [0, 1]$.

Proof. Since c is a regular value and Φ satisfies (PS)$_c$, there exist $\bar{\varepsilon} > 0$ and $\bar{\delta} > 0$ such that for all $u \in \Phi^{-1}([c - 2\bar{\varepsilon}, c + 2\bar{\varepsilon}])$ we have

$$\|\Phi'(u)\| \geq \bar{\delta}.$$

Because otherwise, there would exist $(u_n)_n \subset X$ such that

$$\begin{cases} c - \frac{1}{n} \leq \Phi(u_n) \leq c + \frac{1}{n} \\ \|\Phi'(u)\| \leq \frac{1}{n} \end{cases}$$

and hence by (PS)$_c$, c would be a critical value of Φ.

So, we can use Theorem 4.2 for $\varepsilon \in\ ]0, \bar{\varepsilon}[$, $S = X$ and $\delta = 4\varepsilon/\bar{\delta}$ to conclude. $\quad\square$

Comments and Additional Notes

◇ *4.1 Quantitative Deformation Lemma of Brezis and Nirenberg*

Willem seems to be the first one who tried to remove the (PS) condition and brought up an interesting result. The others focused in general on trying to obtain particular versions to handle situations that lack regularity (for example, Chang [202], Duc [349], Szulkin [891], Ding [333]).

    Another recent work where a similar task has been successfully carried out is the seminal paper by Brezis and Nirenberg [153] where a linking result is proved twice, first using Ekeland variational principle and the second time using a specific deformation lemma not requiring the (PS) condition. We will describe it here.

    Consider a C^1-functional Φ defined on a Banach space X.

Theorem 4.5 (Quantitative Deformation Lemma, Brézis and Nirenberg). *Let $c \in \mathbb{R}$. For any given $\delta < 1/8$, there exists a continuous deformation $\eta: [0, 1] \times X \to X$ such that*

    i. $\eta(0, u) = u$, for all $u \in X$;
    ii. $\eta(t, .)$ is a homemorphism of X onto X, for all $t \in [0, 1]$;
    iii. $\eta(0, u) = u$, for all $t \in [0, 1]$, if $u \in X \setminus \Phi_{c-2\delta}^{c+2\delta}$, or if $\|\Phi'(u)\| \leq \sqrt{\delta}$;
    iv. $0 \leq \Phi(u) - \Phi(\eta(t, u)) \leq 4\delta$, for all $u \in X$ and all $t \in [0, 1]$;
    v. $\|\eta(t, u) - u\| \leq 16\sqrt{\delta}$, for all $u \in X$ and all $t \in [0, 1]$.
    vi. If $u \in \Phi^{c+\delta}$, then the following alternative holds. Either
        a. $\eta(1, u) \in \Phi^{c-\delta}$, or
        b. for some $t_1 \in [0, 1]$, we have

$$\|\Phi'(\eta(t_1, u))\| < 2\sqrt{\delta}.$$

    vii. More generally, let $\tau \in [0, 1]$ and assume that for all $t \in [0, \tau]$,

$$\eta(t, u) \in \tilde{N} = \{v \in \Phi(v)_{c-\delta}^{c+\delta};\ \|\Phi'(v)\| \geq 2\sqrt{\delta}\}.$$

Then, $\Phi(\eta(t, u)) \leq \Phi(u) - \tau/4$.

As an immediate corollary, we get the following result.

Corollary 4.6. *Suppose that* Φ *also satisfies* (PS)$_c$.

  i. Given $\varepsilon > 0$, *there exists* $\delta < \varepsilon$ *and a deformation* η *as shown earlier that also satisfies*

$$\text{if } u \in \Phi^{c+\delta} \text{ and } \Phi(\eta(1, u)) > c - \delta, \quad \text{then } \|\Phi'(\eta(t, u))\| \leq \varepsilon, \ \forall t \in [0, 1].$$
$$(4.3)$$

  ii. Given $\varepsilon > 0$, *and a neighborhood* $\mathcal{O}$ *of* $\mathbb{K}_c$, *there exists* $\delta < \varepsilon$ *and a deformation* η *as in Theorem 4.5 such that, in addition,*

$$\text{if } u \in \Phi^{c+\delta} \setminus \mathcal{O}, \quad \text{the alternative vi. a } \textit{holds}. \quad (4.4)$$

Proof. As in the proof of the deformation lemma of Willem we saw earlier, the idea is to follow the negative gradient flow. Since Φ is only $\mathcal{C}^1$, we use a pseudo-gradient on the set of regular points $X \setminus \mathbb{K}$.

Consider the set

$$N = \left\{ u \in \Phi^{c+\delta}_{c-\delta}; \ \|\Phi'(u)\| > \sqrt{\delta} \right\}.$$

Since $\tilde{N}$ and $X \setminus N$ are disjoint closed sets, the function

$$g(u) = \frac{\operatorname{dist}(u, X \setminus N)}{\operatorname{dist}(u, X \setminus N) + \operatorname{dist}(u, \tilde{N})}$$

is a locally Lipschitz function $0 \leq g \leq 1$ satisfying

$$\begin{cases} 1 & \text{on } \tilde{N}, \\ 0 & \text{outside } N, \end{cases}$$

Consider the vector field

$$V(u) = \begin{cases} -g(u)\dfrac{v(u)}{\|v(u)\|^2} & \text{on } \tilde{N}, \\ 0 & \text{outside } N, \end{cases}$$

where v is a pseudo-gradient vector field. Then, V is locally Lipschitz on X and $\|V(u)\| \leq 1/\sqrt{\delta}$ for all $u \in X$. The deformation $\eta(t, u)$ is defined as the solution $\eta(t)$ of

$$\begin{cases} \dfrac{d\eta}{dt} = V(\eta), \\ \eta(0) = u. \end{cases}$$

Then, η is defined for $t \in [0, 1]$ and satisfies i, ii, iii, iv, v, and vi.   □

◇ **4.II The Deformation Lemma of Shafrir**

In the note [828], Shafrir presented an interesting form of the deformation lemma (suggested by Brézis) with very general conditions. The deformation is only required to strictly decrease the functional at the noncritical points and not near some regular value as usual.

Theorem 4.7 (Deformation Lemma, Shafrir). *Let Φ be a C^1-functional defined on a Banach space X and let $A \subset X$ be a closed set (possibly empty). Then, there exists a continuous deformation $\eta(t, u): [0, 1] \times X \to X$ satisfying*

    i. $\eta(0, u) = u$ for all $u \in X$,
    ii. $\eta(t, u) = u$ for all $t \in [0, 1]$ if $u \in A$ or if $\Phi'(u) = 0$,
    iii. $\Phi(\eta(t, u)) \leq \Phi(u)$ for all $t \in [0, 1]$ and all $u \in X$,
    iv. $\Phi(\eta(t, u)) < \Phi(u)$ for all $t \in (0, 1]$ if $u \notin A$ and $\Phi'(u) \neq 0$.

This result is used to get some additional information in the MPT when the inf max value c is attained by some the maximum of some competing path $\gamma \in \Gamma$. (See Theorem 7.10 in the Notes of Chapter 7.)

◇ 4.III A Quantitative Deformation Lemma with Weight

Chen and Li [221] give the following quantitative deformation lemma. (Compare to the the extension of Ekeland's variational principle by Zhong and Zhao (Theorem 3.12) in the previous chapter.)

Theorem 4.8 (Quantitative Deformation Lemma, Chen and Li). *Let h be a nonnegative continuous function satisfying $\int_0^{+\infty} dr/(1 + h(r)) = +\infty$, $\Phi \in C^1(X; \mathbb{R})$, where X is a Banach space. Let $c \in]a, b[\subset \mathbb{R}$, ε_0, $\delta > 0$ such that $[c - \varepsilon_0, c + \varepsilon_0] \subset]a, b[$, $\bar{\varepsilon} \leq \min]\varepsilon_0, \delta^2/4[$, and $N_\delta \subset X$ be bounded. If there exists $R > 0$ such that*

    1. $\|\Phi'(x)\|(1 + h(\|x\|)) \geq \delta$ for $|\Phi(x) - c| \leq \bar{\varepsilon}$, $\|x\| \geq R$ and
    2. $\|\Phi'(x)\| \geq \delta$ for $|\Phi(x) - c| \leq \bar{\varepsilon}, d(x, X/N_\delta) \leq \delta/2$ and $\|x\| \leq R$,

then for all ε, $0 < \varepsilon < \bar{\varepsilon}/2$, there exists $\eta \in C(X; X)$ such that

    i. $\eta(x) = x$ for $x \notin \Phi_{c-\bar{\varepsilon}}^{c+\bar{\varepsilon}}$, and
    ii. $\eta(\Phi^{c+\varepsilon} \setminus N_\delta) \subset \Phi^{c-\varepsilon}$.

They also use this result to prove an MPT and apply it to solve a semilinear elliptic equation at resonance.

Many forms of the deformation lemma will be seen in the following chapters. We will notice that they all share the same spirit but in general do not follow the same scheme.

◇ 4.IV Extensions of Deformation Techniques to Nonsmooth Theories

1. For Continuous Functionals. As we will see in Chapter 16, in the recent results [257, 261, 310], Corvellec, Degiovanni, and Marzocchi rely on the notion of weak slope of a continuous functional defined on a metric space (see Chapter 16) and corresponding notions of regular and critical points to prove various deformation lemmata.

For other close approaches, the reader is referred to the papers by Ioffe and Schwartzman [494, 495] and by Ribarska et al. [770] where some quantitative deformation lemmata are proved and some nonsmooth variants of the MPT are presented.

2. For Locally Lipschitz Functionals. In the papers [202, 530, 668], for example, Chang, Kourogenis and Papageorgiou, and Motreanu and Varga extend the deformation lemma to a locally Lipschitz setting in view of its application to a nonsmooth critical point theory for locally Lipschitz functionals (see Chapter 15). The paper by Chang [202] being one of the first references in the subject.

The paper by Ribarska et al. [767] contains a deformation lemma for locally Lipschitz continuous functions Φ on a complete Finsler manifold M of class C^1 (already announced in [766]). This lemma is applied to extend some critical point theorems, including an MPT for M and Φ as indicated, whereas in [765], they extend the quantitative deformation lemma of Willem to locally Lipschitz functions and give a simple proof of an extension of the general mountain pass principle of Ghoussoub and Preiss (Theorem 9.6) for locally Lipschitz functions.

◇ *4.V Deformation Lemma for Set-Valued Mappings*

To use variational methods with elliptic equations with discontinuous nonlinearities in Hilbert spaces, Pascali [704] considers critical points of generalized subdifferential functionals. The treatment is based on a deformation lemma for *set-valued mappings* with a suitable version of the (PS) condition. The interested reader may also consult also the papers [233, 409, 540, 541, 768] by Choulli, Deville, and Rhandi; Frigon; Ribarska et al.; and Kristaly and Varga (see also Chapter 25).

◇ *4.VI Uniqueness of the Steepest Descent Direction*

In general the steepest descent direction is not unique (it depends on the choice of the norm), but Boleslaw [133] shows that in super-reflexive spaces there exists a unique steepest descent direction of a locally Lipschitz functional at any noncritical point. He uses properties of Γ-uniformly convex functionals.

II

Reaching the Mountain Pass
Through Easy Climbs

5

The Finite Dimensional MPT

We suspect that results such as Theorem 2.1 (The MPT) and Theorem 2.4 below
exist in the literature although probably not in the generality below.

> A. Ambrosetti and P.H. Rabinowitz, Dual variational methods in
> critical point theory and applications,
> *J. Funct. Anal.*, **14**, 349–381 (1973)

In this chapter, we present *perhaps the first* MPT. It is a finite dimensional version of the MPT
due to Courant (1950) [279, pp. 223–226]. It contains all the ingredients of the MPT in a finite
dimensional setting. That way, we hope to avoid technicalities and make it easy to understand
the basic ideas in a first contact.

As we will discover through the different chapters, critical point theorems of the
MPT type generally require both *geometric* and *compactness* conditions. We can see
this clearly in the theorem of Rolle.[1]

Rolle of Theorem. *Let x_1 and x_2 be two distinct real numbers and $f \in C^1([x_1, x_2]; \mathbb{R})$.
If $f(x_1) = f(x_2)$, then there exists $x_3 \in]x_1, x_2[$ such that*

$$f'(x_3) = 0.$$

The compactness lies in the fact that $[x_1, x_2]$ is compact and f is continuous, hence
it achieves both its maximum and minimum in $[x_1, x_2]$ by a well-known result of
Weierstrass.

The geometric condition

$$f(x_1) = f(x_2)$$

insures that, when f is not constant in $[x_1, x_2]$, in which case the conclusion holds
trivially, at least one of the extrema of f is attained at some interior point x_3 of $[x_1, x_2]$.
That is, x_3 is a critical point of f.

---

[1] The idea to use the theorem of Rolle to see the duality "compactness-geometric conditions" may be found
for example in [623, 628, 982].

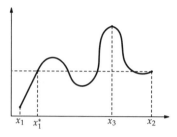

Figure 5.1. The disposition of x_1^*, x_1, x_2, and x_3.

We do not recall the theorem of Rolle only to exhibit the duality geometry-compactness that exists also in the MPT. In fact, it has some links with the MPT. Indeed, consider the intermediate proposition.

Proposition 5.1. *Let $x_1 < x_3 < x_2$ be distinct real numbers and $f \in C^1(\mathbb{R}; \mathbb{R})$. If*

$$\max\{f(x_1), f(x_2)\} < f(x_3), \tag{5.1}$$

then there is a critical point ζ of f in $]x_1, x_2[$ characterized by

$$f(\zeta) = \max_{[x_1, x_2]} f = \inf_{[a,b] \in \Gamma} \max_{x \in [a,b]} f(x), \tag{5.2}$$

where

$$\Gamma = \big\{[a, b]; \ \ such \ that \ a \leq x_1 < x_2 \leq b\big\}.$$

This is an MPT in the real line without the (PS) condition, and it is a consequence of the theorem of Rolle. This will be made clearer in the sequel when the reader would be familiarized with the statement of the MPT. Indeed, we can suppose without loss of generality that $f(x_1) \leq f(x_2)$. Then, by the mean value theorem,

$$\text{there is some point } \overline{x_1^*} \in [x_1, x_3[\text{ such that } f(\overline{x_1^*}) = f(x_2). \tag{5.3}$$

Let us denote by x_1^* the smallest point with this property (see Figure 5.1). Then, it suffices to apply the same idea as in the proof of the theorem of Rolle to get a *maximum* in $[x_1^*, x_2]$. The fact that the maximum is attained is guaranteed by (5.1) and by our choice of x_1^*. Indeed, by (5.1),

$$\max_{[x_1, x_2]} f \geq \max_{[x_1^*, x_2]} f \geq f(x_3) > \max\{f(x_1), f(x_2)\}$$

and because $[x_1, x_2] \in \Gamma$, we have that

$$\max_{[x_1, x_2]} f = \inf_{[a,b] \in \Gamma} \max_{x \in [a,b]} f(x);$$

that is, this maximum also satisfies (5.2).   □

We can easily check that the set Γ can be written also as

$$\Gamma = \big\{\Sigma \subset \mathbb{R}; \ \Sigma \text{ compact and connected and } x_1, x_2 \in \Sigma\big\}.$$

This will help us to understand the generalization that constitutes the principal result of this chapter: the finite dimensional MPT.

In condition (5.1), we must see x_3 as a "set" that *separates* topologically the points x_1 and x_2 in the sense that any connected set that contains both x_1 and x_2 intersects the "set" $\{x_3\}$.

Now, we can extend the last result to higher dimensions, that is, to real functions defined on the Euclidean space $\mathbb{R}^N$ with $N \geq 1$ arbitrary.

Since in higher dimensions we cannot affirm that the value

$$\inf_{\Sigma \in \Gamma} \max_{\Sigma} f$$

is actually a maximum, we cannot say that this value is critical. And in fact *it is not*, unless we require the additional compactness condition (PS) seen in the former chapter, as it was shown by some counterexamples by Brézis and Nirenberg. (See Section 10.2 consecrated to the geometry of the MPT.)

5.1 Finite Dimensional MPT

A finite dimensional version of the MPT was discovered by Courant before 1950, with even some additional information on the topological nature of the critical point obtained by this minimax procedure – a quite impressive thing. In fact, according to Courant, this result was stated before, simultaneously by Morse and Tompkins and by Shiffman in the *1940s* as a general result of critical point theory in functional spaces, and that is really PRODIGIOUS!

Theorem 5.2 (Finite Dimensional MPT, Courant). *Suppose that a continuous function* $\Phi \in C^1(\mathbb{R}^N; \mathbb{R})$ *is coercive and possesses two distinct* strict *relative minima* x_1 *and* x_2. *Then* Φ *possesses a third critical point* x_3 *distinct from* x_1 *and* x_2, *characterized by*

$$\Phi(x_3) = \inf_{\Sigma \in \Gamma} \max_{x \in \Sigma} \Phi(x),$$

where

$$\Gamma = \left\{ \Sigma \subset \mathbb{R}^N; \ \Sigma \text{ is compact and connected and } x_1, x_2 \in \Sigma \right\}.$$

Moreover, x_3 **is not a relative minimizer;** *that is, in every neighborhood of* x_3, *there exists a point* x *such that* $\Phi(x) < \Phi(x_3)$.

Remark 5.1. In fact, Courant proves that

$$\Phi(x_3) > \max\{\Phi(x_1), \Phi(x_2)\}$$

as we will see in the following. Hence, x_3 is different from both x_1 and x_2.

Notice also that, by Theorem 2.1 and since we are in a finite dimensional space, the *coercivity* of Φ implies that Φ satisfies (PS).

In [279], the original theorem is also stated for a domain $\mathcal{O} \subset \mathbb{R}^N$ and instead of the coercivity of Φ, Courant supposes in that case that Φ becomes infinite on $\partial \mathcal{O}$, the boundary of $\mathcal{O}$.

Proof. Let us denote by d_Σ the value $\Phi(x_\Sigma)$ of a point x_Σ in Σ where the maximum of Φ is attained. Note that since x_1 and x_2 are strict relative minimizers, we have

$$d_\Sigma > \max\{\Phi(x_1), \Phi(x_2)\}.$$

Denote by

$$d = \inf_{\Sigma \in \Gamma} \max_\Sigma \Phi = \inf_{\Sigma \in \Gamma} d_\Sigma$$

the greatest lower bound of all possible maximum values d_Σ, and consider a minimizing sequence of such connected compact sets $(\Sigma_n)_n$, that is,

$$d_{\Sigma_n} \to d \qquad \text{as } n \to \infty.$$

Obviously, $d \geq \max\{\Phi(x_1), \Phi(x_2)\}$. In fact, we will see that the inequality is strict because the infimum d is attained for some $\Sigma \in \Gamma$.[2] Indeed, consider the set of all accumulation points of $(\Sigma_n)_n$ defined by

$$\Sigma = \bigcap_{m \in \mathbb{N}} \overline{\bigcup_{i \geq m} \Sigma_i}.$$

This set Σ is compact and connected because it is the intersection of a decreasing sequence of compact connected sets. Moreover, Σ contains both x_1 and x_2, because these are in Σ_i for each i. Therefore, Σ belongs to Γ. A suitable sequence of points $(x_{\Sigma_n})_n$ has then a limit point

$$\lim_{n \to \infty} x_{\Sigma_n} = x_\Sigma \in \Sigma,$$

where $\Phi(x_\Sigma) = d_\Sigma$.

By continuity, we deduce that

$$\Phi(x_\Sigma) \leq \limsup \Phi(x_{\Sigma_n})$$
$$\leq \limsup d_{\Sigma_n}$$
$$\leq d$$

while $\max_\Sigma \Phi \geq d$ because $\Sigma \in \Gamma$. Thus, $d_\Sigma = d$.

Courant calls Σ a *minimizing connecting set joining x_1 and x_2*. So, we have found a set Σ in Γ where the infimum of Φ over Γ is achieved.

Write

$$\mathfrak{M} = \Big\{ x \in \Sigma; \ \Phi(x) = d_\Sigma = \max_\Sigma \Phi = d \Big\}.$$

The set $\mathfrak{M}$ is compact because of the continuity of Φ and the fact that Σ is compact.

We shall show that there exists a point $x_3 \in \mathfrak{M}$ for which $\Phi'(x_3) = 0$. The proof uses the following deformation (See [279, p. 224], we quote Courant):

---

[2] This is something very strong and not true in general in the (classical) MPT as we will see. But here, the situation is favorable.

Transform the minimizing set Σ into a set Σ' by deforming $P \in \Sigma$ into $P' \in \Sigma'$ such that

$$\Phi(P') < \Phi(P)$$

except for the images of arbitrary small neighborhoods of stationary points, along the direction of the gradient of Φ. Hence, if $\mathfrak{M}$ did not contain a stationary point. We could replace Σ by an other connecting set Σ' for which $d_{\Sigma'} < d$, contrary to the minimizing character of Σ.

More precisely, if there were no critical points in the compact set $\mathfrak{M}$, then there would exist a positive constant α such that, for all $x \in \mathfrak{M}$,

$$|\nabla \Phi(x)| > \alpha.$$

By continuity, there exists $\varepsilon > 0$ such that

$$\mathfrak{M}_\varepsilon = \left\{ x \in \mathbb{R}^N; \exists y \in \mathfrak{M} \text{ such that } |x - y| < \varepsilon \right\}$$

is a neighborhood of $\mathfrak{M}$ where $|\nabla \Phi| > \dfrac{\alpha}{2}$. Of course x_1 and x_2 are not in $\mathfrak{M}_\varepsilon$ because they are relative minimizers and, hence, critical points.

Consider now a cutoff function ρ that satisfies the following:

$$\begin{cases} 0 \le \rho \le 1, \\ \text{supp } \rho \subset \mathfrak{M}_\varepsilon, \\ \rho \equiv 1 \text{ in } \mathfrak{M}. \end{cases}$$

The deformation [3] $\eta \colon \mathbb{R}^N \times \mathbb{R} \to \mathbb{R}^N$ is defined by

$$\eta(x, t) = x - t.\rho(x).\nabla \Phi(x).$$

Then η is continuously differentiable and

$$\frac{d}{dt} \Phi(\eta(x, t))|_{t=0} = -\rho(x).|\nabla \Phi(x)|^2.$$

Moreover, we have $|\nabla \Phi(x)|^2 \ge \dfrac{\alpha^2}{4} > 0$ on supp $\rho \subset \mathfrak{M}_\varepsilon$.

By continuity, there exists $T > 0$ such that for all $t \in [0, T]$, the following inequality holds:

$$\frac{d}{dt} \Phi(\eta(x, t))|_{t=0} \le -\frac{\rho(x)}{2} |\nabla \Phi(x)|^2.$$

Thus, if we set

$$\Sigma_T = \eta(\Sigma, T) = \left\{ \eta(x, T); \ x \in \Sigma \right\},$$

---

[3] Under this form, η is taken from [882]. We preferred this one to that of Courant because we care at this stage to present a result as elementary, clear, and elegant as possible. The original one is given later, in the final notes on page 54.

for any point $\eta(x, T) \in \Sigma_T$ we have

$$\Phi(\eta(x, T)) = \Phi(x) + \int_0^T \frac{d}{dt} \Phi(\eta(x, t)) \, dt$$
$$\leq \Phi(x) - \frac{T}{2} \rho(x) |\nabla \Phi(x)|^2.$$

Therefore, either $\Phi(x) < d$ if $x \notin \mathfrak{M}$ or $d - \frac{T}{2} \alpha^2$ if $x \in \mathfrak{M}$. Hence,

$$\max_{x \in \Sigma_T} \Phi(x) < d.$$

But, by the continuity of η, the set Σ_T is compact and connected while $x_i = \eta(x_i, T) \in \Sigma_T$ for $i = 1, 2$ by the choice of $\mathfrak{M}_\varepsilon$ and ρ. Hence, Σ_T is in Γ – a contradiction with the definition of d as a minimum on Γ, thus Σ contains a critical point. $\square$

Lemma 5.3. *In any neighborhood $\mathcal{N}$ of x_3, there exists a point x for which the strict inequality*

$$\Phi(x) < \Phi(x_3)$$

holds.

Proof. Let K be the set of stationary points contained in $\mathfrak{M}(\subset \Sigma)$. Let y be a point lying both in the neighborhood $\mathcal{N}$ of x_3 and $\Sigma \backslash K$; then $\Phi(x_3) = d \geq \Phi(y)$. If $\Phi(y)$ was equal to d, using the deformation defined earlier, there would be another point x in a neighborhood of y contained in $\mathcal{N}$, for which $\Phi(x) < \Phi(y) = d$ since y is not a stationary point. $\square$

5.2 Application

As a very simple application of the finite dimensional MPT, we report a proof of a *global homeomorphism* theorem by the finite dimensional MPT, by Katriel [516]. A functional $f \in C^1(X; Y)$ where X and Y are Banach spaces such that $f'(x)$ is invertible is a local homeomorphism from a neighborhood of x into its image by the local inversion theorem. But it is not in general a global homeomorphism. Think of the real valued function arctan in the real line as an example.

Some global homeomorphism theorems are well known like the Lax-Milgram theorem and Hadamard theorem. (See the Notes for more.) The following version of the Hadamard theorem is an easy consequence of the finite dimensional MPT.

Theorem 5.4. *Let X and Y be finite dimensional Euclidean spaces, and let $\Phi: X \to Y$ be a C^1-function that satisfies*

  i. *$\Phi'(x)$ is invertible for all $x \in X$,*
  ii. *$\|\Phi(x)\| \to \infty$ as $\|x\| \to \infty$.*

Then Φ is a diffeomorphism of X onto Y.

Proof. By i and the inverse function theorem, Φ is an open mapping, then the range of Φ is open in Y. Using ii and the fact that a bounded closed set in a finite dimensional space is compact, we verify easily that $\Phi(X)$ is closed. Indeed, let $(\Phi(x_n))_n$ be a convergent sequence in Y. Then it is bounded in Y and hence $(x_n)_n$ is bounded in X. Thus, it admits a subsequence converging to some $\overline{x} \in X$ and $\Phi(x_n) \to \Phi(\overline{x})$. It remains then to show that Φ is one-to-one, to show that Φ is a diffeomorphism. By contradiction, suppose that $\Phi(x_1) = \Phi(x_2) = y$ for two points x_1 and x_2 in X, and consider the C^1-function $f : X \to \mathbb{R}$ defined by

$$f(x) = \frac{1}{2} \|\Phi(x) - y\|^2.$$

By ii, $f(x) \to \infty$ as $\|x\| \to \infty$. It is easy to see that x_1 and x_2 are global minima of f by the inverse function theorem $\Phi(x) \neq \Phi(x_i)$ in a neighborhood of x_i (for $i = 1, 2$), which implies that x_1 and x_2 are *strict* local minima. Therefore, by the finite dimensional MPT, there exists a third critical point x_3 for f with $f(x_3) > 0$. So $\|\Phi(x_3) - y\| > 0$, and thus $\Phi(x_3) \neq y$. But the fact that x_3 is a critical point of f means that $\Phi^*(x_3)(\Phi(x_3) - y) = 0$, which contradicts the invertiblity of $\Phi'(x_3)$ expressed in i.
□

This proof is due to Katriel [516], who used it as a preamble to motivate his topological version of the MPT, the subject of our next chapter.

Comments and Additional Notes

1. As reported in [50], Ambrosetti and Rabinowitz imagined the possibility of the existence of a result similar to the finite dimensional MPT in the literature. (See the quotation of Ambrosetti and Rabinowitz at the beginning of the chapter.) Of course, their version is much stronger and their paper contains, among other results, both a dual form and a symmetric version of the MPT as well as applications to semilinear partial differential equations.
2. The existence of a connected compact set for which the inf max is attained may also be proved using the fact that the space of compact connected subsets of a compact metric space is compact when endowed with the *Haussdorff distance*:

$$\mathfrak{h}_X(A, B) = \sup\{\sup_{a \in A} \text{dist}\,(a, B), \sup_{b \in B} \text{dist}\,(b, A)\},$$

as in [122, Theorem 2.25, p. 59].

◇ 5.1 On Rolle's Theorem and the MPT

The approach consisting of going from the theorem of Rolle to an MPT in dimension one and finally to the finite dimensional version of the MPT was adopted to make a transition and to exhibit some continuation in the ideas carried by these results.

Recently, Silva and Teixeira [846] proved a version of Rolle's theorem that implies the classical form of the MPT (of Ambrosetti and Rabinowitz).

Let E be a real Banach space and $f : E \to \mathbb{R}$ a C^1-mapping. Denote by Λ the set of nonincreasing, locally Lipschitz continuous functions $\phi : (0, \infty) \to (0, \infty)$ such that $\int_0^\infty \phi(t)\,dt = \infty$. Then, f is said to satisfy the *generalized (PS) condition with respect* to $\phi \in \Lambda$ at $c \in \mathbb{R}$ if every sequence $(u_m)_m$ of points in E satisfying $f(u_m) \to c$ and $\|f'(u_m)\|/\phi(\|u_m\|) \to 0$ $(m \to \infty)$ has a convergent subsequence. The value $c \in \mathbb{R}$ is an *admissible level* for f if c is a regular value of f or c is an isolated critical value with discrete critical points over $f^{-1}(c)$.

The main theorem in [846] implies that, for an admissible critical level c, $f^{-1}(c)$ is arcwise connected or f has a critical point with different critical value from c. It also implies the classical MPT.

◇ 5.II The Deformation that Appears in the Original Proof of Courant

The proof of the finite dimensional MPT is taken from Courant [279], except for the expression of the deformation η, which appears under this form in [882]. In the original work of Courant it was defined (with few changes in the notations but with almost the same words), as follows.

The deformation is effected by moving P into P' along the gradient of the function Φ. Let $U = (u_1, \ldots, u_N)$ be a point in C (in our case $\mathbb{R}^N$) and $V = (v_1, \ldots, v_N)$ an arbitrary vector at the point U. The first derivatives of $\Phi(U)$ were supposed to be continuous in the domain C. Consequently, for any fixed closed subdomain C^* of C and arbitrarily small δ, we can find a positive constant ε_δ so small that, for all points U in C^*,

$$\Phi(U - \varepsilon V) = \Phi(U) - \varepsilon(V \cdot \nabla\Phi(U)) + \varepsilon\tau(\varepsilon) \qquad (*)$$

with $|\tau(\varepsilon)| < \delta$, whenever $|\varepsilon| < \varepsilon_\delta$. We choose C^* so that it contains $\mathfrak{M}$. (The set $\mathfrak{M}$ is defined on page 50.)

Assume that for some positive constant α, everywhere in the closed set $\mathfrak{M}$

$$|\nabla\Phi(u)| \geq \alpha.$$

Then we can choose a constant $a < 1$ so small that

$$|\nabla\Phi(u)| < 2\alpha$$

holds in the larger domain $\mathfrak{M}^* = \{U \in \mathbb{R}^N; \ \mathrm{dist}\,(U, \mathfrak{M}) < a\}$. We may assume that x_1 and x_2 do not belong to $\mathfrak{M}^*$, and that $C^* \supset \mathfrak{M}^*$. We apply $(*)$ for $V = \nabla\Phi(U)$ to the points of $\mathfrak{M}^*$ and obtain

$$\Phi(U - \varepsilon V) = \Phi(U) - 4\varepsilon\alpha^2 + \varepsilon\delta, \ 0 < \varepsilon < \varepsilon_\delta;$$

if we choose $\delta = \alpha^2$ and ε_δ accordingly small, we get

$$\Phi(U - \varepsilon V) = \Phi(U) - 3\varepsilon\alpha^2.$$

Denote by P_0 a fixed point in $\mathfrak{M}$, and by r its distance from P_0. Consider all points U in $\mathfrak{M}^*$ whose distance from P_0 does not exceed a. In this sphere $r \leq a$ we replace each point U by U' according to

$$U' = U - \eta\nabla U,$$

with $\eta = \varepsilon_\delta(a - r)$. Then, in the interior of the sphere

$$\Phi(U') < \Phi(U) < d,$$

while outside the sphere we set $U' = U$ and have $\Phi(U') = \Phi(U)$.

By a finite number of such spheres we can cover $\mathfrak{M}$. Performing a succession of corresponding transformations $U \mapsto U'$ we arrive at a transformation of C into itself in which all points outside $\mathfrak{M}^*$ remain unchanged. The set Σ is transformed into a compact connected set Σ' containing x_1 and x_2, and $\Phi(U) < d$ everywhere in Σ' – a contradiction with the assumption that Σ is a minimizing set. Hence the hypothesis $\alpha > 0$ in $\mathfrak{M}$ is absurd and Σ contains a stationary point of Φ.

◇ 5.III Ancestors of the Finite Dimensional MPT

The finite dimensional MPT was established to investigate the existence of unstable solutions to minima surface problems. Courant quotes in his book [279] the existence of an anterior version of the MPT for abstract functional spaces, first obtained simultaneously by Shiffman [837–839] and by Morse and Tompkins [664, 666]:

> In general, one might expect that the existence of two relative minima in Plateau's problem would guarantee the existence of another minimal surface of unstable character, just as the existence of two distinct relative minima of a differentiable function of a finite number of variables implies the existence of a stationary "mini-maximum." Theorems about the existence of such unstable minimal surfaces were first proved by M. Shiffman and at the same time by M. Morse and C. Tompkins. In addition, these papers propose classification of unstable minimal surfaces by considering the individual features of the problem from within the framework of an abstract theory of critical points in function spaces. The discussions are based on Douglas' explicit expressions from the Dirichlet integral of Harmonic vectors in terms of their boundary values. A different approach due to the author [Courant], using the method of Dirichlet's principle, proceeds by reduction of the problem to that of a finite number of independent variables, provided that the boundary contour is polygonal. This theory was essentially extended by Shiffman. He observed that a passage to the limit from polygons to more general (rectifiable) closed curves is possible on the basis of the continuity theorem of Morse and Tompkins.

Shiffman reports in [837] that his methods were mainly inspired by the work of Courant. We even found in the recent review of the paper [925], for example, a reference to what the reviewer calls the mountain-pass lemma of Morse-Tompkins and in that of [927] that the author obtains as an application a strong version of the theorem of Morse-Shiffman-Tompkins [660, 664] on the existence of a "third" unstable minimal surface if there are two minima being isolated in some strong topology.

◇ 5.IV Hadamard Global Inversion Theorem

The following *global version* of the inverse mapping theorem, also called *monodromy theorem*, is due to Hadamard (1906) in the finite dimensional case. (See, for example

[121, Theorem 5.1.5, p. 222].) Forms for general Banach spaces are due to Cacciopoli [163] and Levy [561].

Theorem 5.5 (Hadamard Theorem). *Let X and Y be two Banach spaces and $\Phi: X \to Y$ a C^1-functional. Suppose that $\Phi': X \to \mathbb{R}$ is an isomorphism. Denote for each $R > 0$*

$$\Psi(R) = \sup_{B(0,R)} \| (\Phi'(x))^{-1} \|.$$

Assume that

$$\int_0^\infty \frac{dr}{\Psi(r)} = \infty.$$

Then, Φ is a diffeomorphism of X onto Y.

In particular, this is the case if, for some $K > 0$,

$$\| (\Phi'(x))^{-1} \| \le K \qquad \text{for all } x \in X.$$

Chapter 3 in Ambrosetti and Prodi's book [49] is devoted to global inversion theorems.

6

The Topological MPT

It is with the help of the methods of topology, that we shall seek answers to the fundamental question connected with the study of nonlinear equations. However, this last assertion does not imply a negative role for other methods of investigations. In fact, topological methods become powerful only by virtue of their combination with other approaches.

> M.A Krasnosel'skii, *Topological methods in the theory of nonlinear integral equations*, Pergamon Press, 1964.

This chapter is devoted to a pure topological version of the MPT due to Katriel [516]. It constitutes a natural "upgrade" of the finite dimensional MPT presented in the former chapter to locally compact topological spaces. It will also help us to clarify more our vision of the situation and will make us *see* the MPT under another angle. In fact, the topological considerations constitute an inherent part of the different faces of the MPT as will be clarified later. This is the case for all variational analysis results.

The MPT we will see is *topological*, in the sense that *no differential structure on X is needed or used.* So, we do not get critical points since this notion does not (yet) make any sense in such spaces. But we get some *particular points* from a pure topological point of view that should be critical in the presence of a differential structure. The theorem can be interpreted as a translation in a "topological language" of the finite dimensional MPT. It seems inspired by the early work of Hofer [480] and Pucci and Serrin [728] on the structure of the critical set in the MPT (see Chapter 12).

This version of the MPT has a great similitude with the finite dimensional MPT. This appears both in the assumptions and in the proof. Both results form an appropriate prelude to the standard MPT.

6.1 Some Preliminaries

First of all, we introduce some notions to adapt the ideas used earlier to this new context and legitimate the fact that we adopt a similar language. All the topological spaces used in this chapter are supposed to be *regular*.

Definition 6.1. A topological space X is said to be *compactly connected* if for each x_1, x_2 in X there exists a compact connected set containing both x_1 and x_2.

We can check immediately that any separated topological space (in particular any normed vector space) is compactly (arcwise) connected.

Definition 6.2. Let X be a topological space and $f: X \to \mathbb{R}$ be a functional. A point $x \in X$ is called a *mountain pass point* (in the sense of Katriel) if for every neighborhood $\mathcal{N}$ of x, the set defined by

$$\mathcal{N} \cap \{y \in X; \ f(y) < f(x)\}$$

is *disconnected*.

This definition was given first by Hofer [1] in [481]. The idea to distinguish the different critical points by the *topological nature of level sets* near the values they take is the essence of both Ljusternik-Schnirelman and Morse approaches to critical point theory.

Now, we introduce a *topological coercivity condition* playing the role of the coercivity that appears in the finite dimensional MPT.

Definition 6.3. Let X be a topological space. A functional $f: X \to \mathbb{R}$ is said to be *increasing at infinity* if for all $x \in X$, there is a compact subset $K \subset X$ such that

$$f(z) > f(x), \qquad \text{for all } z \notin K.$$

Remark 6.1.
1. It is easy to check from the definition that a *continuous* function increasing at infinity is bounded from below.
2. It is also easy to see that a topological space that admits a *continuous* increasing function at infinity f is *locally compact*. Indeed, the family of sets

$$\left(\{x \in X; \ f(x) < c\} \right)_c,$$

where c is such that

$$\inf_{x \in X} f(x) \leq c < \sup_{x \in X} f(x),$$

forms an open covering of X of *precompact sets*.
3. A function increasing at infinity defined on a normed vector space, even finite dimensional, is not necessarily coercive (see Example 6.1). Whereas a coercive functional in a finite dimensional vector space is increasing at infinity.
4. More important is the fact that a function increasing at infinity defined on a normed vector space does not necessarily satisfy the (PS) condition as we see in the same example.

---

[1] In fact, Hofer supposes that the intersection is neither empty nor path-connected. See Chapter 12 on the structure of the critical set of the MPT for more details.

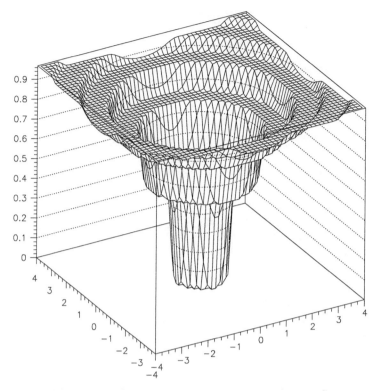

Figure 6.1. The aspect of the function increasing at infinity $\tilde{\psi}$.

Example 6.1. Let X be the real plane $\mathbb{R}^2$, fix $\delta > 0$ and define

$$\psi(x, y) = \begin{cases} 0 & \text{if} \quad (x, y) \text{ is in the ball } B(0, 1 - \delta), \\ 1 - \dfrac{1}{2^n} & \text{if} \quad (x, y) \text{ is in the annulus } A(n + \delta, n + 1 - \delta), \ n \geq 1. \end{cases}$$

Consider a C^1-extension $\tilde{\psi}$ of ψ such that, on each annulus of the form $A(n - \delta, n + \delta)$,

$$1 - \frac{1}{2^{n+1}} \leq \tilde{\psi}(x, y) \leq 1 - \frac{1}{2^n}.$$

Then we can easily see that $\tilde{\psi}$ is increasing at infinity (see Figure 6.1). Moreover, in this example, we can also see that $\tilde{\psi}$ does not verify $(PS)_1$.

In connection with the second point of the former remark, notice that 1 does not belong to $\tilde{\psi}(X)$. Otherwise, the level set $\{x \in X; \ \tilde{\psi}(x) < 1\}$ would be relatively compact and hence $(PS)_1$ would hold. The value 1 in our example is the supremum of $\tilde{\psi}$ and is not attained.

Always, by the second point in the remark above, we can check that a function increasing at infinity f satisfies $(PS)_c$ for all $c < \sup_{x \in X} f(x)$, while we can say nothing about its supremum. This depends essentially on the topology of the underlying space X.

6.2 Topological MPT

We will need the following technical result in the sequel.

Lemma 6.1. *Let X be compactly connected, locally connected, and locally compact, and let C be an open and connected subset of X. Then, C is compactly connected.*

Proof. Let $x_0 \in C$ and

$$B = \{K \subset C; \ K \text{ compact connected containing } x_0\}.$$

We will show that $B = C$. Hence, any two points x and y contained in C will be contained into two compact connected sets containing x_0 whose union forms a compact connected set containing both x and y and therefore is in C. Since C is connected, it suffices to show that B is open and closed in C.

1. The set B is open. Indeed, for $x \in B$, let N_0 be a compact neighborhood of x (recall X is locally compact). Let $\mathcal{O}_1$ and $\mathcal{O}_2$ be nonintersecting open sets containing x and $X \setminus C$, respectively. (Here we use the regularity of X.) Set $N_1 = N_0 \cap \overline{\mathcal{O}_1}$. Then, N_1 is a compact neighborhood of x contained in C. By the local connectedness of X, there is a connected neighborhood N of x contained in N_1. So, $\overline{N}$ is a compact connected neighborhood of x in C. Since $x \in B$, by definition of B there is a compact connected set $K \subset C$ containing x_0 that also contains x. And since $K \cup \overline{N}$ is compact, connected, and contained in C, we see that $N \subset B$; so B is open.

2. The set B is closed in C. Indeed, consider a point $x \in \overline{B} \cap C = \overline{B} \cap \overline{C} = \overline{B}$. ($C$ is connected, so $C = \overline{C}$ and also $B \subset C$, so $\overline{B} \subset \overline{C}$.) As before, let $\overline{N}$ be a compact connected neighborhood of x contained in C. Since $x \in \overline{B}$, there is some y in $B \cap N$. Let $K \subset C$ be a compact connected set containing both x_0 and y. Then $K \cup \overline{N}$ is compact, connected, contains x_0, and is x and is contained in C, so $x \in B$; hence, B is closed in C. $\square$

Now we can state the main result in this chapter.

Theorem 6.2 (Topological MPT, Katriel). *Let X be a topological space locally connected, compactly connected and admitting a continuous function Φ that is increasing at infinity. Let $S \subset X$ be a set that separates x_1 and x_2; that is, x_1 and x_2 lie in different (connected) components of $X \setminus S$, and suppose that*

$$\max \left\{ \Phi(x_1), \Phi(x_2) \right\} < \inf_{x \in S} \Phi(x) = p. \tag{6.1}$$

Then, there is a third point x_3 which is either a local minimum or a mountain pass point of Φ with $\Phi(x_3) = c \geq p > \max\{\Phi(x_1), \Phi(x_2)\}$. Moreover, the value c is characterized by the minimax argument

$$c = \inf_{\gamma \in \Gamma} \max_{x \in \gamma} \Phi(x),$$

where Γ is, as in the previous chapter, the set of all connected compact subsets of X containing both x_1 and x_2.

Proof. Since X is compactly connected, the set Γ is nonempty. Let $\Psi : \Gamma \to \mathbb{R}$ be the functional defined by

$$\Psi(\gamma) = \max_{\gamma \in \Gamma} \Phi(x).$$

It is clear that since S separates x_1 and x_2, every set $\gamma \in \Gamma$ intersects S, so $c \geq p$.

Claim 6.1. The infimum c of Ψ is attained; that is, $\Psi(B) = c$ for some $B \in \Gamma$.

Indeed, consider a minimizing sequence $(A_n)_n$ for Ψ in Γ, such that

$$\begin{cases} \Psi(A_1) \geq \Psi(A_2) \geq \Psi(A_3) \geq \cdots \\ \Psi(A_n) \to c. \end{cases}$$

Since Φ is increasing at infinity, there is a compact set K such that

$$\Phi(z) > \max_{x \in A_1} \Phi(x) \geq \Psi(A_n),$$

for all n and all $z \notin K$.

Then, $A_n \subset K$ for all n. This implies that the sets

$$B_k = \overline{\bigcup_{n \geq k} A_n}$$

are contained in K for all positive integers k. They are also connected since each A_n is connected and $x_0 \in A$ for all n. Hence, $(B_k)_k$ forms a descending sequence of closed connected subsets all contained in a same compact set. The set

$$B = \bigcap_{k \geq 1} B_k = \bigcap_{k \geq 1} \overline{\bigcup_{n \geq k} A_n}$$

is compact and connected[2], and contains both x_1 and x_2. So $B \in \Gamma$, and we can easily see that $\Psi(B) = c$. Indeed, since $B \in \Gamma$, we have that $\Psi(B) \geq c$. But $B = \bigcap_{k \geq 1} B_k$, so we have also that

$$\Psi(B) \leq \Psi(B_k), \qquad \text{for any } k,$$

and

$$\begin{aligned} \Psi(B_k) &= \sup \left\{ \Phi(x); \; x \in \overline{\bigcup_{n \geq k} A_n} \right\} \\ &= \sup \left\{ \Phi(x); \; x \in \bigcup_{n \geq k} A_n \right\} \\ &= \sup \left\{ \Phi(x); \; x \in A_k \right\} \\ &= \Psi(A_k). \end{aligned}$$

Then, since the sequence $(A_n)_n$ is minimizing for Ψ, we get that $\Psi(B) \leq c$. So, $\Psi(B) = c$.

Now, we will show that $\Phi^{-1}(c)$ contains either a local minimum or a mountain pass point of Φ.

---

[2] The same process was used in the proof of the finite dimensional MPT.

By contradiction, suppose that this is not true. Let C be the (connected) component of $X \backslash \Phi^{-1}(c)$ containing x_1. Then C is open because X is locally connected.

Claim 6.2.

$$\overline{C} \subset C \cup \Phi^{-1}(c). \tag{6.2}$$

Suppose that $x \in \overline{C}$ and $x \notin \Phi^{-1}(c)$. Since the set $C \cup \{x\}$ is connected, because it lies between C and $\overline{C}$, and C is a connected component, we shall have $C \cup \{x\} \subset C$, that is, $x \in C$. This proves Claim 6.2.

Claim 6.3.

$$B \subset \overline{C}. \tag{6.3}$$

Assuming Claim 6.3 holds true, Claim 6.2 implies that

$$B \subset C \cup \Phi^{-1}(c).$$

But using (6.1) and the fact that $c \geq p$ we get that $x_2 \notin \Phi^{-1}(c)$; thus, $x_2 \in C$.

So, we have proved that C is open and connected in X, which is compactly connected. The space X is also locally compact because it admits a function increasing at infinity. Then, by Lemma 6.1, C is compactly connected. Therefore, there is a compact connected set $K \subset C$ containing both x_1 and x_2. And since $\Phi(x) < c$ for all $x \in C$, we get that

$$\Psi(K) < c,$$

which is a contradiction with the definition of c.

So, to complete the proof of the topological MPT, we have only to show that Claim 6.3 is true.

To prove that $B \subset \overline{C}$, it is enough to prove that $B \cap \overline{C} = B$. Clearly, $B \cap \overline{C}$ is relatively closed in B. And since B is connected, it suffices to show that $B \cap \overline{C}$ is relatively open in B. So, consider $x \in B \cap \overline{C}$. It suffices to construct a neighborhood $\mathcal{N}$ of x such that

$$\mathcal{N} \cap B \subset B \cap \overline{C}.$$

If $x \in C$, we can take $\mathcal{N} = C$, because C is open. Otherwise, by (6.2) we have $x \in \Phi^{-1}(c)$.

Since x is not a mountain pass point, there is a neighborhood $\mathcal{N}$ of x such that the set

$$M = \{y \in X; \ \Phi(y) < c\} \cap N$$

is connected. And since $x \in \overline{C}$ (recall that $x \in B \cap \overline{C}$), there exists a point $u \in C \cap N$. Thus $\Phi(u) < c$ and then $u \in M$. Hence, the two connected sets C and M intersect, but since C is a maximal connected set in $X \setminus \Phi^{-1}(c)$, we must have

$$M \subset C. \tag{6.4}$$

Take now $w \in \Phi^{-1}(c) \cap \mathcal{N}$. By assumption, w is not a local minimum of Φ. So, any neighborhood of w intersects M. Thus $w \in \overline{M}$ and

$$\Phi^{-1}(c) \cap \mathcal{N} \subset \overline{M}. \qquad (6.5)$$

By (6.4) and (6.5), we have

$$\mathcal{N} \cap B \subset \mathcal{N} \cap \{y \in X; \; \Phi(y) \le c\}$$
$$= \left(N \cap \{y \in X; \; \Phi(y) < c\}\right) \cup \left(N \cap \{y \in X; \; \Phi(y) = c\}\right)$$
$$\subset C \cup \overline{M}$$
$$\subset \overline{C}$$

So

$$\mathcal{N} \cap B \subset B \cap \overline{C}.$$

This proves Claim 6.3. $\square$

In the real line $\mathbb{R}$, if $x_1 < x_3 < x_2$, the set $\{x_3\}$ separates x_1 and x_2 as in the statement of the result described as an MPT in one dimension in the earlier chapter. In higher dimensions, as an easy consequence of the topological MPT, we have the following result.

Corollary 6.3. *Let X be a locally connected and compactly connected topological space and let $\Phi \colon X \to \mathbb{R}$ be continuous and increasing at infinity. If x_1 and x_2 are strict local minima of Φ, then there exists a point $x_3 \in X$ different from x_1 and x_2 such that it is either a local minimum or a mountain pass point of Φ and $\Phi(x_3) > \max\{\Phi(x_1), \Phi(x_2)\}$.*

Proof. Without loss of generality, we can assume that $\Phi(x_1) \ge \Phi(x_2)$. Since X admits a function increasing at infinity, it is locally compact. Let $\mathcal{N}$ be a compact neighborhood of x_1 such that

$$\Phi(x) > \Phi(x_1) \qquad \text{for all } x \in \mathcal{N} \setminus \{x_1\}.$$

This is possible because x_1 is a strict local minimum.

Let $p = \min_{\partial \mathcal{N}} \Phi$, then $p > \Phi(x_1) \ge \Phi(x_2)$; thus, $x_2 \notin \partial \mathcal{N}$. Hence, taking $S = \partial \mathcal{N}$, we have that S separates x_1 and x_2 and we can apply Theorem 6.2 to get x_3. $\square$

As we remarked before, a coercive function on a finite dimensional vector space is increasing at infinity, since a mountain pass point for a smooth[3] function is a critical point.

The finite dimensional MPT (Theorem 5.2) is a consequence of Corollary 6.3.

Remark 6.2. The two cases wherein x_3 is a *mountain pass point* and x_3 is a *local minimum* occur as we can see easily on simple examples of coercive functionals with two strict minima x_1 and x_2 defined on the real line $\mathbb{R}$. The reader is referred to

---

[3] There, *smooth* means smooth enough that the notion of critical point makes sense, otherwise a function will be said to be nonsmooth.

Chapter 12, which is devoted to the study of the critical set in the situation of the MPT, for more details.

Remark 6.3. The topological MPT was used by Katriel to study some criteria permitting affirmation that a local homeomorphism is a global one. These will not be given here. The interested reader can consult [516].

Comments and Additional Notes

1. Topology is a very important ingredient in variational methods. The terminology "variational methods," methods based on critical point theorems and variations, is used in general in opposition to "topological methods," the methods that use some kind of topological degree or index. But in fact, these methods are also topological methods. The topological MPT is *unique* in the sense that no kind of nonsmooth analysis is used, but rather topology, and as a result, the local minimum or the mountain pass point obtained are critical any time this makes sense. We think that the assumptions required are somewhat severe, in particular the local compactness condition, and we believe they can be weakened by reformulating the definition of a function increasing at infinity.

2. As in the finite dimensional MPT, the inf max value c is also attained by the maximum value of some path in Γ. See a result by Shafrir and another one by Taubes in the final Notes of Chapter 7.

3. We saw in the earlier chapter that Katriel applied the finite dimensional MPT of Courant to prove a global homeomorphism theorem. He also used his topological variant to prove some new global homeomorphism theorems. He is able to deal with continuous functions between topological spaces. These are purely topological problems with no apparent relation to nonlinear boundary value problems which constitute the classical field of application of different kinds of critical point theorems, including the different versions of the MPT.

7

The Classical MPT

The mountain pass lemma of Ambrosetti and Rabinowitz is a result of great intuitive appeal as well as practical importance in the determination of critical points of functionals, particularly those which occur in the theory of ordinary and differential equations.

> P. Pucci and J. Serrin, Extensions of the mountain pass theorem,
> *J. Funct. Anal.,* **59**, 185–210 (1984)

One would expect that it is virtually impossible to find critical points which are not extrema. The first to show that this is not the case are Ambrosetti and Rabinowitz.

> M. Schechter, *Linking methods in critical point theory*, Birkhäuser, 1999

This chapter is devoted to *the* MPT, the source of inspiration for all the results that constitute the material of this monograph. We will also see two of its direct applications to boundary value problems: a superlinear Dirichlet problem and a problem of Ambrosetti-Prodi type.

A very interesting situation that occurs when treating nonlinear problems by variational methods is the following. The "energy" associated with the problem, whose critical points are the weak solutions, is *indefinite*, in the sense that it is bounded neither from above nor from below. In such cases, there are of course no absolute extrema, and the direct method of the calculus of variations that looks for absolute minimizers fails to apply. It may also happen that, in addition to being indefinite, the energy functional does not possess any local maxima or minima; therefore, some of its critical points with some *different nature*[1] than minima and maxima should be found.

7.1 The Name of the Game

The MPT is one of the most popular theorems in critical point theory that do not seek local extrema but characterize a critical value of the functional by a *minimax argument*. (Such results are known in the literature as *minimax theorems*.) This

---

[1] Mountain pass points; for example, see Chapter 6 devoted to the topological MPT.

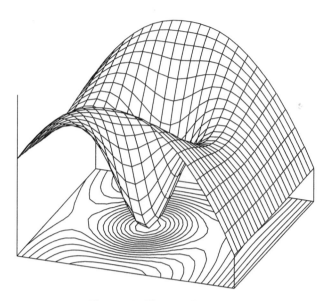

Figure 7.1. The MPT landscape.

theorem is known as the *mountain pass theorem*, (also called, in some references, the mountain pass lemma). It owes this name to its *geometric interpretation*: (see Figure 7.1)

> Consider two valleys A and B such that A is surrounded by a mountain ridge that separates it from B. To go from A to B, we must cross the mountain chain. If we want to climb as little as possible, we would have to consider the maximal elevation of each path. The path with the minimal one (of these maximal elevations) will cross a *mountain pass*.

7.2 The MPT

In fact the preceding statement is not correct. We have to require an additional compactness assumption.

Theorem 7.1 (The Classical MPT, Ambrosetti and Rabinowitz). *Let X be a Banach space and $\Phi: X \to \mathbb{R}$ a C^1-functional satisfying* (PS). *Suppose that there is $e \in X$, $\|e\| > r > 0$ and*

$$\alpha = \max\{\gamma(0), \gamma(e)\} < \inf_{u \in S(0,\rho)} \Phi(u) = \beta. \tag{7.1}$$

Then, Φ possesses a critical value $c \geq \beta$ characterized by

$$c = \inf_{\gamma \in \Gamma} \max_{u \in \gamma([0,1])} \Phi(u), \tag{7.2}$$

where

$$\Gamma = \big\{\gamma \in \mathcal{C}([0,1]; X);\ \gamma(0) = 0 \qquad and \qquad \gamma(1) = e\big\}.$$

Remark 7.1. The set Γ of compact connected sets containing 0 and e used in the two preceding chapters is *different* from the set Γ that appears there. Nevertheless, we will see below that it is still possible to use the former set Γ and that both of them yield exactly the same critical value!

A first proof using the deformation lemma. It is obvious that $c < \infty$. For all paths $\gamma \in \Gamma$, $\gamma([0, 1])$ is connected, $\gamma(0) = 0$ and $\gamma(1) = e$, so

$$\gamma([0, 1]) \cap S(0, \rho) \neq \varnothing.$$

Hence,

$$\max_{u \in \gamma([0,1])} \Phi(u) \geq \inf_{v \in S(0,\rho)} \Phi(v) \geq \beta,$$

and therefore $c \geq \beta$.

Suppose now by contradiction that c is a regular value; that is, $\mathbb{K}_c = \varnothing$. Then, by Lemma iv there exist $\varepsilon \in]0, (\beta - \alpha)/2[$ and a deformation η such that

$$\eta(1, \Phi^{c+\varepsilon}) \subset \Phi^{c-\varepsilon}. \tag{7.3}$$

By definition of c, there is some $\gamma \in \Gamma$ such that

$$\max_{u \in \gamma([0,1])} \Phi(u) \leq c + \varepsilon. \tag{7.4}$$

The function $\gamma^*(t) = \eta(1, \gamma(t))$ clearly belongs to $\mathcal{C}([0, 1]; X)$. Moreover, $\gamma^*(0) = \eta(1, 0) = 0$ and $\gamma^*(1) = \eta(1, e) = e$ by (7.3) and also because $\max\{\gamma(0), \gamma(e)\} = \alpha < c - 2\varepsilon$; then, γ^* belongs to Γ, whereas, by (7.4), the set $\gamma([0, 1]) \subset \Phi^{c+\varepsilon}$; So that, by (7.3) we get

$$\gamma^*([0, 1]) = \eta(1, \gamma([0, 1])) \subset \Phi^{c-\varepsilon};$$

that is,

$$\max_{u \in \gamma^*([0,1])} \Phi(u) \leq c - \varepsilon, \tag{7.5}$$

Which is a contradiction because $\gamma^* \in \Gamma$. □

Remark 7.2. The aforementioned form of the MPT and its proof contain the ingredients used in the bulk of minimax theorems that use the compactness condition (PS). Their proofs are very similar to this one, as can be checked, for example, from [628, 748, 882] and the references appearing therein. The MPT is their (young) grandfather to all.

Remark 7.3. Notice that only the condition (PS)$_c$ is needed as it was pointed out by Brézis, Coron, and Nirenberg in [155], the paper where the condition (PS)$_c$ was introduced for the first time. The former proof is the original one; it is attributed to Ambrosetti and Rabinowitz.

Now we will give another *nice* proof which, instead of using the deformation lemma, uses Ekeland's variational principle.

A second proof with Ekeland's variational principle. Consider the set Γ, defined earlier, as a normed space for the uniform topology generated by the norm

$$\|\gamma\|_\Gamma = \max_{t \in [0,1]} |\gamma(t)| \qquad \text{for } \gamma \in \Gamma.$$

It is known (we can easily check) that $(\Gamma, \|.\|_\Gamma)$ is a Banach space. Consider the functional $\Psi : \Gamma \to \mathbb{R}$ defined by

$$\Psi(\gamma) = \max_{t \in [0,1]} \Phi(\gamma(t)).$$

Then, Ψ is lower semicontinuous (l.s.c.) as the upper bound of a family of (lower semi-) continuous functions.

We have that

$$c = \inf_\Gamma \Psi \geq \max \{\Phi(0), \Phi(e)\}.$$

Thus, Φ is bounded below. So, we are in the context of the variational principle. Then, for every $\varepsilon > 0$, there exists a path $\gamma_\varepsilon \in \Gamma$ such that

$$\begin{cases} \Psi(\gamma_\varepsilon) \leq c + \varepsilon, \text{ and} \\ \Psi(\gamma) \geq \Psi(\gamma_\varepsilon) - \varepsilon \|\gamma - \gamma_\varepsilon\|_\Gamma \qquad \text{for all } \gamma \in \Gamma. \end{cases} \tag{7.6}$$

Set

$$M(\varepsilon) = \left\{ t \in [0,1]; \ \Phi(\gamma_\varepsilon(t)) = \max_{s \in [0,1]} \Phi(\gamma_\varepsilon(s)) \right\}.$$

Using (7.6), we can deduce that there exists $t_\varepsilon \in M(\varepsilon)$ such that

$$\|\Phi'(\gamma_\varepsilon(t_\varepsilon))\| \leq \varepsilon.$$

It suffices then to consider the (PS) sequence $x_n = \gamma_{1/n}(t_{1/n})$ and to use the $(PS)_c$ condition to conclude.   □

Remark 7.4. So, without assuming the *compactness* condition of $(PS)_c$, we were able to prove the existence of a (PS) sequence.

> The geometric condition (7.1) suffices to get a sequence of almost critical points as close to c as we want.

This is also the case when using the elegant quantitative form of the deformation lemma of Willem seen in Chapter 4. Nevertheless, it is important to note that the geometric condition (7.1) does not suffice to guarantee that c is a critical value of Φ, even when X is finite dimensional, as we can see in two examples attributed, respectively, to Nirenberg [683] and to Brézis and Nirenberg [153] given in Chapter 10.

Using Ekeland's approach, we can even get additional information.

Proposition 7.2 (Taubes [359]). *For any neighborhood N of $\mathbb{K}_c$ and any $\varepsilon > 0$, there exists $\varepsilon_0 \in]0, \varepsilon[$ and $\gamma \in \Gamma$ such that*

$$\Phi(\gamma(t)) > d - \varepsilon_0 \qquad \textit{implies that} \qquad \gamma(t) \in N.$$

As Φ' is small in N, we can interpret the proposition as follows.

When $\Phi(\gamma(t))$ is close to c, Φ is *flat*.

In the next chapter, we will present, among other results, some of the *immediate* consequences and variants of the MPT.

We present now two *typical* applications of the MPT to partial differential equations. These show clearly the way it is used in the treatment of the numerous applications to boundary value problems you may find in the literature.

7.3 A Superlinear Problem

The reader who is not familiarized with applications of variational methods to boundary value problems is strongly encouraged to get a look at Appendix A for the necessary material on Nemytskii operators, Sobolev spaces, and so forth.

Consider the semilinear elliptic Dirichlet problem[2]:

$$(\mathcal{P}) \quad \begin{cases} -\Delta u(x) = f(x, u(x)) & \text{in } \Omega, \\ u(x) = 0 & \text{on } \partial\Omega, \end{cases}$$

where Ω is a bounded smooth domain of $\mathbb{R}^N$. The function $f: \mathbb{R} \to \mathbb{R}$ is supposed to be a Carathéodory function satisfying the growth condition

$$|f(x, s)| \le a(x) + b|s|^{p-1}, \tag{7.7}$$

where $a(x) \in L^{p'}(\Omega)$ with $\frac{1}{p} + \frac{1}{p'} = 1$ and $1 \le p \le \frac{2N}{N-2}$ if $N \ge 3$, and $1 \le p < \infty$ if $N = 2$.

So the energy functional Φ associated to $(\mathcal{P})$, defined by

$$\Phi(u) = \frac{1}{2} \int_\Omega |\nabla u(x)|^2 \, dx - \int_\Omega F(x, u(x)) \, dx,$$

is well defined on $H_0^1(\Omega)$, is of class $\mathcal{C}^1$, and its critical points are weak solutions of $(\mathcal{P})$ (cf. Appendix A).

Applications based on the MPT require conditions on the behavior of f at 0 and $|s| = \infty$.

Theorem 7.3 (Ambrosetti-Rabinowitz). *Suppose that f satisfies:*

f1. $f(x, s) = o(|s|)$ at $s = 0$ uniformly in $x \in \overline{\Omega}$.
f2. There are constants $\mu > 2$ and $r > 0$ such that for $|s| \ge r$,

$$0 < \mu F(x, s) \le s.f(x, s).$$

Then, $(\mathcal{P})$ possesses a nontrivial solution.

Remark 7.5.
   1. Notice that, by $f1$, $u \equiv 0$ is a trivial solution of $(\mathcal{P})$.

---

[2] The operator $-\Delta$ in $(\mathcal{P})$ can be replaced by a more general second-order uniformly elliptic operator. The choice of $-\Delta$ has been made for simplicity.

2. After integrating $f2$, we get constants $a, b > 0$ such that

$$F(x, s) \geq a|s|^{\mu} - b \qquad (7.8)$$

for almost every $x \in \Omega$ and all $s \in \mathbb{R}$. Hence, since $\mu > 2$, the potential $F(x, s)$ has a *superquadratic* growth rate in s and $f(x, s)$ has a *superlinear* growth in s.

Proof. The proof will be done in two steps. We first show that Φ has the right geometry and then that it satisfies the (PS) condition.

1. The Geometry of the MPT

Clearly $\Phi(0) = 0$. And by (7.8), for all $u \in H_0^1(\Omega)$ we have

$$\int_{\Omega} F(x, u(x)) \, dx \geq a \int_{\Omega} |u(x)|^{\mu} \, dx - b \cdot \text{meas}\,(\Omega). \qquad (7.9)$$

Then, for u non-null, equation (7.9) gives that

$$\Phi(tu) = \frac{t^2}{2} \int_{\Omega} |\nabla u|^2 \, dx - \int_{\Omega} F(x, tu(x)) \, dx,$$

$$\leq \frac{t^2}{2} \|u(x)\|_{H_0^1(\Omega)}^2 - t^{\mu} a \int_{\Omega} |u(x)|^{\mu} \, dx + b \cdot \text{meas}\,(\Omega).$$

Hence,

$$\Phi(tu) \to -\infty \text{ as } t \to \infty.$$

To complete the proof of the geometric condition, it suffices to prove that 0 is a local minimum of Φ.

By $f1$, for any $\varepsilon > 0$ there exists $\delta > 0$ such that for almost every $x \in \Omega$,

$$|F(x, s)| \leq \frac{1}{2} \varepsilon |s|^2 \qquad \text{for all } s \in B(0, \delta).$$

And by the growth condition, there is a constant $A = A(\delta) > 0$ such that for almost every $x \in \Omega$

$$|F(x, s)| \leq A|s|^p \qquad \text{for each } s \in \mathbb{R}^N \setminus B(0, \delta).$$

Then, combining these two estimates, we have for almost every $x \in \Omega$,

$$|F(x, s)| \leq \frac{1}{2} \varepsilon |s|^2 + A|s|^p \qquad \text{for every } s \in \mathbb{R}.$$

Using the Poincaré inequality and the Sobolev embedding theorem, it follows that

$$\left| \int_{\Omega} F(x, u(x)) \, dx \right| \leq \int_{\Omega} |F(x, u(x))| \, dx,$$

$$\leq \frac{\varepsilon}{2} \int_{\Omega} |u(x)|^2 \, dx + A \int_{\Omega} |u(x)|^p \, dx,$$

$$\leq C \cdot \left(\frac{\varepsilon}{2} + A\|u\|_{H_0^1(\Omega)}^{p-2} \right) \|u\|_{H_0^1(\Omega)}^2.$$

Thus, for $\|u\|_{H_0^1(\Omega)} = \rho$ sufficiently small,

$$\Phi(u) \geq \|u\|_{H_0^1(\Omega)}^2 - C\varepsilon \|u\|_{H_0^1(\Omega)}^2.$$

Then, the fact that $\varepsilon > 0$ was taken arbitrary implies that

$$\Phi(u) \geq \alpha > 0. \tag{7.10}$$

This proves that 0 is a local minimum of Φ.

2. The Palais-Smale Condition

Proving that (PS) holds in our case is made by proving that any (PS) sequence is bounded. Indeed, by the growth condition the functional $u(x) \mapsto g(x, u(x))$ takes bounded sets of $L^p(\Omega)$ into $L^{p'}(\Omega) \subset H^{-1}(\Omega)$. By the Rellich-Kondrachov theorem, since $p < \frac{2N}{N-2}$, $H_0^1(\Omega)$ embeds compactly into $L^p(\Omega)$.

Then the functional

$$K: \mathbf{H}_0^1(\Omega) \to L^p(\Omega) \to L^{p'}(\Omega) \to \mathbf{H}^{-1}(\Omega)$$
$$\mathbf{u}(\mathbf{x}) \mapsto u(x) \mapsto f(.,u(.)) \mapsto \mathbf{f}(.,\mathbf{u}(.))$$

is compact.

And since $\Phi'(u(x)) = -\Delta u(x) - f(x, u(x))$, we get by Proposition 2.2 that it is enough to show that any (PS) sequence of Φ is bounded in $H_0^1(\Omega)$ to conclude that (PS) holds for Φ.

Let $(u_n)_n$ be a (PS) sequence; that is,

$$\begin{cases} |\Phi(u_n)| \leq M, \\ \Phi'(u_n) \to 0 \qquad \text{as } n \to \infty. \end{cases}$$

Then, it follows that

$$\mu \Phi(u_n) - \langle \Phi'(u_n), u_n \rangle \leq M\mu + \|u_n\|_{H_0^1(\Omega)}. \tag{7.11}$$

But, if we denote $T_n = f(x, u_n(x))u_n(x) - \mu F(x, u_n(x))$, we have that

$$\mu \Phi(u_n) - \langle \Phi'(u_n), u_n \rangle$$
$$= \frac{\mu - 2}{2} \|u_n\|_{H_0^1(\Omega)} + \int_\Omega T_n \, dx \geq \frac{\mu - 2}{2} \|u_n\|_{H_0^1(\Omega)}$$
$$+ \underbrace{\int_{\{x \in \Omega; \, |u_n(x)| < r\}} T_n \, dx}_{\text{Term 1}} + \underbrace{\int_{\{x \in \Omega; \, |u_n(x)| \geq r\}} T_n \, dx}_{\text{Term 2}}$$

where Term 2 is positive and Term 1 is bounded by a constant not depending on n. Then, by (7.11) and the fact that $\mu > 2$, the sequence $(u_n)_n$ is bounded in $H_0^1(\Omega)$.

As remarked before, 0 is a critical point of Φ and $\Phi(0) = 0$, while for the critical point u obtained with the MPT, we have by (7.10) that

$$\Phi(u) \geq \alpha > 0.$$

Thus, u is a nontrivial weak solution of Φ. $\square$

Remark 7.6. The condition $f2$ was used to get (PS), whereas only the condition

$$F(x, s) \geq C|s|^{\mu} + C'$$

is needed to obtain the geometry of the MPT.

If instead of requiring $f: \Omega \times \mathbb{R} \to \mathbb{R}$ to be a Carathéodory function, we imposed that f is locally Lipschitz continuous in $\overline{\Omega} \times \mathbb{R}$ and satisfies the standard growth condition (7.7). Then, by the *regularity theory*, any weak solution of $(\mathcal{P})$ would also have been a classical solution.[3] See [5] for more details. Moreover, it has been proved using the maximum principle that $(\mathcal{P})$ has both *a positive and a negative classical solution*. In fact, a third sign-changing solution also exists [96, 187, 949]. When f is odd in u, $(\mathcal{P})$ possesses infinitely many solutions.

7.4 A Problem of Ambrosetti-Prodi Type

Consider the semilinear elliptic Dirichlet problem

$$(\mathcal{P}_{AP}) \quad \begin{cases} -\Delta u = f(u(x)) + g(x) & \text{in } \Omega \\ u = 0 & \text{on } \partial\Omega \end{cases}$$

where Ω is some bounded smooth domain in $\mathbb{R}^N$, g is a given function in $\mathcal{C}^{0,\alpha}(\overline{\Omega})$ and $f: \mathbb{R} \to \mathbb{R}$ is a function of class $\mathcal{C}^2$ such that

i. $f(0) = 0$,
ii. $f''(t) > 0$ for all t, and

$$\lim_{t \to \infty} f'(t) < \lambda_1 < \lim_{t \to +\infty} f'(t).$$

In 1972, Ambrosetti and Prodi [48] proved that there is a connected $\mathcal{C}^1$-manifold M in $\mathcal{C}^{0,\alpha}(\overline{\Omega})$ that disconnects $\mathcal{C}^{0,\alpha}(\overline{\Omega})$ into exactly two open connected sets $\mathcal{O}_1$ and $\mathcal{O}_2$ such that

- if $g \in \mathcal{O}_1$, then $(\mathcal{P}_{AP})$ has no solution,
- if $g \in M$, then $(\mathcal{P}_{AP})$ has a unique solution, and
- if $g \in \mathcal{O}_2$, then $(\mathcal{P}_{AP})$ has exactly two solutions.

Ambrosetti and Prodi were looking for (and obtained) their solutions in $\mathcal{C}_0^{2,\alpha}(\overline{\Omega})$, the set of $\mathcal{C}^{2,\alpha}$-functions vanishing on the boundary of Ω. This is natural in view of the regularity imposed on f, g and Ω.

---

[3] This holds also for the solutions that can be obtained by the direct method of the calculus of variations.

Consider the normalized eigenfunction φ_1 associated to the first eigenvalue λ_1 of $-\Delta$ with Dirichlet boundary data in $H_0^1(\Omega)$ with positive sign everywhere:

$$\varphi_1 > 0 \quad \text{and} \quad \int_\Omega \varphi_1^2(x)\, dx = 1.$$

Then, any function $g \in C^{0,\alpha}(\overline{\Omega})$ can be written uniquely as $g = t.\varphi_1 + h$, where $t \in \mathbb{R}$ and $h \in \varphi_1^\perp = \{u \in C^{0,\alpha}(\overline{\Omega}); \int_\Omega u\varphi = 0\}$.

Consider now the parameterized family of Dirichlet problems

$$(\mathcal{P}_t) \quad \begin{cases} -\Delta u(x) = f(x, u(x)) + t\varphi_1 + h & \text{in } \Omega, \\ u = 0 & \text{on } \partial\Omega. \end{cases}$$

where f is locally Lipschitz continuous in $\overline{\Omega} \times \mathbb{R}$.

Suppose that

$$\limsup_{s \to -\infty} \frac{f(x,s)}{s} < \lambda_1 < \liminf_{s \to +\infty} \frac{f(x,s)}{s}. \tag{7.12}$$

As far as only existence is concerned, the conditions on f are somewhat stringent. What matters seems to be the crossing of the first eigenvalue λ_1, as expressed in (7.12).

Indeed, we can prove the following result using the MPT.

Theorem 7.4. *If* (7.12) *holds true, the functional f satisfies the classical growth condition* (7.7) *and*

$$0 < \mu.F(x, s) \le s.f(x, s), \qquad \forall x \in \Omega, \ \forall s \in \mathbb{R}. \tag{7.13}$$

Then, there exists $t_0 \in \mathbb{R}$ such that, for all $t \le t_0$, the problem $(\mathcal{P}_t)$ has at least two solutions in $C^{2,\alpha}(\overline{\Omega})$.

First, let us prove the following technical result.

Lemma 7.5. *Assume that* (7.12), (7.13), *and the growth condition* (7.7) *hold. Then, the functional*

$$\Phi_t(u) = \frac{1}{2}\int_\Omega \nabla u(x)^2\, dx - \int_\Omega F(x, u(x))\, dx - \int_\Omega (t\varphi_1(x) + h(x)).u(x)\, dx$$

satisfies the (PS) *condition.*

Proof. Fix $t_0 \in \mathbb{R}$ and write Φ for Φ_{t_0} for simplicity.

Let $(u_n)_n$ be a (PS) sequence in $H_0^1(\Omega)$; then there is some constant C such that for all n,

$$|\Phi(u_n)| = \left| \frac{1}{2}\int_\Omega \nabla u_n^2\, dx - \int_\Omega F(x, u_n)\, dx - \int_\Omega (t\varphi_1 + h).u_n\, dx \right| \le C. \tag{7.14}$$

And for all $N \in \mathbb{N}$, there exists $p(N) \in \mathbb{N}$ such that if $n \ge p(N)$, we have for all $v \in H_0^1(\Omega)$

$$|\langle \Phi'(u_n), v\rangle| \le \frac{1}{N}\|v\|, \tag{7.15}$$

where

$$|\langle \Phi'(u_n), v \rangle| = \left| \int_\Omega \nabla u_n . \nabla v \, dx - \int_\Omega f(x, u_n).v \, dx - \int_\Omega (t\varphi_1 + h).v \, dx \right|.$$

Since Φ' has the particular form $I_d - K'$ where K is compact, it suffices, by Theorem 2.2, to show that $(u_n)_n$ is bounded to prove that Φ satisfies (PS).

Consider a (PS) sequence $(u_n)_n$ in $H_0^1(\Omega)$. Notice first that $v \in W^{k,p}$ implies that its positive and negative parts v^+, v^- are also in $W^{k,p}$. We will prove that both $\|u_n^-\|_{H_0^1}$ and $\|u_n^+\|_{H_0^1}$ are bounded and, hence, $\|u_n\|_{H_0^1(\Omega)}$ is bounded.

Using the first inequality in (7.12), we can show easily that there exists $0 < \mu < \lambda_1$ and a constant C such that

$$f(x, s) > \mu s - C \qquad \text{for } s \leq 0. \tag{7.16}$$

Replacing v by u_n^- in (7.15), we obtain that

$$\int_\Omega |\nabla u_n^-|^2 \leq - \int_\Omega f(x, u_n).u_n^- + \varepsilon_n \|u_n^-\|_{H_0^1(\Omega)}.$$

Then, by (7.16), we have

$$\int_\Omega |\nabla u_n^-|^2 \leq \mu \int_\Omega (u_n^-)^2 \, dx + \int_\Omega u_n^- \, dx + \varepsilon_n \|u_n^-\|_{H_0^1(\Omega)}.$$

And by Poincaré and Schwarz inequalities,

$$\|u_n^-\|_{H_0^1(\Omega)} \leq M \qquad \text{for some constant } M > 0, \tag{7.17}$$

to prove that $\|u_n^+\|_{H_0^1(\Omega)}$ is bounded. By the first inequality in (7.12), there are constants $0 < \mu < \lambda_1$ and $C > 0$ such that for $x \in \Omega$, we have

$$F(x, s) \leq \frac{\mu}{2}s^2 - C.s \qquad \text{for } s \leq 0.$$

Then,

$$\int_\Omega F(x, -u_n^-) \leq \frac{\mu}{2} \int_\Omega (u_n^-)^2 - \int_\Omega u_n^- \leq \text{constant}.$$

So, from (7.14), we obtain that

$$\frac{1}{2} \int_\Omega |\nabla u_n^+|^2 - \int_\Omega F(x, u_n^+) \leq \text{constant}. \tag{7.18}$$

From the equation (7.15) with $v = v_n^+$, we get that

$$\left| \frac{1}{2} \int_\Omega |\nabla u_n^+|^2 - \int_\Omega F(x, u_n^+) \right| \leq \varepsilon_n . \|u_n^+\|_{H_0^1}. \tag{7.19}$$

Multiplying (7.18) by μ and subtracting (7.19), we get

$$\left(\frac{\mu}{2} - 1\right)\frac{1}{2} \int_\Omega |\nabla u_n^+|^2 \leq \int_\Omega \left[\mu F(x, u_n^+) - f(x, u_n^+)u_n^+ \right] dx$$

$$+ \varepsilon_n . \|u_n^+\|_{H_0^1(\Omega)} + \text{constant}.$$

Finally, we conclude using (7.13). □

The proof of the theorem consists in showing that, assuming (7.12),

1. For each $t \in \mathbb{R}$, the problem $(\mathcal{P}_t)$ has a classical subsolution $\underline{u_t}$ such that, for any classical supersolution $\overline{u_t}$ of $(\mathcal{P}_t)$,

$$\underline{u_t} < \overline{u_t}, \qquad \forall x \in \Omega,$$

and

$$\frac{\partial \underline{u_t}(x)}{\partial \nu} > \frac{\partial \overline{u_t}(x)}{\partial \nu}, \qquad \forall x \in \partial\Omega.$$

2. There exists $t_0 \in \mathbb{R}$ such that, for all $t \le t_0$, $(\mathcal{P}_t)$ has a classical supersolution $\overline{u_t}$.

3. Set $C = \{u \in H_0^1; \ \underline{u_t} \le u \le \overline{u_t}\}$. Then, C is a closed convex subset of H_0^1, and the rest of the proof is carried in three steps related to the set C.
   a. The functional Φ restricted to C has a minimum u_0 in C that is a critical point of Φ.
   b. The local minimum u_0 of Φ in C is indeed a local minimum of Φ in H_0^1.
   c. And of course the last step is to obtain a second solution using the MPT.

By a *classical subsolution* $\underline{u}$, we mean a $C^{2,\alpha}(\Omega)$-functional that satisfies

$$-\Delta\underline{u} - f(x, \underline{u}) \le t\varphi_1 + h \ \text{in} \ \Omega, \qquad \underline{u} = 0 \ \text{on} \ \partial\Omega.$$

A *supersolution* is defined in a similar way.

Using the MPT, we can also prove the following result concerning the Ambrosetti-Prodi problem.

Theorem 7.6. *If* (7.12) *holds and* f *has a* linear growth,

$$|f(x, s)| \le a|s| + b, \qquad \forall x \in \overline{\Omega}, \ \forall s \in \mathbb{R}, \tag{7.20}$$

where $a, b > 0$, *then there exists* $t_0 \in \mathbb{R}$ *such that for all* $t \le t_0$ *the problem* $(\mathcal{P}_t)$ *has at least two solutions in* $C^{2,\alpha}(\overline{\Omega})$.

Indeed, we can prove that under the assumptions (7.12) and (7.20), the functional

$$\Phi_t(u) = \frac{1}{2}\int_\Omega \nabla u(x)^2 \, dx - \int_\Omega F(x, u(x)) \, dx - \int_\Omega (t\varphi_1(x) + h(x)).u(x) \, dx$$

satisfies the (PS) condition.

The proof of the fact that Φ satisfies (PS) is left to the reader. The proof of the theorem follows the same scheme just described. The results presented here are attributed to De Figueiredo and Solimini [302].

The difference between the proofs of the superlinear problem and the Ambrosetti-Prodi problem is mainly that no obvious local minimum, that is, 0, is available and we have to find one. The Ambrosetti-Prodi problem is famous and has been investigated by many authors.

The last two sections were given with some details to show the way the MPT can be applied to solve semilinear elliptic problems.

Many spectacular applications of the MPT exist in the literature. They are behind the mere existence of this book. Prospect the bibliography and you will not be deceived!

Comments and Additional Notes

The (classical) MPT, if we exclude the finite dimensional MPT, is the first of all the results exposed here. The first papers all used Clark's version of the deformation lemma (modulo some changes occasionally) that require (PS). Ekeland's principle permitted, however, to give more elegant proofs and the strongest results requiring less regularity on the functional and without supposing the (PS) condition a priori. Of course, what we get then is only a (PS) sequence (of almost critical points) as near as we want to $c = \inf \max \Phi$. The same is now possible using the quantitative deformation lemma.

◇ 7.1 A Third Proof of the MPT

Recently, J.P. Aubin and I. Ekeland have found a very elegant proof of the Mountain Pass lemma.

Haïm Brézis, [147, p. 182].

We report a third *nice proof* of the MPT from de Figueiredo [296], who cites Brézis [146]. The critical point theorem is stated in a general form that will be met many times in the sequel. The proof uses some elements of *convex analysis* and *measure theory*.

Let X be a Banach space and $\Phi \colon X \to \mathbb{R}$ a C^1-functional. Let K be a compact metric space and $K_0 \subset K$ a closed subset. Consider the space $E = \mathcal{C}(K; \mathbb{R})$ endowed with the norm of the uniform convergence, that is, the supremum norm

$$\|f\|_\infty = \max \{ f(s); \ s \in K \}.$$

By the Riesz representation theorem, the dual E^* of E is isometric isomorphic to the Banach space $\mathcal{M}(K)$ of all regular countably additive real-valued functions, called *Radon measures*, defined in the σ-algebra of all Borel sets in K, endowed with the *norm of the total variation*:

$$\|\mu\| = \sup \left\{ \sum_{i=1}^{k} |\mu(E_i)|; \ \cup_{i=1}^{k} E_i \subset E, \ E_i \cap E_j = \varnothing, k = 1, 2, \ldots \right\}$$

Lemma 7.7. *We have that* $\| \cdot \|_\infty$ *is continuous and convex and the subdifferential*

$$\partial \|f\|_\infty = \left\{ \mu \in \mathcal{M}(K); \ \mu \ge 0, \int_K d\mu = 1, \operatorname{supp}(\mu) \subset \{t; \ f(t) = \|f\|_\infty\} \right\}$$

the set of positive *Radon measures with* mass one *and support in* $\{t; \ f(t) = \|f\|_\infty\}$.

The support is defined as follows. The measure *vanishes* in an open set U if $\int_U f\, d\mu = 0$ for any $f \in E$ with compact support in U. Using the partition of the unity, we can prove that if μ vanishes in a collection of open sets, it vanishes on their union. So, there exists a largest open set $\tilde{U}$ where μ vanishes. The support of μ is $\operatorname{supp}(\mu) = K \setminus \tilde{U}$.

Lemma 7.8. *Let* $\Phi\colon X \to \mathbb{R}$ *be convex and continuous. Then, for each* $x, y \in X$,

$$\lim_{t \searrow 0} \frac{\Phi(x + ty) - \Phi(x)}{t} = \max_{\mu \in \partial\Phi(x)} \langle \mu, y \rangle.$$

For the proof, see [296], for example.

Denote by

$$\Gamma = \left\{ \gamma \in C(K; X);\ \gamma = \gamma_0 \text{ on } K_0 \right\}$$

where $\gamma_0\colon K_0 \to X$ is a given continuous functional. Then, Γ is a complete metric space for the distance of the uniform convergence $\operatorname{dist}(\gamma_1, \gamma_2) = \|\gamma_1 - \gamma_2\|_\infty$.

Theorem 7.9. *Suppose that*

$$c = \inf_{\gamma \in \Gamma} \max_{s \in K} \Phi(\gamma(s)) > \max_{\gamma_0(K_0)} \Phi = b. \tag{$*$}$$

Then, for all $\varepsilon > 0$ *and all* $\gamma \in \Gamma$ *such that* $\max_{s \in K} \Phi(\gamma(s)) \le c + \varepsilon$, *there exists* $u_\varepsilon \in X$ *such that*

$$c - \varepsilon \le \Phi(u_\varepsilon) \le \max_{s \in K} \Phi(\gamma(s)),$$
$$\operatorname{dist}(u_\varepsilon, \gamma(K)) \le \sqrt{\varepsilon},$$
$$\|\Phi'(u_\varepsilon)\| \le \sqrt{\varepsilon}.$$

Proof. Define the functional $\Psi\colon \Gamma \to \mathbb{R}$ by $\Psi(\gamma') = \max_{s \in K} \Phi(\gamma'(s))$. The assumption $(*)$ means that $c = \inf_{\gamma' \in \Gamma} \Phi(\gamma'(s)) > b$. Moreover, Ψ is l.s.c. as the supremum of a family of (lower semi-) continuous functions. Assume that $0 < \varepsilon < c - b$ and consider $\gamma \in \Gamma$ such that

$$\max_{s \in K} \Phi(\gamma(s)) \le c + \varepsilon.$$

By Ekeland's principle, there exists $\gamma_\varepsilon \in \Gamma$ such that:

i. $\Psi(\gamma_\varepsilon) \le \Psi(\gamma) \le c + \varepsilon$,
ii. $\Psi(\gamma) \ge \Psi(\gamma_\varepsilon) - \sqrt{\varepsilon}\|\gamma - \gamma_\varepsilon\|_\infty,\ \forall \gamma \in \Gamma$,
iii. $\|\gamma - \gamma_\varepsilon\|_\infty \le \sqrt{\varepsilon}$.

Consider $\varphi\colon K \to X$ such that $\varphi(s) = 0$ for all $s \in K_0$. For any $r > 0$, we conclude by ii that

$$\frac{\Psi(\gamma_\varepsilon + r\varphi) - \Psi(\gamma_\varepsilon)}{|r|} \ge -\sqrt{\varepsilon}\|\gamma\|_\infty.$$

By definition of Ψ,

$$\Psi(\gamma_\varepsilon + r\varphi) - \Psi(\gamma_\varepsilon) = \max_K \Phi(\gamma_\varepsilon + r\varphi) - \max_K \Phi(\varphi_\varepsilon)$$
$$= \max_{s \in K} \left\{ \Phi(\gamma_\varepsilon(s)) + r\langle \Phi'(\gamma_\varepsilon(s)), \varphi(s) \rangle + o(r) \right\} - \max_K \Phi(\varphi_\varepsilon).$$

So, by Lemma 7.8, as r tends to 0, we get

$$-\sqrt{\varepsilon} \|\varphi\|_\infty \le \lim_{r \searrow 0} \frac{\Phi(\gamma_\varepsilon + r\varphi) - \Phi(\gamma_\varepsilon)}{r}$$
$$\le \max \left\{ \int_K \langle \Phi'(\gamma_\varepsilon(s)), \varphi(s) \rangle \, d\mu; \ \mu \in \partial \|\Phi(\gamma_\varepsilon)\|_\infty \right\},$$

and

$$\min \left\{ \int_K \langle \Phi'(\gamma_\varepsilon(s)), \varphi(s) \rangle \, d\mu; \ \mu \in \partial \|\Phi(\gamma_\varepsilon)\|_\infty \right\} \le \sqrt{\varepsilon}.$$

Denote

$$\Gamma_0 = \left\{ k \in \mathcal{C}(K; X); \ k \equiv 0 \text{ on } K_0, \|k\|_\infty \le 1 \right\}.$$

Divide by $\|\Phi(\gamma_\varepsilon)\|_\infty$ and take inf on Γ_0:

$$\sup_{k \in \Gamma_0} \min \left\{ \int_K \langle \Phi'(\gamma_\varepsilon(s)), k(s) \rangle \, d\mu; \ \mu \in \partial \|\Phi(\gamma_\varepsilon)\|_\infty \right\} \le \sqrt{\varepsilon}. \qquad (7.21)$$

By Von Neumann min-max theorem [101] applied to $\mathcal{G}: \mathcal{M}(\mathbb{R}) \times \mathcal{C}(K; X) \to \mathbb{R}$, ($\mathcal{M}(\mathbb{R})$ is endowed with the weak-$*$ topology) defined by $\mathcal{G}(\mu, k) = \int_K \langle \Phi'(\gamma_\varepsilon(s)), k(s) \rangle \, d\mu$, we can interchange sup and min in the preceding equation. The function $\mathcal{G}$ is continuous and linear in each variable separately and the sets $\{t \in K; \ k(t) = \max_{t \in K} \Phi(k(t))\}$ and $\{k \in \mathcal{C}(K; X); \ \|k\| \le 1\}$ are convex, the former one being weak-$*$ compact.

Denote $K_1 = \{t \in K; \ \Phi(\gamma_\varepsilon(t)) = \|\Phi(\gamma_\varepsilon)\|_\infty\}$. Then, K_1 is a compact set disjoint from K_0. So, there is $\varphi: K \to \mathbb{R}$ such that $\varphi(t) = 1$ if $t \in K_1$, $\varphi(t) = 0$ if $t \in K_0$, and $0 \le \varphi(t) \le 1$ for all $t \in K$. Then, given $k \in \mathcal{C}(K; X)$ with $\|k\| \le 1$, $k_1 = \varphi \cdot k \in \Gamma_0$, $\|k_1\| \le 1$ and

$$\int_K \langle \Phi'(\gamma_\varepsilon(s)), k(s) \rangle \, d\mu = \int_K \langle \Phi'(\gamma_\varepsilon(s)), k_1(s) \rangle \, d\mu$$

because supp $\mu \subset K_1$. So,

$$\sup_{k \in \Gamma_0} \int_K \langle \Phi'(\gamma_\varepsilon(s)), k(s) \rangle \, d\mu = \sup_{\substack{k \in \mathcal{C}(K;X) \\ \|k\| \le 1}} \int_K \langle \Phi'(\gamma_\varepsilon(s)), k(s) \rangle \, d\mu.$$

We have

$$\sup_{\substack{k \in \mathcal{C}(K;X) \\ \|k\| \le 1}} \int_K \langle \Phi'(\gamma_\varepsilon(s)), k(s) \rangle \, d\mu \le \int_K \langle \sup_{\substack{k \in \mathcal{C}(K;X) \\ \|k\| \le 1}} \Phi'(\gamma_\varepsilon(s)), k(s) \rangle \, d\mu = \|\Phi'(\gamma_\varepsilon)\|_{X'}.$$

The first inequality is true since it is taken on a uniformly bounded family of functions in $C(K; X)$ and $\mu > 0$. So, (7.21) with sup and min interchanged gives

$$\min_{\mu \in \partial \|\Phi'(\gamma_\varepsilon)\|_\infty} \int_K \|\Phi'(\gamma_\varepsilon)\| \, d\mu \leq \sqrt{\varepsilon}.$$

Let $\overline{\mu} \in \partial \|\Phi'(\gamma_\varepsilon)\|_\infty$ that realizes the minimum. Since it has mass one and is supported in K_1 by Lemma 7.7, there exists $\overline{s} \in K_1$ such that

$$\|\Phi'(\gamma_\varepsilon)\| \, d\mu \leq \sqrt{\varepsilon}.$$

Set $u_\varepsilon = \gamma_\varepsilon(\overline{s})$, then by i and iii,

a. $c \leq \Phi(u_\varepsilon) = \max_{s \in K} \Phi(\gamma_\varepsilon(s)) = \Psi(\gamma_\varepsilon) \leq \inf_\Gamma \Psi + \sqrt{\varepsilon} = c + \sqrt{\varepsilon}$,
b. $\|\Phi'(u_\varepsilon)\| \leq \sqrt{\varepsilon}$,
c. $\mathrm{dist}(u_\varepsilon, \gamma(K)) \leq \sqrt{\varepsilon}$.

$\square$

A *homological proof* (that relies on a homological approach theory) of the MPT may be consulted in Chang's book [205] (see Chapter 25).

◇ 7.II The Situation of a Path $\gamma \in \Gamma$ Achieving the inf max

For c being the inf max value in the MPT statement, we saw that by Ekeland's variational principle, for all $\varepsilon > 0$, there exists $\gamma = \gamma_\varepsilon \in \Gamma$ and $t \in [0, 1]$ such that

$$\begin{cases} d \leq \Phi(\gamma(t)) = \max_s \Phi(\gamma(s)) \leq d + \varepsilon, \\ \|\Phi'(\gamma(t))\| \leq \varepsilon. \end{cases}$$

An ideal situation would be to have some $\gamma \in \Gamma$ achieving the inf max as in the finite dimensional and the topological versions of the MPT; that is,

$$\max_{t \in [0,1]} \Phi(\gamma(t)) = c.$$

This is in "general hopeless," to quote Ekeland [359] (counterexamples exist). So, the following natural question comes to mind. *Do we get any supplementary information in such situations?*

As expected, the answer is *yes*. Indeed, we have the following additional information.

Theorem 7.10 (Shafrir [828]). *Assume that*

$$c = \inf_{\gamma \in \Gamma} \max_{t \in [0,1]} \Phi(\gamma(t)) > \max \{\gamma(x_0), \gamma(x_1)\}$$

and that the infimum is attained. Then, there exists $\gamma_1 \in \Gamma$ such that

$$\begin{cases} \max_{t \in [0,1]} \Phi(\gamma_1(t)) = c \qquad and \\ \Phi(\gamma_1(t)) = c \qquad implies \ that \qquad \Phi'(\gamma_1(t)) = 0. \end{cases}$$

This result was obtained by Shafrir [828] using his deformation lemma (Theorem iv in the Notes of Chapter 4). Compare to Proposition 7.2 of Taubes.

◇ *7.III Some Surveys on the MPT*

Many surveys on the MPT and its symmetric variants have been written during the past years in the framework of a treatment of minimax methods. We cite in particular the following ones by Ambrosetti, Rabinowitz, and Chang.

[744] by Rabinowitz. Detailed treatment of the MPT, index theory, and the symmetric MPT (see Chapter 11); the use of the MPT in bifurcation problems.

[748] by Rabinowitz. Monograph, one of the oldest surveys on modern minimax methods in critical point theory, discusses the MPT, linking in the sense of Benci-Rabinowitz (see Chapter 19), index theories, symmetric MPT (see Chapter 11), perturbation from symmetry, and the MPT in bifurcation problems (see Chapter 24).

[749] by Rabinowitz. Considers the MPT, symmetric versions ($\mathbb{Z}_2$ and S^1 symmetries), and perturbation.

[753] by Rabinowitz. Surveys the developments of critical point theory and its applications to differential equations in the past 20–25 years. The main results are stated, discussed, and related bibliographic references are also given.

[43] by Ambrosetti. Monograph, discusses Ljusternik-Schnirelman theory, the MPT, and the linking theorem. Many applications to differential equations (partial and ordinary) are also given. Particular attention is devoted to discontinuous nonlinearities.

[44] by Ambrosetti. Gives a short account of the Ljusternik-Schnirelman theory, reviews the MPT, its symmetric version and gives some applications.

[209] by Chang. A *homological approach* to minimax theorems is used to give a *homological proof* of the MPT with some information on the critical groups, which describes the local behavior near critical point. (See Chang's book [205] for more details.)

◇ *7.IV On Applications*

The applications we considered are simple enough. They have been given for illustration. Practically, discussing the different applications of the MPT is a huge work that would have enlarged this book considerably. A specific survey about this subject is in preparation [497].

8

The Multidimensional MPT

Fortunately, there are existence results for saddle taylor-made for applications. These are the famous (infinite-dimensional) mountain pass lemma and its variants, due to Ambrosetti and Rabinowitz.

> M. Struwe, *Variational methods, applications to nonlinear partial differential equations and Hamiltonian systems*, Springer-Verlag, 1990

In this chapter, we present some of the earliest variants of the MPT. We will focus our attention on a generalization by Rabinowitz [737], with a more general geometric condition that involves a splitting of the underlying space into two topologically supplementary spaces, one of them being finite dimensional. This extension of the MPT to higher dimensions was essentially destined to provide a more adapted tool than the MPT to treat particular kinds of semilinear elliptic equations.

The *multidimensional* MPT of Rabinowitz [737], which is the main result in this chapter, dates from 1978 and constitutes in some sense the beginning of the fame for the MPT. It allows us to treat semilinear elliptic partial differential equations where the nonlinearity contains a linear part at 0, as we will see later in Section 8.2, in contrast with the superlinear problem seen in Chapter 7 where the nonlinearity had the form $f(s) = o(|s|)$ in a neighborhood of 0.

Some early variants that contributed, with the multidimensional MPT, to give to the (classical) MPT a special status in the family of main results of nonlinear analysis are recalled briefly in the Notes and final comments at the end of the chapter.

8.1 The Multidimensional MPT

The novelty in this extension of the MPT, attributed to Rabinowitz [737], is a more general geometric condition with a splitting of the underlying space into two topologically supplementary spaces. One of them is finite dimensional, technically because of the use of the topological degree that is defined for general functions only in finite dimensions. The passage to a splitting involving infinite dimensional spaces requires that we restrict ourselves to functionals that are compact perturbations of the identity.

We begin first by stating the abstract result.

Theorem 8.1 (Multidimensional MPT, Rabinowitz). *Let X be a real Banach space with $X = X_1 \oplus X_2$, where X_1 is finite dimensional. Suppose that $\Phi \in C^1(X; \mathbb{R})$ satisfies* (PS) *and*

i. *there are constants $\rho, \alpha > 0$ such that $\Phi|_{\partial B_\rho \cap X_2} \geq \alpha$ and that*
ii. *there is $e \in \partial B_1 \cap X_2$ and $R > \rho$ such that $\Phi|_{\partial Q} \leq 0$, where*

$$Q = (\overline{B_R} \cap X_1) \oplus \{re; \ 0 < r < R\}.$$

Then, Φ possesses a critical value $c \geq \alpha$ characterized by

$$c = \inf_{\gamma \in \Gamma} \max_{u \in Q} \Phi(\gamma(u)),$$

where

$$\Gamma = \{\gamma \in C^1(\overline{Q}; X); \ \gamma = I_d \text{ on } \partial Q\}.$$

Remark 8.1.
i. The set ∂Q refers to the boundary of Q relative to $X_1 \oplus \text{span} \{e\}$; span $\{e\}$ refers to the space spanned by e.
ii. When $X_1 = \{0\}$, the space $X = X_1$ and if $\Phi(0) = 0$ we get the standard MPT (Theorem 7.1).

Proof. The proof is done in two steps. Suppose the following claim to be true.

Claim 8.1.

$$c \geq \alpha. \tag{8.1}$$

Then, by contradiction, if c was not a critical value, by the deformation lemma, for $\bar{\varepsilon} = \alpha/2$, there would be some $\varepsilon \in]0, \bar{\varepsilon}[$ and a deformation $\eta \in C([0, 1] \times X; X)$ such that

$$\eta(1, \Phi^{c+\varepsilon}) \subset \Phi^{c-\varepsilon}. \tag{8.2}$$

Choose $\gamma \in \Gamma$ such that

$$\max_{u \in Q} \Phi(\gamma(u)) \leq c + \varepsilon. \tag{8.3}$$

Since $\Phi(u) \leq 0$ on the boundary ∂Q and by our choice of $\bar{\varepsilon}$, we have that $\eta(1, \gamma(u)) = u$ for all $u \in \partial Q$, which means that the mapping $u \mapsto \eta(1, \gamma(u))$ is in Γ. But (8.3) and the definition of c imply that

$$c \leq \max_{u \in Q} \Phi(\eta(1, \gamma(u))) \leq c - \varepsilon,$$

a contradiction.

We have to prove (8.1). To do so, it suffices to establish the following *linking* property.

Claim 8.2. For any path $\gamma \in \Gamma$, we have

$$\gamma(Q) \cap \partial B_\rho \cap X_2 \neq \varnothing.$$

Supposing that Claim 8.2 is true. For $\gamma \in \Gamma$ and $w \in \gamma(Q) \cap \partial B_\rho \cap X_2$, we have by i that

$$\max_{u \in Q} \Phi(\gamma(u)) \geq \Phi(w) \geq \inf_{v \in \partial B_\rho \cap X} \Phi(v) \geq \alpha, \tag{8.4}$$

which, once combined with the definition of c, implies (8.1).

Therefore, we have only to verify Claim 8.2. This will be done using some arguments from the degree theory (see, for example, [402, 748]).

Denoting by P the projector of X onto X_1, the affirmation of the claim is equivalent to

$$\begin{cases} P\gamma(u) = 0, \\ \|(I_d - P)\gamma(u)\| = \rho, \end{cases} \tag{8.5}$$

for some $u \in Q$ (depending on γ). Write $u \in \overline{Q}$ as $u = v + re$, where $v \in \overline{B_R} \cap X_1$ and $0 \leq r \leq R$, and

$$\Phi(r, v) = \big(\|(I_d - P)\gamma(v + re)\|, P\gamma(v + re) \big).$$

Then, $\Phi \in \mathcal{C}(\mathbb{R} \times X_1; \mathbb{R} \times X_1)$. Since $\gamma|_{\partial Q} = I_d$, when $u \in \partial Q$ we get

$$\Phi(r, v) = \big(\|re\|, v \big) = (r, v).$$

This means that $\Phi \equiv I_d$ on ∂Q. In particular, $\Phi(r, v) \neq (\rho, 0)$ for $u \in \partial Q$, because by ii we have that $(\rho, 0) \in Q$. So, identifying $\mathbb{R} \times X$ with $\mathbb{R}^n$ for some n, we get that the Brouwer degree $d(\Phi, Q, (\rho, 0))$ is well defined. And then, by the homotopy property of the degree,

$$d(\Phi, Q, (\rho, 0)) = d(I_d, Q, (\rho, 0)) = 1.$$

So, by the existence property, there exists $u \in Q$ such that $\Phi(u) = (\rho, 0)$, i.e., (8.5). □

Remark 8.2.
- Claim 8.2 expresses the fact that ∂Q, the boundary of Q relative to $X_1 \oplus \text{span} \{e\}$, and the sphere $S_R = S(R; X_2)$ of X_2 with center 0 and radius R *link* in the sense that any hypersurface modeled on ∂Q intersects S_R.
- In applications, when the classical MPT is used (as we saw in Chapter 7), we have to show that the origin 0 is a local minimum; then we get another critical point different from the origin, while for the multidimensional MPT, it may be that the origin is a local minimum for $\Phi|_{X_2}$, the restriction of Φ to X_2, without being a local minimum of Φ (with respect to the whole space X).

This will be more clear in the next section.

8.2 Application

Consider the following semilinear problem:

$$(\mathcal{P}) \begin{cases} -\Delta u(x) = \lambda a(x) u + f(x, u(x)) & \text{in} \quad \Omega, \\ u(x) = 0 & \text{on} \quad \partial\Omega, \end{cases}$$

where Ω is a bounded smooth domain of $\mathbb{R}^N$. The function $f : \mathbb{R} \to \mathbb{R}$ is supposed to be a Carathéodory function that satisfies the growth condition

$$|f(x, s)| \leq a(x) + b|s|^{p-1}, \tag{8.6}$$

where $a(x) \in L^{p'}(\Omega)$, $\frac{1}{p} + \frac{1}{p'} = 1$, and $1 \leq p \leq \frac{2N}{N-2}$ if $N \geq 3$ and $1 \leq p < \infty$ if $N = 2$.

So, the energy functional associated to $(\mathcal{P})$,

$$\Phi(u) = \int_\Omega \left[\frac{1}{2} |\nabla u(x)|^2 - \frac{\lambda}{2} a(x) u^2(x) - F(x, u(x)) \right] dx,$$

is well defined on $H_0^1(\Omega)$, is of class $\mathcal{C}^1$, and its critical points are weak solutions of $(\mathcal{P})$. Suppose also that f satisfies the following:

$f1.$ $f(x, s) = o(|s|)$ at $s = 0$ uniformly in $x \in \overline{\Omega}$.
$f2.$ There are constants $\mu > 2$ and $r > 0$ such that for $|s| \geq r$

$$0 < \mu F(x, s) \leq s.f(x, s).$$

The situation here differs from the case treated by the classical MPT by the presence of a linear term.

If $\lambda < \lambda_1$ where λ_1 is the first eigenvalue of the problem

$$(\mathcal{EP}) \begin{cases} -\Delta u(x) = \lambda a(x).u & \text{in} \quad \Omega, \\ u(x) = 0 & \text{on} \quad \partial\Omega, \end{cases}$$

then the functional

$$\left[\int_\Omega \left(|\nabla u|^2 - \lambda a u^2 \right) dx \right]^{1/2}$$

can be taken as a norm on the space $X = H_0^1(\Omega)$, and the energy Φ associated to the problem has the right geometry of the (classical) MPT.

Notice that if a is assumed to be positive and Lipschitz continuous in $\overline{\Omega}$, $(\mathcal{EP})$ possesses a sequence of eigenvalues $(\lambda_j)_j$ such that

$$0 < \lambda_1 < \lambda_2 \leq \lambda_3 \leq \cdots \leq \lambda_j \leq \cdots \to \infty \text{ as } j \to \infty.$$

However, if $\lambda > \lambda_1$, this is no longer the case and we can no longer apply the MPT. Nevertheless, we can use the multidimensional MPT.

So, for any λ, the following result holds.

Theorem 8.2. *Suppose that f satisfies the above assumptions $f1$, $f2$ and also*

$f3$. $sf(x, s) > 0$ for all $s \in \mathbb{R}$.

Then, for any $\lambda \in \mathbb{R}$, the problem $(\mathcal{P})$ admits a nontrivial weak solution.

Proof. If $\lambda < \lambda_1$, the result follows by applying the (classical) MPT as explained above.

So, suppose that $\lambda > \lambda_1$, say $\lambda \in [\lambda_k, \lambda_{k+1}]$. As remarked before, the functional Φ is of class C^1. Set $X_1 = \mathrm{span}\,\{v_1, \ldots, v_k\}$ where v_j is the jth eigenvalue of $(\mathcal{EP})$ associated to λ_j and set $X_2 = (X_1)^{\perp}$.

For $u \in X_2$, we have

$$\int_{\Omega} (|\nabla u(x)|^2 - \lambda a u^2)\,dx \geq \left(1 - \frac{\lambda}{\lambda_{k+1}}\right) ||u(x)||^2.$$

As in the treatment of the superlinear semilinear problem by the classical MPT, by $f1$, it holds that

$$\int_{\Omega} F(x, u)\,dx = o(||u(x)||^2) \text{ as } u \to 0.$$

Hence, condition i of Theorem 8.1 is satisfied. To verify ii, it suffices to show that

a. the restriction $\Phi|_{X_1} \leq 0$ and that
b. there is some $e \in \partial B_1 \cap X_2$ and $\overline{R} > \rho$ such that $\Phi(u) \leq 0$ for $u \in X_1 \oplus \mathrm{span}\,\{e\}$ and $||u(x)|| \geq \overline{R}$.

The assumption a follows from $f3$, and the assumption b follows by setting $e = v_{k+1}$ and noting that the argument used to show that $\Phi(tu) \to -\infty$ as $t \to \infty$ in the part (geometry of the MPT) in the proof of Theorem 7.3 holds *uniformly* for finite dimensional subspaces of X.

It remains only to show that Φ satisfies (PS) by proving that any (PS) sequence is bounded to prove that. This follows using $f2$ in the standard way seen in the proof of Theorem 7.3. $\square$

The multidimensional MPT was *improved* by Benci and Rabinowitz in [113] where the geometric condition was sharpened and the *finite dimension* condition removed for Hilbert spaces. This latter result was also behind some more general minimax principles that unified the approach of a great number of minimax theorems requiring the (PS) condition extending the MPT (see Chapter 19).

Comments and Additional Notes

◇ 8.1 Some Extensions of the Multidimensional MPT

The multidimensional MPT is important for two reasons. It permits the treatment of more applications and it makes the existence of some *linking* more clear in the geometric conditions. This was exploited by Benci and Rabinowitz, who formulated explicitly a notion of linking that permits use of the same language to describe the geometric conditions of the (classical) MPT, the multidimensional MPT, and the saddle point theorem of Rabinowitz. In [113], the finite dimension assumption that appears in the statement of the classical MPT is not required. This is discussed in detail in Chapter 19. Going a little further, Ding showed in [333], using a modified version of the deformation lemma, that the limiting case also holds true for the theorem of Benci and Rabinowitz. (We will see in the next chapter what the limiting case means.)

◇ 8.II Different Minimaxing Sets

We will consider other sets that proved to be appropriate candidates to play the role of the minimaxing set Γ in the MPT.

The set Γ of continuous paths joining 0 and e, appearing in the MPT, has a particular interest but is by no means privileged. Indeed, as remarked by Rabinowitz in [748], we could *minimax* Φ over the sets

$$\Gamma_0 = \left\{ K = \gamma([0, 1]); \ \gamma \in \mathcal{C}([0, 1]; X), \ \gamma \text{ is one-to-one}, \ \gamma(0) = 0 \text{ and } \gamma(1) = e \right\},$$

$$(8.7)$$

or as for the finite dimensional and topological versions of the MPT seen before, on

$$\Gamma_1 = \left\{ K \subset X; \ K \text{ is compact, connected, and } 0, e \in K \right\},$$

or even on

$$\Gamma_2 = \left\{ K \subset X; \ K \text{ is closed, connected, and } 0, e \in K \right\}.$$

Set, for $i = 0, 1$, and 2,

$$c_i = \inf_{K \in \Gamma_i} \sup_{u \in K} \Phi(u). \qquad (8.8)$$

Since the following inclusions hold,

$$\Gamma_0 \subset \left\{ \gamma([0, 1]); \ \gamma \in \Gamma \right\} \subset \Gamma_1 \subset \Gamma_2,$$

we conclude that, for the c_i's appearing in (8.8), the inequality

$$c_2 \leq c_1 \leq c \leq c_0$$

holds. And using the fact that $\eta(1, .)$ is a homeomorphism of X onto X, where η is the deformation that appears in the deformation lemma, we see that the proof of the standard MPT (Theorem 7.1) can also be used to show that the c_i's for $i = 0, 1, 2$ are also critical values of Φ.

In fact, $c_1 = c = c_0$ because, if $K \in \Gamma_1$, for all $\varepsilon > 0$ there is $\gamma([0, 1])$ in Γ_0 such that K is a uniform ε-neighborhood of $\gamma([0, 1])$.

The first equality shows us the similitude with the finite dimensional and topological versions seen before. But as reported by Rabinowitz [748, p. 19], "we do not know if c_2 can differ from this common value."

◇ *8.III The Dual MPT*

We will now see a version of the MPT where c is not a "inf sup" value but it is rather a "sup inf" value. We expose what was done on dual families by Ambrosetti and Rabinowitz in [50]. Fang and Ghoussoub arrived to formulate a consistent critical point theory only from a study of duality and perturbation principles. A detailed approach to the duality theory and the MPT can be found in the book by Ghoussoub [425].

The critical values of Φ were obtained in the results seen until now as

$$\inf_{K \in \Gamma} \sup_{u \in K} \Phi(u)$$

over some set Γ. But in the situation of the MPT, critical values can also be characterized in a *dual* sense to those already seen. Let

$$W = \{B \subset X;\ B \text{ is open},\ 0 \in B \text{ and } e \notin \overline{B}\}.$$

The "duality" meant there is expressed by the fact that

$$B \cap K \neq \varnothing \text{ for all } K \in \Gamma_0, \Gamma_1 \text{ and all } B \in W.$$

Setting

$$W_1 = \{h(S(0, \rho));$$
$$h \colon X \to X \text{ is a homeomorphism and } h(0) = 0, h(1) = e\}, \quad (8.9)$$

and

$$W_2 = \{K \in X;\ K \cap \gamma([0, 1]) \neq \varnothing \text{ for all } \gamma \in \Gamma\},$$

we have the relation

$$W_2 \supset \{\partial B;\ B \in W\} \supset W_1.$$

Then, for

$$\begin{cases} b_i = \sup_{K \in W_i} \inf_{u \in K} \Phi(u) & i = 1, 2, \\ b = \sup_{K \in W} \inf_{u \in K} \Phi(u), \end{cases}$$

we have

$$c \geq b_2 \geq b \geq b_1.$$

There too, it is not difficult to show that b, b_1, and b_2 are critical values using the following more appropriate form of the deformation lemma (the interested reader may consult [748] for the details).

Theorem 8.3 ("Ascending Deformation" Lemma). *Let $c \in \mathbb{R}$ and consider $\Phi \colon X \to \mathbb{R}$ a C^1-functional satisfying* (PS)$_c$. *If c is a regular value of Φ, then for every ε*

sufficiently small, there exists $\eta \in C([0, 1] \times X; X)$ such that

> i. $\eta(0, u) = u$ *for all u in X,*
> ii. $\eta(t, u) = u$ *for all* $u \notin \Phi^{-1}([c - 2\varepsilon, c + 2\varepsilon])$ *and all* $t \in [0, 1]$,
> iii. $\eta(t, .)$ *is a homeomorphism of X onto X for every* $t \in [0, 1]$,
> iv. $\Phi(\eta(t, x)) \leq \Phi(x)$ *for all* $x \in X$, $t \in [0, 1]$, *and*
> v. $\eta(1, \Phi^{c-\varepsilon}) \subset \Phi^{c+\varepsilon}$.

There too, we have the equality

$$c = b_2,$$

because for any $\gamma \in \Gamma$, the maximum of Φ over $\gamma([0, 1])$ is attained at some point $\zeta = \zeta(\gamma) \in X$. Then, setting

$$K = \bigcup_{\gamma \in \Gamma} \zeta(\gamma),$$

we have that

$$K \cap \gamma([0, 1]) \neq \varnothing, \qquad \text{for any } \gamma \in \Gamma.$$

Thus, $K \in W_2$ and $c = \inf_{u \in K} \Phi(u)$.

A strong form of the dual MPT established by Rabinowitz is treated in the next chapter consecrated to the limiting case of the MPT.

◇ 8.IV On a More General Minimax Principle

For the history, we report a little paragraph from a text by Ljusternik [591], whose translation to English goes back to 1966.

Let $\mathcal{B}$ be a family of nonempty subsets of X, $\Phi: X \to \mathbb{R}$ be a functional and $\eta: X \times [0, 1] \to X$ such that $\eta(., 1)$ maps elements of $\mathcal{B}$ into $\mathcal{B}$. Suppose also that for all $c \in \mathbb{R}$ and any $\varepsilon > 0$ small enough,

$$\text{if } \Phi(u) \leq d + \varepsilon, \text{ then } \Phi(\eta(n, 1)) \leq d - \varepsilon.$$

Set

$$c = \inf_{A \in \mathcal{B}} \sup_{u \in A} \Phi(u).$$

Then, if c is finite, it is a critical value of Φ.

Proof. Suppose by contradiction that c is regular. For $\varepsilon > 0$ small enough, there exists $A \in \mathcal{B}$ such that

$$c \leq \sup_{v \in A} \leq c + \varepsilon.$$

But then $B = \eta(A, 1) \in \mathcal{B}$ satisfies $\Phi|_B \leq c - \varepsilon$ which is absurd. □

This general principle is very simple but a *pertinent choice* of B in practice is not obvious. In general, we use some topological invariants (genus, homotopy classes, homology and cohomology classes) that can guarantee some stability by η when crossing regular values (in the sense $A \in B$ implies that $\eta(A, 1) \in B$). *Finding η requires the use of some form of the (PS) condition. And linking theorems are now the most general form of* usable theorems *belonging to this category.*

III

A Deeper Insight in Mountains Topology

9

The Limiting Case in the MPT

Even as the finite encloses an infinite series
And in the unlimited limits appear,
So the soul of immensity dwells in minutia
And in the narrowest limits no limit in here.
What joy to discern the minute in infinity!
The vast to perceive in the small, what divinity!
Jacob Bernoulli, Ars Conjectandi

Only those who risk going too far
can possibly find out how far they can go.
Anonymous

In this chapter we give a detailed account of the results from studying what is known as the *limiting case* of the MPT from the very beginning until a complete treatment was given. We focus on the development of the ideas about what should be the right geometry to enable such an extension.

The Situation

In the statement of the MPT, the strict inequality in the geometric condition

$$\inf_{S(0,\rho)} \Phi > \max\{\Phi(0), \Phi(e)\}, \tag{9.1}$$

meaning that the *mountain ridge* $(S(0, \rho))$ separating the two *valleys* (0 and e) has an altitude strictly higher than that of both 0 and e, plays an essential role in the proof. This condition is a topological *separation* property in the sense that the set

$$\{x \in X; \ \Phi(x) \geq c\}$$

separates 0 and e in the following sense.

Definition 9.1. A subset $S \subset X$ of a topological space X is said to *separate* two points u and v of X if u and v belong to two disjoint components of $X \setminus S$.

Early, Rabinowitz wondered whether the conclusion of the MPT remains true in the case of *zero altitude*, that is, when the aforementioned strict inequality becomes large. This situation is known as the *limiting case*.

We will use ideas and papers by Rabinowitz [748], Pucci and Serrin [727], Willem [952], and Ghoussoub and Preiss [427], who have all contributed to shed the light on this problem.

9.1 Limiting Case

So what happens when we have equality in (9.1)? There were many attempts to answer this question. The first ones brought only partial answers but contributed to the comprehension of the whole situation.

9.1.1 Some Tentatives

First, Pucci and Serrin [727] proved that the MPT still holds true with the *zero altitude* assumptions provided that the *mountain ridge* has a *nonzero thickness*. Moreover, in the case of equality, the pass occurs precisely on the mountain ridge.

Theorem 9.1 (Limiting Case, Pucci and Serrin). *Let* $\Phi \in C^1(X; \mathbb{R})$ *satisfying the* (PS) *condition, and let c and* Γ *be as in Theorem 7.1.*

Suppose that there are numbers a, r, and R such that $0 < r < R < \|e\|$ *and*

$$\inf_{r < \|u\| < R} \Phi(u) \geq a = \max\{\gamma(0), \gamma(e)\}.$$

Then, the value $c \geq a$*, defined by*

$$c = \inf_{\gamma \in \Gamma} \max_{t \in [0,1]} \Phi(\gamma(t)),$$

where Γ *is the set of all continuous paths* γ *joining 0 and e, is a critical value for* Φ*. Moreover, if* $c = a$*, then there is a critical point in the open ring* $\{u \in X; r < \|u\| < R\}$*.*

Remark 9.1. Notice that when $c = a$, Theorem 9.1 implies in fact the existence of an infinite number of critical points in the ring $\{u \in X; r < \|u\| < R\}$, since the preceding arguments apply in any of its "subrings."

We will not report Pucci and Serrin's proof here, because we will see next "another version" of this result due to Willem, whose proof is very simple.

Indeed, practically the same result was proved under the *unusual but acceptable requirement* that

$$\Phi \text{ satisfies both (PS)}_c \text{ and (WPS)}$$

by Willem [952] to treat a forced pendulum equation. Note that requiring two conditions of (PS) type is not necessary to reach the final form of the limiting case in the MPT. The exact statement of the result of Willem is the following.

Theorem 9.2 (Limiting Case, Willem). *Let $\Phi \in C^1(X; \mathbb{R})$, the minimaxing set Γ, the point e, and the inf max value c as above. If*

1. *Φ satisfies (PS)$_c$ and (WPS), and*
2. *there is $0 < r < R < \|e\|$ such that*

$$r \leq \|u\| \leq R \quad \text{implies that} \quad \Phi(u) \geq a = \max\{\Phi(0), \Phi(e)\},$$

then $c \geq a$ is a critical value of Φ. Moreover, if $c = a$, there is a critical point w such that $\Phi(w) = a$ and $\|w\| = (R + r)/2$.

We will report here the original proof of Willem. It is based on his quantitative deformation lemma seen in Chapter 3.

Proof. If $c > a$, it suffices to follow the proof of the classical MPT. Indeed, even if the assumptions are different, only the fact that $c > a$ is used. Hence, the same proof applies.

Let us assume that $c = a$. Let $n \in \mathbb{N}^*$ such that $1/\sqrt{n} \leq (R - r)/2$. By the definition of $c(= a)$, there is some $\gamma \in \Gamma$ such that

$$\max_{0 \leq t \leq 1} \Phi(\gamma(t)) \leq a + \frac{1}{n}.$$

For some $t \in [0, 1]$, we have that $\|\gamma(t)\| = (R + r)/2$. Let us write $u_n = \gamma(t)$. Then $\Phi(u_n) \leq a + 1/n$ and $\|u_n\| = (R + r)/2$. Using the quantitative deformation lemma with $S = \{u_n\}$, $c = a$, $\varepsilon = 1/n$, and $\delta = 1/\sqrt{n}$, we get that if for every $u \in \Phi^{-1}([c - 2/n, c + 2/n]) \cap S_{2/\sqrt{n}}$, $\|\Phi'(u)\| \geq 4/\sqrt{n}$, then

$$v_n = \eta(1, u_n) \in \Phi^{c - 1/n} \cap S_{1/\sqrt{n}}.$$

We obtain then that

$$\|v_n\| \leq \|v_n - u_n\| + \|u_n\| \leq \frac{R - r}{2} + \frac{R + r}{2} = R,$$

and

$$\Phi(v_n) \leq a - \frac{1}{n}.$$

This contradicts condition 2. Hence, there is some w_n such that $c - 2/n \leq \Phi(w_n) \leq c + 2/n$, $\|u_n - w_n\| \leq 2/\sqrt{n}$, and $\|\Phi'(w_n)\| < 4/\sqrt{n}$.

By condition 1, $(w_n)_n$ contains a subsequence converging to some w. Clearly, $\Phi(w) = c$, $\|w\| = (R + r)/2$, and $\Phi'(w) = 0$. $\square$

Remark 9.2. When $c = a$, Theorem 9.2 implies the existence of a critical point on every sphere with center 0 and radius $\rho \in]r, R[$.

Answering the question, "*Is Theorem 9.1 still valid without assuming that the mountain ridge has a nonzero thickness, i.e., when passing from a ring to a sphere ($r = R$)?*" would give a complete generalization of the MPT containing the limiting case.

9.1.2 The Finite Dimensional Case

The answer was known to be *yes* in *finite dimensions* since the work of Pucci and Serrin [727] who indeed located a critical point of level c on the sphere $S(0, R)$. Pucci and Serrin proved the existence of a critical point in the closure of an open ring A around $S(0, R)$ such that the distance from the boundary of A to $S(0, R)$ can be taken arbitrarily small. Then by the fact that the dimension of X is finite and using a standard compactness argument, they showed that there is a critical point $x_0 \in S(0, R)$ with $\Phi(x_0) = a$.

9.1.3 The Infinite Dimensional Case

In [746], Rabinowitz proved a stronger form of the dual MPT that includes the limiting case providing, in some sense, an improvement of the result of Pucci and Serrin (Theorem 9.1).

Theorem 9.3 (Limiting Case, Rabinowitz). *Let $\Phi \in C^1(X; \mathbb{R})$ satisfying the (PS) condition. Suppose that $\Phi(0) = 0$ and that*

$\Phi 1.$ *there is an open neighborhood B, of 0 such that*

$$\Phi|_{\partial B} \geq 0,$$

$\Phi 2.$ *and there is $e \notin \overline{B}$ such that*

$$\Phi(e) \leq 0.$$

Then there exists a critical value b of Φ at the level characterized by

$$b = \sup_{B \in W} \inf_{u \in \partial B} \Phi(u),$$

where

$$W = \left\{ B \subset X; \ B \text{ is open}, \ 0 \in B \text{ and } e \notin \overline{B} \right\}.$$

Moreover, if $b = 0$ there is a critical point of Φ on ∂B.

Proof. If $b > 0$, the proof of Theorem 7.1 carries over this situation as remarked in Chapter 8. So, we have to treat only the situation $b = 0$. Since $0 \in B$ and $e \notin \overline{B}$, we can assume without loss of generality that

$$\min(\text{dist}(e, \overline{B}), \text{dist}(0, \partial B)) > 1. \tag{9.2}$$

Suppose by contradiction that Φ has no critical point on ∂B. Since Φ satisfies (PS), the intersection $\Theta \cap \partial B = \varnothing$ where Θ is a neighborhood of $\mathbb{K}_0$, the set of critical points of level 0 of Φ, which is *compact*. By the ascending deformation lemma (Theorem 8.3 on page 88), for $\varepsilon > 0$ sufficiently small, there is a deformation $\eta \in C([0, 1] \times X; X)$ such that

$$\eta(1, \Phi^{-\varepsilon} \setminus \Theta) \subset \Phi^{\varepsilon}. \tag{9.3}$$

Then, by $\Phi 1$, we have that $\partial B \subset \Phi^{-\varepsilon} \setminus \Theta$. Therefore, by (9.3),

$$\eta(1, \partial B) = \partial(\eta(1, B)) \subset \Phi^{\varepsilon}.$$

And then,

$$\inf_{\eta(1,\partial B)} \Phi \geq \varepsilon > 0. \tag{9.4}$$

But since η is a homeomorphism, there is a point $x \in X$ such that $\eta(1, x) = 0$. And because

$$\| \overbrace{\eta(1, x)}^{0} - x \| = \|x\| \leq 1,$$

by (9.2), we get that $x \in B$; that is, $\eta(1, B)$ is a neighborhood of 0. Also, $e \notin \eta(1, B)$. Indeed, because η is a homeomorphism, $\eta(1, y) = e$ for some $y \in X$ and $\|\eta(1, y) - y\| = \|e - y\| \leq 1$. Hence, by the inequality (9.2), $y \notin \overline{B}$. So, we have proved that $\eta(1, B)$ is an open neighborhood of 0 such that $e \notin \eta(1, B)$. In other words, it is in W. But then, (9.4) would contradict the assumption $b = 0$. $\quad\square$

But, *is it possible to still get the same conclusion using the set Γ and the* inf sup *argument as in the classical MPT?*

The answer is yes again. Indeed, This has been done by many authors recently, beginning with Ghoussoub and Preiss [427].

But before that and without passing by the dual sets, it is worth noticing that de Figueiredo and Solimini gave a proof of the MPT in [302], also including the limiting case arguing by contradiction. Their proof uses only properties of local minima of a functional satisfying the (PS) condition and Ekeland's variational principle. But the corresponding critical value *is not characterized by a minimax argument.* The proof relies on the following property.

Lemma 9.4. *Let $\Phi \in C(X; \mathbb{R})$ be a functional satisfying* (PS). *Suppose that u_0 is a local minimum of Φ, that is, there is some $\varepsilon > 0$ such that*

$$\Phi(u_0) \leq \Phi(u) \qquad \text{for all } u \in B(u_0, \varepsilon).$$

Then, for some $0 < \varepsilon_0 \leq \varepsilon$, the following alternative holds. Either

i.   *there exists α, with $0 < \alpha < \varepsilon_0$ such that*

$$\inf_{S(u_0,\alpha)} \Phi(u) > \Phi(u_0),$$

   or

ii. *for each α, with $0 < \alpha < \varepsilon_0$, Φ admits a local minimum $u_\alpha \in S(u_0, \alpha)$ such that $\Phi(u_0) = \Phi(u_\alpha)$.*

Proof. Take ε_0 with $0 < \varepsilon_0 \leq \varepsilon$ and suppose that condition i does not hold. Then, for any fixed α with $0 < \alpha < \varepsilon_0$,

$$\inf_{S(u_0,\alpha)} \Phi(u) = \Phi(u_0). \tag{9.5}$$

Let δ be a positive real number such that $0 < \alpha - \delta < \alpha + \delta < \varepsilon_0$ and consider the restriction of Φ to the ring

$$R = \{u \in X; \ \alpha - \delta \le \|u - u_0\| \le \alpha + \delta\}.$$

By (9.5), there is u_n such that

$$u_n \in S(u_0, \alpha) \qquad \text{and} \qquad \Phi(u_n) \le \Phi(u_0) + \frac{1}{n}.$$

By Ekeland's variational principle, there exists a $v_n \in R$ such that

$$\Phi(v_n) \le \Phi(u_n), \quad \|u_n - v_n\| \le \frac{1}{n}, \tag{9.6}$$

and

$$\Phi(v_n) \le \Phi(u) + \frac{1}{n}\|u - v_n\| \qquad \text{for any } u \in R. \tag{9.7}$$

Then, v_n belongs to the interior of R for large n. Take $u = v_n + tw$ in (9.7), where $w \in X$ has norm 1 and t is sufficiently small. Tending t to 0, we get that $\|\Phi'(v_n)\| \le 1/n$. Using the first assertion in (9.6) and (PS) we conclude that there exists a subsequence of $(v_n)_n$ converging to some v_α. So that $\Phi(v_\alpha) = \Phi(u_0)$, $\Phi'(v_\alpha) = 0$, and $\|v_\alpha - u_0\| = \alpha$. □

The exact statement of the version of the MPT attributed to De Figueiredo and Solimini is the following.

Theorem 9.5 (Limiting Case, De Figueiredo and Solimini). *Let $\Phi \in C^1(X; \mathbb{R})$ satisfying the* (PS) *condition. Suppose that*

$$\inf_{u \in S(0,R)} \Phi(u) \ge \max\{\Phi(0), \Phi(e)\}, \tag{9.8}$$

where $0 < r < \|e\|$. Then, Φ has a critical point $u_0 \ne 0$.

Proof. When the inequality (9.8) is strict, we just apply the classical MPT. Therefore, let us assume equality in (9.8). If e is not a local minimum, we may assume that there exists a point e' near e with $\Phi(e') < \Phi(e)$. Therefore, by replacing e with e', one of the following two possibilities occurs.

- Either we gain strict inequality in (9.8) and again the classical MPT applies and we are done,
- or equality persists and we have

$$\inf_{u \in S(0,r)} \Phi(u) = \Phi(0) > \Phi(e).$$

We may also assume that

$$\inf_{u \in B(0,r)} \Phi(u) = \Phi(0) > \Phi(e), \tag{9.9}$$

because otherwise, the classical MPT would apply again. But (9.9) means that 0 is a local minimum. So, we can apply the former lemma (Lemma 9.4) to conclude. □

9.2 Mountain Pass Principle

The MPT proved to be a special case of a more general principle due to Ghoussoub and Preiss [427], which also includes the limiting case and carries some information on the location of the (PS) sequence found by the MPT. You may think of it as the counterpart of Ekeland's variational principle in the one-dimensional minimax setting. It was also used, as we will see in Chapter 12, to derive some results that can help in understanding the structure of the critical set in the situation of the MPT if appropriately exploited.

For this result, we need the following weaker form of the (PS) condition.

Definition 9.2. We will say that a C^1-functional Φ satisfies the (PS) *condition around a set F at the level c if*

$$(PS)_{F,c} \quad \begin{cases} \text{every sequence } (u_n)_n \text{ in } X \text{ that satisfies conditions i, ii, and iii,} \\ \text{has a convergent subsequence.} \end{cases}$$

Theorem 9.6 (General Mountain Pass Principle, Ghoussoub and Preiss). *Let $\Phi \in C^1(X; \mathbb{R})$. Consider the number*

$$c = \inf_{\gamma \in \Gamma} \max_{t \in [0,1]} \Phi(\gamma(t)),$$

where Γ is the set of all continuous paths joining two points u and v in X. Suppose that F is a closed subset of X such that

$$F \cap \{u \in X; \ \Phi(u) \geq c\}$$

separates u and v. Then, there is a sequence $(u_n)_n$ in X such that

   i. $\lim_n \text{dist}(u_n, F) = 0$,
   ii. $\lim_n \Phi(u_n) = c$, *and*
   iii. $\lim_n \|\Phi'(u_n)\| = 0$.

If we suppose, moreover, that Φ satisfies $(PS)_{F,c}$, then Φ has a critical point of level c on F.

The proof uses Ekeland's variational principle. We will sketch it here following Ekeland [358, p. 13].

Proof. Instead of Φ, we will consider the perturbation Φ_η defined by

$$\begin{cases} \Phi_\eta(u) = \Phi(u) & \text{if } \text{dist}(u, F) \geq 2\eta, \\ \Phi_\eta(u) \leq \Phi(u) + \eta^2 & \text{for all } u \in X, \\ \Phi_\eta(u) = \Phi(u) + \eta^2 & \text{if } \text{dist}(u, F) \leq \eta, \\ \|\Phi'(u) - \Phi'_\eta(u)\| \leq \eta. \end{cases}$$

Then,

$$\inf_{\gamma \in \Gamma} \max_{t \in [0,1]} \Phi_\eta(\gamma(t)) = c + \eta^2.$$

Thus, for $\gamma_\varepsilon \in \Gamma$ such that

$$\max_{t\in[0,1]} \Phi(\gamma_\varepsilon(t)) \leq c + \varepsilon,$$

we have

$$\max_{t\in[0,1]} \Phi_\eta(\gamma_\varepsilon(t)) \leq c + \varepsilon + \eta^2.$$

Moreover, if $\bar{t}$ is such that $\Phi_\eta(\gamma_\varepsilon(\bar{t})) = \max_{t\in[0,1]} \Phi_\eta(\gamma_\varepsilon(t))$, we have

$$\Phi'(\gamma_\varepsilon(\bar{t})) \leq \varepsilon$$

and

$$\mathrm{dist}\,((\gamma_\varepsilon(\bar{t})), F) \leq 2\eta,$$

provided that $\varepsilon < \eta^2$.   □

In the standard MPT (Theorem 7.1), the geometric condition (7.1) means that $\{u \in X; \ \Phi(u) \geq c\}$ separates 0 and e. So, Theorem 9.6 applies with $F = X$.

The *limiting case* corresponds to the situation where u and v are separated by a sphere on which Φ is larger or equal to c. Thus there, with

$$F = S = S \cap \{u \in X; \ \Phi(u) \geq c\},$$

one can apply Theorem 9.6 and find a critical point of Φ of level c on the sphere S.

Brézis and Nirenberg [153] also proved an MPT that includes the limiting case. They used Ekeland's variational principle and a perturbation different from that of Ghoussoub and Preiss. We describe it in the Notes that follow.

Comments and Additional Notes

1. There; the question is *closed* and we got a *happy ending*. This is the reason for the historical approach adopted. There are two things that must be kept in mind after having read this chapter. First, in the case of a zero-altitude situation, not only do we get a critical point corresponding to the "inf sup" value but we have additional information on its location. (We know it is on S.) Second, we could clearly see a beginning of distinction between the situations of the finite and infinite dimensions. This will prove to be essential in the study of the critical set in the situation of the MPT.

2. In the original statement of the mountain pass principle by Ghoussoub and Preiss, the functional Φ was supposed to be continuous and Gâteaux differentiable on a Banach space X such that $\Phi': X \to X^*$ is continuous from the norm topology of X to the weak-* topology of X^*. By particular choices of F, this principle was used to extend and simplify some results of Hofer [481] and Pucci and Serrin [726, 727] concerning the structure of the critical set in the MPT (see Chapter 12).

◇ 9.I The Perturbation of Brézis and Nirenberg

In [153], Brézis and Nirenberg also proved a form of the MPT that includes the limiting case (in a more general form), once using a quantitative deformation lemma (Theorem 4.5 seen in Chapter 4) and a second time using Ekeland's variational principle and a perturbation argument different from that of Ghoussoub and Preiss. We describe here the second approach to let the reader make the comparison.

Sketch of the proof of Brézis and Nirenberg [153] adapted to the case of the MPT. For $t \in [0, 1]$, set

$$\rho(t) = \min\{\text{dist}(t, \{0, 1\}), 1\} = \min\{1 - t, t\}$$

and consider, for any fixed $\varepsilon > 0$ and $\gamma \in \Gamma$, the perturbation

$$G(\gamma, t) = \Phi(\gamma(t)) + \varepsilon\rho(t).$$

Set

$$\Psi_\varepsilon(\gamma) = \max_{t \in [0,1]} G(\gamma, t),$$

$$c_\varepsilon = \inf_{\gamma \in \Gamma} \Psi_\varepsilon(\gamma).$$

The functional $\gamma \mapsto \Psi_\varepsilon(\gamma)$ is continuous on Γ, and by Ekeland's principle, there is a $\gamma \in \Gamma$ and there exists $t_0 \in \{t \in K;\ G(\gamma, t) = \Psi_\varepsilon(\gamma)\}$ such that

$$c \leq c_\varepsilon \leq c + \varepsilon$$

and

$$\|\Phi'(\gamma(t_0))\| \leq 2\varepsilon.$$

$\square$

From the general mountain pass principle, we also obtain the following result that already appears in an earlier paper [726] by Pucci and Serrin. It was used later to solve some semilinear problems in a search of unstable solutions.

Corollary 9.7. *Let $\Phi \in C^1(X; \mathbb{R})$ be a functional satisfying* (PS). *If Φ has a pair of local minima (or maxima), then Φ possesses a third critical point.*

Proof. Let u_1, u_2 be the two critical points and suppose that $\Phi(u_1) \geq \Phi(u_2)$. We may suppose $u_1 = 0$ and $\Phi(u_1) = 0$. Then, with $e = u_2$, and B a small neighborhood of 0, we conclude that Φ possesses a critical value $c \geq \Phi(u_1)$. Moreover, when $c = \Phi(u_1)$, Φ has a critical value on ∂B. $\square$

◇ 9.II Higher Dimensional Links and the Limiting Case

In [423, 424], Ghoussoub considers the more general setting of the higher dimensional links and the limiting case using Ekeland's variational principle.

Theorem 9.8 (MPT with Higher Dimensional Links and Limiting Case, Ghoussoub). *Let Φ be a C^1-functional on a complete connected C^1-Finsler manifold X, B be a closed subset of X, and F be a class of compact subsets of X such that*

 a. *every set in F contains B and*
 b. *for any set A in F and any $\eta \in C([0, 1] \times X; X)$ satisfying $\eta(t, x) = x$ for all $(t, x) \in (\{0\} \times X) \cup ([0, 1] \times B)$, one has $\eta(\{1\} \times A) \in F$.*

Define $c = \inf_{A \in \mathcal{F}} \sup_{x \in A} \Phi(x)$ and suppose there exists a closed subset F of X such that $A \cap F \supset B \neq \varnothing$ for every $A \in \mathcal{F}$ and $\inf \Phi(F) \geq c$. Then, for any sequence of sets $(A_n)_n$ in $\mathcal{F}$ with $\limsup_n \Phi(A_n) = c$, there exists a sequence $(x_n)_n$ in X such that

 i. $\lim_n \Phi(x_n) = c$,
 ii. $\lim_n \|d\Phi(x_n)\| = 0$,
 iii. $\lim_n \operatorname{dist}(x_n, F) = 0$, *and*
 iv. $\lim_n \operatorname{dist}(x_n, A_n) = 0$.

◇ 9.III Other MPT Versions with the Limiting Case

In the first part of [460], Guo et al. proved, in 1988, the following form of the MPT.

Theorem 9.9 (Guo et al.). *If $\Phi: E \to \mathbb{R}$ is a C^1-functional that satisfies the (PS) condition on a real Banach space E, $x_0, x_1 \in E$ and $\mathcal{D}$ is an open neighborhood of x_0 such that $x_1 \notin \mathcal{D}$,*

$$\inf \Phi(\partial \mathcal{D}) \geq \max\{\Phi(x_0), \Phi(x_1)\}.$$

Then,

$$c = \inf_{\gamma \in \Gamma} \max_{t \in [0,1]} \Phi(\gamma(t))$$

where

$$\Gamma = \left\{ \gamma \in C([0, 1]; E),\ \gamma(0) = x_0, \gamma(1) = x_1 \right\}$$

is a critical value of Φ. Moreover, if $c = \inf \Phi(\partial \mathcal{D})$, then $\partial \mathcal{D} \cap \mathbb{K}_c \setminus \{x_0, x_1\}) \neq \varnothing$.

The paper [730] (1987) by Qi is earlier than Ghoussoub and Preiss's paper, and Qi also proved an MPT for a continuously differentiable functional (satisfying the (PS) condition) with the limiting case using a deformation lemma. See also [153, 333, 347, 348, 407].

10

Palais-Smale Condition versus Asymptotic Behavior

One of the most influential ideas in the modern era of variational calculus is probably the new belief that the failure of the Palais-Smale condition is not always the final word and that a finer analysis of the behavior of non-convergent (PS) sequences may require new variational methods that would prevent such an eventuality.

> I. Ekeland and N. Ghoussoub, New aspects of the calculus of variations in the large. *Bull. Am. Math. Soc.*, **39**, no. 2, 207–265 (2001)

This chapter is a continuation of Chapter 2 wherein some *introductory* material on the Palais-Smale condition was given. In particular, we will see that (PS) implies a particular asymptotic behavior on the functional when some control is imposed on its level sets. We will also see some examples of functionals where the functional has the geometry of the MPT but does not satisfy (PS) and the inf max value is not critical. In the particular situation of the MPT, the geometric conditions give some second-order information on (PS) sequences.

The Palais-Smale condition is considered by many authors to be quite stringent. For example, Schechter gave a considerable effort trying to weaken or avoid its use in the particular situation of the MPT and got many interesting results in many directions. We will try to get an idea of its influence through the study of particular situations where we have some information of the asymptotic behavior on the functional.

10.1 (PS), Level Sets, and Coercivity

We recall first that a real-valued functional Φ defined on a Banach space X is *coercive* if $\Phi(u) \to +\infty$ as $||u|| \to \infty$. As one can check immediately, Φ is coercive if and only if for any real number $d \in \mathbb{R}$, the set

$$\Phi^d = \{u \in X; \ \Phi(u) \le d\}$$

is bounded.

In [565], S. Li used a deformation lemma to prove the coercivity of functionals that satisfy (PS) and are bounded from below.

Proposition 10.1. *If a C^1-functional $\Phi\colon X \to \mathbb{R}$ is bounded from below and satisfies* $(PS)_c$ *for all $c \in \mathbb{R}$, then Φ is coercive.*

Later, the same result was proved independently by Willem using his quantitative deformation lemma, and again by Caklovic, Li, and Willem [164] using Ekeland's variational principle.

It was proved again another time by Costa and Silva [265] as a part of their study of some aspects relating the (PS) condition, level sets, and coercivity.

We report the two nice and short proofs of this Proposition by Willem [954, Appendix] and by Caklovic et al. [164].

Proof of Proposition 10.1 by Willem. By contradiction, if Φ was not coercive, then the value

$$c = \sup \{d \in \mathbb{R};\ \Phi^d \text{ is bounded}\}$$

would be finite. Let $\mathcal{U}$ be an open neighborhood of the compact set $\mathbb{K}_c$, $\varepsilon \in\,]0, 1[$, and η be given by the quantitative deformation lemma. It follows from the definition of c that the set $(\Phi^{c+\varepsilon} \setminus \mathcal{U})$ is unbounded and that $\Phi^{c-\varepsilon}$ is bounded. On the other hand, $\eta(1, \Phi^{c+\varepsilon} \setminus \mathcal{U}) \subset \Phi^{c-\varepsilon}$ and $\eta(1, .)\colon X \to X$ maps unbounded sets into unbounded sets; thus, we get a contradiction. $\square$

The second proof from [164] is obtained as an immediate consequence of Ekeland's principle.

Proof of Proposition 10.1 by Caklovic et al. By contradiction, suppose that

$$c = \liminf_{\|u\|\to\infty} \Phi(u) \in \mathbb{R}.$$

Therefore, for every $n \in \mathbb{N} \setminus \{0\}$, there would exist $u_n \in X$ such that

$$\Phi(u_n) \le c + \frac{1}{n} \qquad \text{and} \qquad \|u\| \ge 2n.$$

Ekeland's principle with $\varepsilon = c + (1/n) - \inf_X \Phi$ and $\lambda = 1/n$ implies then the existence of $v_n \in X$ such that

$$\Phi(v_n) \le \Phi(u_n) \le c + \tfrac{1}{n},$$

$$\|v_n\| \ge \|u_n\| - \|u_n - v_n\| \ge \|u_n\| - n \ge n, \quad \text{and}$$

$$\|\Phi'(v_n)\| \le \tfrac{1}{n}(c + \tfrac{1}{n} - \inf_X \Phi).$$

Since $\|v_n\| \to \infty$, it follows that $\Phi(v_n) \to c$. But $\Phi'(v_n) \to 0$, which is a contradiction with the (PS) condition. $\square$

Remark 10.1. Note that the converse (coercivity implying (PS)) is valid in finite dimension only, as we saw in Proposition 2.1.

In [164], even the case of a functional Φ not bounded below is treated.

Proposition 10.2. *Let X be a Banach space and let $\Phi\colon X \to \mathbb{R}$ be a C^1-functional satisfying* (PS). *If there is some $d \in \mathbb{R}$ such that $\Phi^{-1}(d)$ is bounded, then $|\Phi|$ is coercive.*

Proof. Without loss of generality, we can assume that $d = 0$. By the assumption $\Phi^{-1}(d)(= \Phi^{-1}(0))$ is bounded, there is n_0 such that $\Phi(u) \neq 0$ when $||u|| \geq n_0$.

Suppose that

$$c = \liminf_{||u|| \to \infty} |\Phi(u)| \in \mathbb{R}.$$

For every $n > n_0$, there exists $u_n \in X$ such that $|\Phi(u_n)| \leq c + 1/n$ and $||u_n|| \geq 2n$. By Ekeland's principle, with $\varepsilon = c + (1/n) - \inf_X |\Phi|$ and $\lambda = 1/n$, there exists $v_n \in X$ such that

$$|\Phi(v_n)| \leq |\Phi(u_n)| \leq c + \frac{1}{n},$$

$$||v_n|| \geq ||u_n|| - ||u_n - v_n|| \geq ||u_n|| - n \geq n > n_0, \quad \text{and}$$

$$||\Phi'(v_n)|| \leq \tfrac{1}{n}(c + \tfrac{1}{n} - \inf_X |\Phi|).$$

(The fact that $\Phi(w) \neq 0$ in a neighborhood of v_n is used.) As shown earlier, $||v_n|| \to \infty$, $\Phi(v_n) \to c$, and $\Phi'(v_n) \to 0$ and we again get a contradiction with the (PS) condition. $\qquad \square$

The approach of Costa and Silva in [265] is somewhat different and relies on the properties of the level sets. It is illustrated by the following result.

Proposition 10.3. *Let* $\Phi \in C^1(X; \mathbb{R})$ *be bounded from below and set* $a = \inf_X \Phi$. *If* Φ *satisfies* $(PS)_a$, *then the set* $\Phi^{a+\alpha}$ *is bounded for some* $\alpha > 0$.

Proof. By contradiction, suppose that $\Phi^{a+\alpha}$ is unbounded for all $\alpha > 0$. Then, there would exist $(v_n)_n \subset X$ such that

$$a \leq \Phi(v_n) \leq a + \frac{1}{n} \qquad \text{and} \qquad ||v_n|| \geq n.$$

Ekeland's variational principle (with $\varepsilon = \frac{1}{n}$ and $\delta = 1/\sqrt{n}$) implies then that there exists a sequence $(u_n)_n \subset X$ satisfying

$$a \leq \Phi(u_n) \leq \Phi(v_n) \leq a + \frac{1}{n},$$

$$\Phi(u_n) \leq \Phi(u) + \frac{1}{\sqrt{n}}||u - u_n|| \qquad \text{for all } u \in X, \text{ and} \qquad (10.1)$$

$$||u_n - v_n|| \leq \frac{1}{\sqrt{n}},$$

which implies that

$$\Phi(u_n) \to a, \qquad ||\Phi'(u_n)|| \leq \frac{1}{\sqrt{n}} \to 0,$$

and

$$||u_n|| \geq n - \frac{1}{\sqrt{n}} \to +\infty.$$

That is a contradiction with $(PS)_a$. $\qquad \square$

They also prove the slightly more general result.

10.2 On the Geometry of the MPT and (PS)

We know that by using Ekeland's principle and supposing only that a functional satisfies the geometric conditions of the MPT, we can affirm that there exists a sequence of almost critical points $(u_n)_n \subset X$ such that

$$\Phi(u_n) \to c = \inf_{\gamma \in \Gamma} \max_{t \in [0,1]} \Phi(\gamma(t)),$$

where Γ is the set of all continuous paths joining 0 and e.

Without supposing $(PS)_c$, the sequence $(u_n)_n$ may fail to have an accumulation point and we cannot affirm that c is a critical value because Φ may *lack compactness* at that level. Indeed, counterexamples were given by Nirenberg and Brézis.

Example 10.1. Consider the functional $\Phi \colon \mathbb{R}^2 \to \mathbb{R}$ (see Figure 10.1) defined by

$$\Phi(x, y) = |\exp(x + iy) - 1|^2$$

where i is an imaginary number such that $i^2 = -1$. Then it can be checked easily that F achieves its minimum 0 at $(0, 0)$ and $(0, 2\pi)$, and that for $r > 0$ small enough

$$\Phi(x, y) \geq c_0 > 0 \qquad \text{for} \qquad x^2 + y^2 = r.$$

On the other hand, 0 is the only critical value of Φ.

Example 10.2. In $\mathbb{R}^2$, consider the function (see Figure 10.2)

$$\Phi(x, y) = x^2 - y^2(x - 1)^3.$$

Choose R in a way that

$$\Phi(x, y) > 0 \qquad \text{for} \qquad 0 < x^2 + y^2 < R^2$$

and

$$\Phi(x_0, y_0) \leq 0 \qquad \text{for some } (x_0, y_0) \text{ with } x_0^2 + y_0^2 > R^2.$$

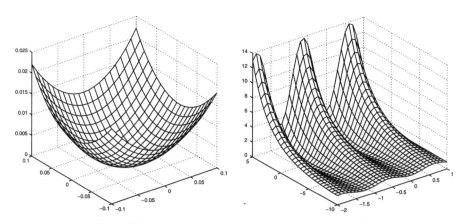

Figure 10.1. A mountain paysage lacking compactness (I)

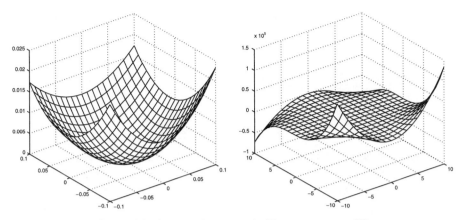

Figure 10.2. A mountain paysage lacking compactness (II).

There too, one checks easily that $(0, 0)$ is the only critical point of Φ, while the real number c given by the "inf sup" argument (7.2) in this case, in both examples, is positive and then cannot be a critical value.

In the case of a finite dimensional space X, we saw that coercivity implies (PS) while, even when X is infinite dimensional, there is still a close relation between the asymptotic behavior of a functional Φ and the fact that it satisfies the (PS) condition or not.

So, exploiting the fact that the geometric condition that appears in the MPT (in the preceding examples in particular) are local, occurring in some ball $B(0, R)$, and modifying Φ outside $B(0, R)$ for some $R' > R$ is crucial for the behavior of (PS) and can raise up critical points of level $c = \inf \max \Phi$.

10.3 Second-Order Information on (PS) Sequences in the MPT

Let $\Phi \colon H \to \mathbb{R}$ be a functional defined on a Hilbert space H.

We saw in Chapter 3 that when Φ is bounded from below, Ekeland's principle yields the existence of a minimizing sequence tending to the infimum of Φ,

$$\begin{cases} x_n \to \inf_H \Phi, \\ \Phi(x_n) \to 0. \end{cases} \tag{10.11}$$

When Φ is of class C^2, Borwein and Preiss proved the existence of a sequence $(x_n)_n$ which, in addition to satisfying (10.11), also satisfies

$$\liminf_n \langle \Phi''(x_n)w, w \rangle \geq 0 \qquad \text{for all } w \in H.$$

When the functional Φ satisfies the geometric conditions of the MPT, Fang and Ghoussoub [392] proved the existence of a sequence $(x_n)_n$ in H which, in addition to being a (PS) sequence, has an additional property involving the Hessian of Φ, which implies that when this one is nondegenerate, any cluster point for $(x_n)_n$ is a *critical point of*

Morse index of at most one. This additional information in both results (minimization and MPT) proves to be of primordial importance in applications (see [275, 583], for example). The main result in [392] is the following.

Theorem 10.7. *Let* $\Phi\colon H \to \mathbb{R}$ *be a* C^2*-functional and let* $u, v \in H$ *be such that the value*

$$c = \inf_{\gamma \in \Gamma} \max_{t \in [0,1]} \Phi(\gamma(t))$$

is finite, where, as usual, $\Gamma = \big\{\gamma \in C([0, 1]; H);\ \gamma(0) = u \text{ and } \gamma(1) = v\big\}$.

Suppose that Φ' *and* Φ'' *are Hölder continuous in a neighborhood of* $\{\Phi = c\}$ *and let* F *be a closed subset of* Φ_c *that separates* u *and* v.

Then, for any minimaxing sequence $(\gamma_n)_n \subset \Gamma$, *there exists a sequence* $(x_n)_n$ *in* H *such that*

    i. $\lim_n \Phi(x_n) = c$;
    ii. $\lim_n \Phi'(x_n) = 0$;
    iii. $\lim_n \mathrm{dist}(x_n, \gamma_n([0, 1])) = 0$;
    iv. $\lim_n \mathrm{dist}(x_n, F) = 0$;
    v. for each n, *if* $\langle \Phi''(x_n)u, u\rangle < -\frac{1}{n}\|u\|^2$ *for all* u *in a subspace* E *of* H, *then* $\dim(E) \le 1$.

And in the case where $F = \Phi_c$, we obtain the following result in the situation of the MPT.

Corollary 10.8. *Let* $\Phi\colon H \to \mathbb{R}$ *be a* C^2*-functional and let* $u, v \in H$ *be such that*

$$\max\big\{\Phi(u), \Phi(v)\big\} < c = \inf_{\gamma \in \Gamma} \max_{t \in [0,1]} \Phi(\gamma(t)).$$

Suppose that Φ' *and* Φ'' *are Hölder continuous in a neighborhood of* $\{\Phi = c\}$. *Then, there exists a sequence* $(x_n)_n$ *in* H *such that*

    i. $\lim_n \Phi(x_n) = c$;
    ii. $\lim_n \Phi'(x_n) = 0$;
    iii. for each n, *if* $\langle \Phi''(x_n)u, u\rangle < -\frac{1}{n}\|u\|^2$ *for all* u *in a subspace* E *of* H, *then* $\dim(E) \le 1$.

In the situation of the finite dimensional MPT, we get as a corollary the more precise result.

Corollary 10.9. *Suppose that the situation of Corollary 10.8 holds with* $H = \mathbb{R}^m$ $(2 \le m < \infty)$. *Let* F *be a closed subset of* $\{\Phi = c\}$ *that separates* u *and* v. *Then, there exists a sequence* $(x_n)_n$ *in* H *such that*

    i. $\lim_n \Phi(x_n) = c$;
    ii. $\lim_n \Phi'(x_n) = 0$;
    iii. $\lim_n \mathrm{dist}(x_n, F) = 0$;

iv. for each n, there exists at least $m - 1$ *eigenvalues of* $\lambda_n^1, \ldots, \lambda_n^{m-1}$ *of* $\Phi''(x_n)$
    such that

$$\liminf_n \lambda_n^i \geq 0 \qquad \text{for all } i = 1, \ldots, m - 1.$$

Comments and Additional Notes

Schechter [793] noticed that the (PS) condition at a level c has the drawback that it requires sets of the form $|G(u) - c| < \varepsilon$, $||G'(u)|| < \varepsilon$, to be bounded for $\varepsilon > 0$ sufficiently small.

The condition $(PS)_c$ requires (PS) sequences of level c to be bounded. (In fact, this is true for perturbations of the identity whose derivatives are compact, as we have seen.) So, some substitutes must be proposed. We will see in Chapter 13 some that have been formulated by Bartolo, Benci, and Fortunato, Cerami, Amrouss and Moussaoui, and some other strategies proposed by Schechter. There is a large and valuable work by Schechter on the subject.

Brézis [148] gives a brief summary of the current state of the treatment of variational problems that lack compactness using critical point theory, with an extensive bibliography, arranged by topic. He focuses in particular on Dirichlet problems for second-order elliptic problems with critical exponents and on the Yamabe problem. He also discusses weaker versions of the (PS) conditions and results on the best constants in the Sobolev inequalities.

◇ 10.1 The Concentration Compactness Principle

In some situations, there is a lack of compactness, in the sense that $(PS)_c$ fails at certain levels and, hence, not all (PS) sequences contain some convergent subsequence, but it may still be possible to say when the problem possesses solutions. This happens, for example, with the loss of compactness of the Sobolev embedding in unbounded domains. Some tools used include special functions spaces where the compactness is preserved or use of weighted Sobolev spaces. Another interesting technique developed for this aim, proved to be useful in a certain number of situations to find solutions when they exist by performing a careful study of the behavior of (PS) sequences, is the *concentration compactness principle* of Lions [579–582]. It was used for minimization problems and it was also *combined* with the MPT to study a certain number of problems as in [65, 76, 102, 174, 193, 197, 200, 237, 251, 263, 395, 416, 443, 564, 595, 643, 762, 819, 854, 855, 890, 961, 1000, 1005]. The concentration-compactness principle is also treated in some textbooks on variational methods, for example in [544, Chapter 3], where the MPT together with the concentration compactness method are applied to noncoercive elliptic problems on $\mathbb{R}^N$, as well as in [517, 882].

In [192], Chabrowski formulated a *concentration compactness principle at infinity*, which he used instead of the principle of Lions to treat some nonlinear elliptic equations. More recently, Schindler and Tintarev [821] proved a more abstract formulation of the original form of Lions' principle, which is stated for specific functional spaces. Their result is expressed in terms of a noncompact group of bounded operators on a Banach space.

◇ 10.II Hartree-Fock Equations for Coulomb Systems

Consider the Hartree-Fock equations

$$- \Delta\varphi_i + V\varphi_i + \left(\rho * \frac{1}{|x|}\right)\varphi_i - \int_{\mathbb{R}^3} \rho(x, y)\frac{1}{|x - y|}\varphi_i(y)\, dy$$
$$+ \varepsilon_i \varphi_i = 0, \qquad \text{for } 1 \leq i \leq N,$$

where $V(x) = -\sum_{j=1}^{m} z_j |x - \bar{x}_j|^{-1}$, $m \geq 1$, $z_j > 0$, $\bar{x}_j \in \mathbb{R}^3$ are fixed, $(\varphi_1, \ldots, \varphi_N) \in H^1(\mathbb{R}^3)^N$, $\int_{\mathbb{R}^3} \varphi_i \varphi_j^* \, dx = \delta_{ij}$ for $1 \leq i, j \leq N$. Write $Z = \sum_{j=1}^{m} z_j$ for the total charge of the nuclei.

The solutions of this problem are the critical points of the functional

$$\mathcal{E}(\varphi_1, \ldots, \varphi_N) = \sum_{i=1}^{N} \int_{\mathbb{R}^3} |\nabla\varphi_i|^2 + V|\varphi_i|^2 \, dx$$
$$+ \frac{1}{2} \iint_{\mathbb{R}^2 \times \mathbb{R}^3} \rho(x)\frac{1}{|x - y|}\rho(y)\, dx dy - \frac{1}{2}\int_{\mathbb{R}^3 \times \mathbb{R}^3} \frac{1}{|x - y|}|\rho(x, y)|^2 \, dx dy$$

on the manifold

$$F = \left\{(\varphi_1, \ldots, \varphi_N) \in H^1(\mathbb{R}^3)^N; \int_{\mathbb{R}^3} \varphi_i \varphi_j^* \, dx = \delta_{ij} \text{ for } 1 \leq i, j \leq N\right\}$$

where z^* is the conjugate of the complex number z, $\rho(x) = \sum_{i=1}^{n} |\varphi_i|^2(x)$ is the density, and $\rho(x, y) = \sum_{i=1}^{N} \varphi_i(x)\varphi_j^*(y)\, dx dy$ is the density matrix associated to $(\varphi_1, \ldots, \varphi_n)$.

In [583], Lions showed that the functional $\mathcal{E}$ restricted to F does not verify the (PS) condition. However, if $Z > N$, then $\mathcal{E}$ satisfies the following form of the (PS) condition:

Any sequence $(x_n)_n \subset H$ such that

$$\mathcal{E}(x_n) \to c, \quad \mathcal{E}'(x_n) \to 0$$

and $\dim(E) \leq k$ whenever E is a subspace of H such that $\langle \mathcal{E}''(x_n)u, u \rangle < -\frac{1}{n}\|u\|^2$ for every $u \in E$ admits a convergent subsequence

for every $c \in \mathbb{R}$ and any $k \in \mathbb{N}$. This allows him to prove that the Hartree-Fock equations have an infinite number of solutions. He approximates $\mathcal{E}$ with the functionals that are associated with the same problem but restricted to a suitable approximating bounded domain and obtains (PS) sequences with approximate second-order information using Morse theory.

◇ *10.III Coercivity versus (PS) and Nonsmooth Critical Point Theory*

In [256] (a note on coercivity of lower semicontinuous functions and nonsmooth critical point theory), Corvellec uses the notion of the weak slope of a lower semicontinuous function defined on a metric space (see Chapter 16) to extend the result of coercivity of a functional that satisfies (PS) and is bounded from below to a very general version in this paper, such that almost all known results in this direction are included as special cases. A kind of "uniform Γ-convergence" and the notion of F bounded from below are introduced in the statement.

Using Ekeland's principle, Chang and Shi proved the following result [215].

Theorem 10.10. *Suppose that E is a closed convex set of a Banach space. Assume that $\Phi: E \to R$ is l.s.c., bounded from below, and satisfies the following form of the (PS) condition:*

Any sequence $\{x_n\} \subset E$, along which $\Phi(x_n) \to \alpha \in \mathbb{R}$ and

$$\inf_{h \in E - x_n} \liminf_{t \to 0} [\Phi(x_n + th) - \Phi(x_n)]/t \|h\| \to \beta \geq 0,$$

possesses a convergent subsequence,

Then, $\Phi(x) \to \infty$ as $\|x\| \to \infty$.

11

Symmetry and the MPT

In the present section, we shall develop methods, employing ideas contained in some of L.A. Lyusternik's work, which allow us to establish the existence of a denumerable number of stable critical values of an even functional – they do not disappear under small perturbations by odd functionals.

> M.A. Krasnosel'skii, *Topological methods in the theory nonlinear integral equations*, 1956.

This chapter is devoted to the study of the *symmetric* MPT and its subsequent extensions. It is a multiplicity result asserting the existence of multiple critical points, when the functional is invariant under the action of a group of symmetries. It has been stated in the same time as the classical MPT by Ambrosetti and Rabinowitz [50]. This theorem can be seen as an extension of older multiplicity results of Ljusternik Schnirelman type. We will also review two other ways of obtaining multiplicity results; a procedure that inductively uses the (classical) MPT and does not pass by any Index theory, and a generalization of the symmetric MPT, the fountain theorem of Bartsch and its dual form by Bartsch and Willem.

Some basic references for the material presented here include [93, 734, 748, 882] and of course [50]. The lecture notes [93] by Bartsch discuss very nicely and exhaustively the role of symmetry in variational methods.

When a functional is invariant under a group of symmetries, we expect the existence of infinitely many solutions. This was remarked on by Ljusternik and Schnirelman in 1930. The symmetric MPT constitutes a twofold extension of the Ljusternik-Schnirelman theorem. It deals with *unbounded functionals* on *Banach spaces*.

11.1 Ljusternik-Schnirelman Theory

In the early 1930s, Ljusternik and Schnirelman developed a critical point theory for differentiable functions on finite dimensional Hilbert manifolds. They adopted the same technique as in Morse theory; that is, the critical points are obtained by deforming the manifold along gradientlines.

We begin by stating the finite dimensional result of Ljusternik and Schnirelman [592]. It is probably the earliest multiplicity result for symmetric functionals.

Theorem 11.1 (Ljusternik-Schnirelman). *Let* $\Phi \in C^1(\mathbb{R}^N; \mathbb{R})$ *be an even function. Then, the restriction of* Φ *to the unit sphere* S^{N-1} *of* $\mathbb{R}^N$ *possesses at least* N *distinct pairs of critical points.*

This result has many infinite dimensional parents that naturally require some additional compactness. As usual, this is the (PS) condition. In a chronological order, the first extensions are due to Krasnoselskii [534] and Schwartz [824] who considered C^2-functionals on C^2 Hilbert manifolds. Then, Palais [694] and Browder [156] considered C^1-functionals on C^2 Finsler manifolds. More recently, Szulkin removed the condition on the space in [891] where he supposes only that the functional is of class C^1. He also proved a yet more recent extension to functions of the form $I = \Phi + \psi$, where $\Phi \in C^1(X, \mathbb{R})$ and $\psi : X \to (-\infty, +\infty]$ is convex and lower semicontinuous. (See the final Notes at the end of this chapter.) We present here the following infinite dimensional extension of the Ljusternik-Schnirelman theorem.

Theorem 11.2 (Rabinowitz). *Let* X *be an infinite dimensional Hilbert space, and consider an even functional* $\Phi \in C^1(X; \mathbb{R})$. *If* $\Phi|_{\partial B_r}$ *satisfies* (PS) *and is bounded below for some* $r > 0$, *then* Φ *possesses infinitely many pairs of critical points.*

Remark 11.1. Theorem 11.2 is also valid with the (PS) condition replaced by $(PS)_{c_j}$ for each $1 \leq j \leq N$ where the c_j's will be defined in the proof. This is more convenient for the treatment of applications in which requiring (PS) may be stringent.

Theorem 11.2 has some interesting applications (see, for example, Rabinowitz [748]). We will prove these two results as a direct consequence in application of the index theory (as formulated by Rabinowitz [744]).

11.1.1 Index Theory

Suppose that X is a Banach space with a compact *group action* $\mathfrak{G}$. Let

$$\mathcal{S} = \{A \subset X; A \text{ is closed and } \mathfrak{g}(A) = A \text{ for all } \mathfrak{g} \in \mathfrak{G}\}$$

be the set of $\mathfrak{G}$-*invariant* subsets of X. Consider also the class of all $\mathfrak{G}$-*equivariant* mappings of X:

$$\Gamma = \{\gamma \in C(X; X); \gamma(\mathfrak{g}(x)) = \mathfrak{g}(\gamma(x)) \text{ for all } \mathfrak{g} \in \mathfrak{G} \text{ and all } x \in X\}.$$

If $\mathfrak{G} \neq \{e\}$, denote the *set of fixed points of* $\mathfrak{G}$ by

$$\text{Fix } \mathfrak{G} = \{x \in X; \mathfrak{g}x = x \text{ for all } \mathfrak{g} \in \mathfrak{G}\}.$$

The definition of the general concept of *index* is the following.

Definition 11.1. An *index* for the triple $(\mathfrak{G}, \mathcal{S}, \Gamma)$ is a mapping $\operatorname{Ind} : \mathcal{S} \to \mathbb{N} \cup \{\infty\}$ such that for all $A, B \in \mathcal{S}$ and $\gamma \in \Gamma$, the following properties hold:

1. *Definiteness*: $\operatorname{Ind}(A) = 0 \iff A = \varnothing$.
2. *Monotonicity*: $A \subset B \Rightarrow \operatorname{Ind}(A) \leq \operatorname{Ind}(B)$.
3. *Subadditivity*: $\operatorname{Ind}(A \cup B) \leq \operatorname{Ind}(A) + \operatorname{Ind}(B)$.
4. *Supervariance*: $\operatorname{Ind}(A) \leq \operatorname{Ind}\left(\overline{\gamma(A)}\right)$.
5. *Continuity*: If A is compact and $A \cap \operatorname{Fix} \mathfrak{G} = \varnothing$, then $\operatorname{Ind}(A) < \infty$ and there is a $\mathfrak{G}$-invariant neighborhood N of A such that $\operatorname{Ind}(\overline{N}) = \operatorname{Ind}(A)$.
6. *Normalization*: If $u \notin \operatorname{Fix} \mathfrak{G}$, then $\operatorname{Ind}\left(\bigcup_{g \in \mathfrak{G}} gu\right) = 1$.

Property 4 is also known as *the mapping property*.

The Krasnoselskii Genus

The Krasnoselskii genus is a particular index that will be used to prove the symmetric MPT. The notion of index will not be needed in all its generality. This simple example of index notion is a tool that measures the "size" of symmetric sets, to quote Rabinowitz [748].

It is defined for the *symmetric group* $\mathfrak{G} = \mathbb{Z}_2 = \{I_d, -I_d\}$,

$$\mathcal{S} = \big\{A \subset X; A \text{ closed and } A = -A\big\},$$

the set of closed symmetric subsets of X, and $\operatorname{Fix} \mathfrak{G}$ being the singleton $\{0\}$. This particular index is called the (Krasnoselskii) genus and is denoted by γ.

Definition 11.2. Let $A \in \mathcal{S}$. We denote $\gamma(A) = N$ and say that A has genus N if N is the smallest integer such that there is an odd continuous function $f \in C(A; \mathbb{R}^N \setminus \{0\})$. Otherwise, $\gamma(A)$ is set to ∞.

We can easily check that γ satisfies the properties of an index. We also have the following useful information that exhibits some similarity between the notion of genus and that of *dimension of a linear space*.

Proposition 11.3. *Let A be a symmetric bounded neighborhood of 0 in $\mathbb{R}^N$. Then, $\partial A \in \mathcal{S}$ and $\gamma(A) = N$.*

The fact that $\gamma(A) \leq N$ is trivial. And if we suppose that the inequality is strict, we get a contradiction with the Borsuk-Ulam theorem. (The details may be found, for example, in the book [315] by Deimling.)

This result has also a converse.

Proposition 11.4. *Suppose that X is a Hilbert space, $A \subset X$ is compact, symmetric, and $\gamma(A) = m < \infty$. Then A contains at least m mutually orthogonal vectors $(u_k)_{k=1}^m$, i.e., $(u_k, u_l) = \delta_{kl}$ where $\delta_{kl} = 1$ if $k = l$ and $\delta_{kl} = 0$ if not (Kronecker index).*

Proof. Let $\{u_1, \ldots, u_\ell\}$ be a maximal set of the mutually orthogonal vectors in A, and denote $W = \operatorname{span}\{u_1, \ldots, u_\ell\} \simeq \mathbb{R}^\ell$. The orthogonal projection $\pi : X \to W$ is an odd continuous mapping and $\pi|_A: A \to \mathbb{R}^\ell \setminus \{0\}$. Since by assumption $\gamma(A) = m$, we conclude that $\ell \geq m$. $\quad \square$

Remark 11.2. If $A = \{u_1, \ldots, u_k, -u_1, \ldots, -u_k\}$ is *finite*, *symmetric*, and $0 \notin A$, then $\gamma(A) = 1$. More generally, if $B \subset X$ is closed and $B \cap -B = \varnothing$, then $\gamma(B \cup (-B)) = 1$. It suffices to consider the odd function $\varphi \in C^1(B \cup (-B); \mathbb{R} \setminus \{0\})$ defined by $\varphi(x) = 1$ if $x \in B$ and $\varphi(x) = -1$ if $x \in (-B)$.

Now, we can prove *Ljusternik-Schnirelman theorem* (Theorem 11.1).

Proof of Theorem 11.1. The idea of the proof consists of minimaxing Φ over N families of sets Γ_j, $1 \leq j \leq N$, to obtain N critical points. Consider

$$\Gamma_j = \{K \in S^{N-1}; \ \gamma(K) \geq j\}, \qquad 1 \leq j \leq N.$$

Using the properties of the index, we easily obtain the following relations.

The set $\Gamma_j \neq \varnothing$, for all $1 \leq j \leq n$.
Monotonicity: $\Gamma_{j+1} \subset \Gamma_j$.
Excision: If $B \in \Gamma_j$ and $A \in S$ with $\gamma(A) \leq s < j$, then $\overline{B \setminus A} \in \Gamma_{j-s}$.
Invariance: If $f \in C(S^{N-1}; S^{N-1})$ is odd, then $f(\Gamma_j) \subset \Gamma_j$.
Then set

$$c_j = \inf_{B \in \Gamma_j} \max_{x \in B} \Phi(x), \qquad \text{for } 1 \leq j \leq n. \tag{11.1}$$

We will show that the c_j's are critical values of $\Phi|_{S^{N-1}}$ for $1 \leq j \leq N$, but this is not sufficient to prove our theorem, because some of the c_j's may coincide and then we would not be able to certify that there are n pairs of critical points of $\Phi|_{S^{N-1}}$. So, we must prove the following. If we denote

$$\mathbb{K}_c = \left\{x \in S^{N-1}; \ \Phi(x) = c \text{ and } \left(\Phi|_{S^{N-1}}\right)'(x) = 0\right\},$$

then we have the property that

$$(c_j = \ldots = c_{j+k-1} \equiv c) \qquad \text{implies that} \qquad \gamma(\mathbb{K}_c) \geq k. \tag{11.2}$$

Notice that by the former remark, if $k > 1$ in (11.2), $\mathbb{K}_c$ contains infinitely many critical points.

To prove (11.2), suppose by contradiction that $\gamma(\mathbb{K}_c) < k$. By the continuity property, there is a $\delta > 0$ such that $\gamma(N_\delta(\mathbb{K}_c)) = \gamma(\mathbb{K}_c)$. Notice that $N_\delta(\mathbb{K}_c) \not\subset S^{N-1}$. So, denoting by

$$\mathcal{O} = N_\delta(\mathbb{K}_c) \cap S^{N-1},$$

we get that

$$\mathbb{K}_c \subset \mathcal{O} \subset N_\delta(\mathbb{K}_c).$$

Hence, by monotonicity, $\gamma(\mathcal{O}) < k$.

Now, by the deformation lemma and for ε small enough, there is some continuous deformation $\eta \in \mathcal{C}([0, 1] \times S^{N-1}; S^{N-1})$ such that

$$\eta(1, \Phi^{c+\varepsilon} \setminus \mathcal{O}) \subset \Phi^{c-\varepsilon}.$$

Choose $B \in \Gamma_{j+k-1}$ such that $B \subset \Phi^{c+\varepsilon}$; that is,

$$\max_{x \in B} \Phi(x) \le c + \varepsilon.$$

Then,

$$\max_{x \in \eta(1, \overline{B \setminus \mathcal{O}})} \Phi(x) \le c - \varepsilon. \tag{11.3}$$

But by the excision property, $\overline{B \setminus \mathcal{O}} \in \Gamma_j$, and by invariance, $\eta(1, \overline{B \setminus \mathcal{O}}) \in \Gamma_j$. Then, the definition of $c = c_j$ implies that

$$\max_{x \in \eta(1, \overline{B \setminus \mathcal{O}})} \Phi(x) \ge c$$

and contradicts (11.3).   □

Historical Remark – First Index

A variety of indices have been introduced for locally compact Lie groups [104, 385, 386] to characterize critical points of functionals with some particular symmetries. The *Ljusternik-Schnirelman category* [592] is the first known example of index; it was introduced in 1934.

Let X be a *topological space* and consider a closed subset $A \subset X$. A is said to have category k relative to X, which is denoted

$$\operatorname*{cat}_{X} A = k$$

if A is covered by k closed *contractible* sets in X and k is minimal with this property. We recall that a subset $C \subset X$ is *contractible* in X if it can be mapped continuously in X into a single point of X.

The notions of genus, category, Morse-Conley index, and so forth are all easily and clearly *defined* but are difficult to *compute* in practice except when the group acting on X is S^1 or $\mathbb{Z}/p\mathbb{Z}$ for some prime number p (cf. [93]).

11.2 Symmetric MPT

The symmetric MPT may be considered an extension of the Ljusternik-Schnirelman theorem to Banach spaces and *unbounded* (from below or from above) functionals. Its exact statement is the following.

Theorem 11.5 (Symmetric MPT, Ambrosetti-Rabinowitz). *Let X be a real infinite dimensional Banach space and $\Phi \in C^1(X; \mathbb{R})$ a functional satisfying (PS). If Φ satisfies*

i. $\Phi(0) = 0$ *and there are constants $\rho, \alpha > 0$ such that*

$$\Phi|_{\partial B_\rho} \geq \alpha,$$

ii. Φ *is even, and*

iii. *for all finite dimensional subspaces $\tilde{X} \subset X$, there exists $R = R(\tilde{X}) > 0$ such that*

$$\Phi(u) \leq 0 \qquad for\ u \in \tilde{X} \setminus B_R(\tilde{X}),$$

then Φ possesses an unbounded sequence of critical values (as usual) characterized by a minimax argument.

Condition iii is the natural generalization to our new context of the condition

$$\Phi(e) < \max_{u \in S(0,R)} \Phi(u), \qquad \|e\| > R.$$

Proof. We begin with an intermediate result.

Lemma 11.6. *There is a sequence $(\Gamma_j)_j$ satisfying the following properties:*

Each set $\Gamma_j \neq \varnothing$, for all $1 \leq j \leq n$.
Monotonicity: $\Gamma_{j+1} \subset \Gamma_j$.
Excision: If $B \in \Gamma_j$ and $A \in S$ with $\gamma(A) \leq s < j$, then $\overline{B \setminus A} \in \Gamma_{j-s}$.
The following modified version of the invariance property:

$$If\ f \in C(S^{N-1}; S^{N-1})\ is\ odd,\ then\ f(\Gamma_j) \subset \Gamma_j.$$

Moreover, the reals $(c_j)_j$ defined in (11.1) are critical values of Φ for each $j \in \mathbb{N}$ and are unbounded.

We begin by choosing a sequence of unit vectors $(e_m)_m$ such that $e_{m+1} \notin$ span $\{e_1, \dots, e_m\} \equiv X_m$. Set $R_m = R(X_m)$, $D_m \equiv \overline{B_{R_m}} \cap X_m$, and

$$\mathfrak{G}_m = \left\{ h \in C(D_m; X);\ h\ is\ odd\ and\ h = I_d\ on\ \partial B_{R_m} \cap X_m \right\}.$$

Since $I_d \in \mathfrak{G}_m$, the set $\mathfrak{G}_m$ is nonempty.
Set

$$\Gamma_j = \left\{ h(\overline{D_m \setminus Y});\ m \geq j,\ h \in \mathfrak{G}_m,\ Y \subset X\ and\ \gamma(Y) \leq m - j \right\}.$$

The sets Γ_j are nonempty for all $j \in \mathbb{N}$. Likewise, the monotonicity of $(\Gamma_j)_j$, the excision property, and the fact that

if $\psi \in C(X; X)$ is odd and $\psi = I_d$ on $\partial B_{R_m} \cap X_m$ for all $m \in \mathbb{N}$, then ψ maps Γ_j into Γ_j for each $j \in \mathbb{N}$

follow easily.
Now, let c_j as in (11.1) for each $j \in \mathbb{N}$. Since each $B \in \Gamma_j$ is compact, $c_j < \infty$.

Claim 11.1.

$$c_1 \geq \alpha > 0. \tag{11.4}$$

Assume that (11.4) holds true. It follows that c_j is a critical value for each $j \in \mathbb{N}$. The multiplicity assumption

$$(c_j = \ldots = c_{j+k-1} \equiv c) \quad \text{implies that} \quad \gamma(\mathbb{K}_c) \geq k \tag{11.5}$$

also holds. Indeed, notice first that $c = c_j \geq c_1 \geq \alpha$ by the monotonicity and (11.1). Therefore, $0 \notin \mathbb{K}_c$ since $\Phi(0) = 0$. Hence, $\mathbb{K}_c \in S$. Using (PS) and the fact that the genus of a compact set is finite, $\gamma(\mathbb{K}_c) < \infty$. Now, if $\gamma(\mathbb{K}_c) < k$, by the continuity property there is a $\delta > 0$ such that $\gamma(N_\delta(\mathbb{K}_c)) < k$. By the deformation lemma, there is a deformation $\eta \in \mathcal{C}([0,1] \times X; X)$ for $\varepsilon \leq \alpha/2$ such that

$$\eta(1, \Phi^{c+\varepsilon} \setminus \{N_\delta(\mathbb{K}_c)\}) \subset \Phi^{c-\varepsilon}. \tag{11.6}$$

Choose $B \in \Gamma_{j+k-1}$ such that

$$\max_B \Phi \leq c + \varepsilon. \tag{11.7}$$

By (11.6),

$$\max \left\{ \Phi \circ \eta(1, \overline{B \setminus \{N_\delta(\mathbb{K}_c)\}}) \right\} \leq c - \varepsilon. \tag{11.8}$$

But, by excision $\overline{B \setminus \{N_\delta(\mathbb{K}_c)\}} \in \Gamma_j$ and by the choice of ε, the mapping property:

if $A, B \in S$ and $\psi \in \mathcal{C}(A, B)$ is odd, then $\gamma(A) \leq \gamma(B)$, and if $\psi \in \mathcal{C}(A; X)$ is an odd homeomorphism, $\gamma(\psi(A)) = \gamma(A)$,

we have that $\eta(1, .) \in \mathfrak{G}_m$ for all $m \in \mathbb{N}$. Hence, by the modified version of the invariance property defined earlier, $\eta(1, \overline{B \setminus \{N_\delta(\mathbb{K}_c)\}}) \in \Gamma_j$. The definition of c_j then gives

$$\max \left\{ \eta(1, \overline{B \setminus \{N_\delta(\mathbb{K}_c)\}}) \right\} \geq c, \tag{11.9}$$

which is a contradiction with (11.8). So, (11.5) is true.

We now have to verify (11.4). Let $B \in \Gamma_j$, $B = h(\overline{D_m \setminus Y})$ where $Y \in S$ and $\gamma(Y) \leq m - 1$. Since $\Phi \geq \alpha$ on ∂B_ρ by condition i and $\Phi(u) \leq 0$ outside D_m in X_m by condition iii, we have that $\rho < R_m$. Let

$$\widehat{\Omega} = \left\{ x \in D_m; \ h(x) \in B_\rho \right\}$$

and let Ω denote the component of $\widehat{\Omega}$ containing 0. Since h is odd and $h = I_d$ on $\partial B_{R_m} \cap X_m$, the set Ω is a symmetric bounded open neighborhood of 0 in X_m. So, identifying X_m with $\mathbb{R}^m$ we get $\gamma(\partial \Omega) = m$.

Moreover, if $x \in \partial \Omega$, $h(x) \in \partial B_\rho$. Denote

$$W = \left\{ x \in D_m; \ h(x) \in \partial B_\rho \right\}.$$

We have that $\gamma(W) \geq \gamma(\partial \Omega) = m$ because if $\gamma(B) < \infty$, $\gamma(\overline{A \setminus B}) \geq \gamma(A) - \gamma(B)$. So, $\partial \Omega \subset W$ and $\gamma(\overline{W \setminus Y}) \geq m - (m-1) = 1$. Thus $\overline{W \setminus Y} \neq \emptyset$ and there exists

$\zeta \in W$ such that $h(\zeta) \in B \cap \partial B_\rho$. Accordingly, by condition i,

$$\max_B \Phi \geq \Phi(h(\zeta)) \geq \inf_{\partial B_\rho} \Phi \geq \alpha. \tag{11.10}$$

And since this is true for any set B in Γ_1, we have that $c_1 \geq \alpha$.

To complete the proof, we must show that $c_j \to \infty$ as $j \to \infty$. Suppose by contradiction that this is not the case. Since by monotonicity $(c_j)_j$ is a monotone nondecreasing sequence, there is $\bar{c} < \infty$ such that $c_j \to \bar{c}$ as $j \to \infty$. In fact, $\bar{c} > c_j$ for all j. Otherwise, by (11.5) we would have $\gamma(\mathbb{K}_{\bar{c}}) = \infty$. However, the set $\mathbb{K}_{\bar{c}} \in \mathcal{S}$ and is compact, so $\gamma(\mathbb{K}_{\bar{c}}) < \infty$. Set

$$\mathcal{H} = \left\{ u \in X; \ c_1 \leq \Phi(u) \leq \bar{c} \text{ and } \Phi'(u) = 0 \right\} = \bigcup_{c_1 \leq c \leq \bar{c}} \mathbb{K}_c.$$

By (PS), $\mathcal{H}$ is compact and by condition ii and (11.10), $\mathcal{H} \in \mathcal{S}$. Again, the continuity property and the fact that the genus of a compact set is finite, there is a $\delta > 0$ such that $\gamma(\mathcal{N}_\delta(\mathcal{H})) = \gamma(\mathcal{H}) < \infty$. Suppose now that $\gamma(\mathcal{H}) = n$. By the deformation lemma, for $c = \bar{c}$ there is some $\varepsilon < \bar{c} - c_1$ and a deformation $\eta \in \mathcal{C}([0,1] \times X; X)$ with $\eta(1, u)$ odd such that

$$\eta\left(1, \overline{\Phi^{\bar{c}+\varepsilon} \setminus \mathcal{N}_\delta(\mathcal{H})}\right) \subset \Phi^{\bar{c}-\varepsilon}. \tag{11.11}$$

Choose the smallest m such that

$$c_m > \bar{c} - \varepsilon. \tag{11.12}$$

We have that $m > 1$ because $\bar{c} - \varepsilon > \bar{c} - (\bar{c} - c_1) = c_1$. Take $B \in \Gamma_{m+n}$ such that

$$\max_B \Phi \leq \bar{c} + \varepsilon. \tag{11.13}$$

Then $\eta(1, u) = u$ is in $\partial B_{R_n} \cap X_k$ for all $k \in \mathbb{N}$. Hence, by excision and by the modified version of the invariance property, both $\overline{B \setminus \mathcal{N}_\delta(\mathcal{H})}$ and $\eta(1, \overline{B \setminus \mathcal{N}_\delta(\mathcal{H})})$ belong to Γ_m. But then, by (11.11)–(11.13) we have that

$$c_m \leq \max\left\{ \eta\left(1, \overline{B \setminus \mathcal{N}_\delta(\mathcal{H})}\right) \right\} \leq \bar{c} - \varepsilon < c_m,$$

which is absurd. $\quad\square$

Courant-Fischer Minimax Principle

It is interesting to compare the symmetric MPT, which is only a particular case of a more general minimax principle for invariant functionals under the action of a group of symmetries, to the *Courant-Fischer minimax principle* in the linear spectral theory. Let $L : \mathbb{R}^N \to \mathbb{R}$ be a symmetric linear mapping, then the ℓth eigenvalue of L is given by the formula

$$\lambda_\ell = \min_{\substack{X_\ell \subset \mathbb{R}^N \\ \dim X_\ell = \ell}} \max_{\substack{u \in X_\ell \\ \|u\|=1}} (Lu, u).$$

Using a language proper to critical point theory, this problem of determining the eigenvalues of L can be transformed into the following. Consider the functional

$\Phi(u) = (Lu, u)$ on the unit sphere S^{N-1} of $\mathbb{R}^N$ and compute c_ℓ as described in the minimax principle (11.1). We can easily check that Φ satisfies (PS) and that the critical value $c_\ell = \lambda_\ell$.

Notice also the analogy between the facts that

- when $\lambda_\ell = \cdots = \lambda_{\ell+m-1} = \lambda$ coincide, then L has an m-dimensional eigenspace of eigenvectors associated to λ and
- when $c_{\ell+1} = \cdots = c_{\ell+m-1} = c$, then $\gamma(\mathbb{K}_c) \geq m$.

The assumptions of Theorem 11.5 can be weakened a little. But we don't get this for free and the price to pay, to quote Rabinowitz [748] once again, is that the c_j's as defined in (11.1) will no longer be critical values of Φ unless j is sufficiently large.

Theorem 11.7. *Suppose that X is a real Banach space and $\Phi \in C^1(X;\mathbb{R})$ satisfies* (PS). *Denote by $X_j^\perp$ a supplementary subspace to X_j, i.e., $X = X_j \oplus X_j^\perp$. If Φ satisfies assumptions* ii *and* iii *of Theorem 11.5 and*

  i.' *There are constants $\rho, \alpha > 0$ and $n \in \mathbb{N}$ such that*

$$\Phi|_{\partial B_\rho \cap X_n} \geq \alpha,$$

then Φ possesses an unbounded sequence of critical values.

Proof. Define c_j as in the proof of Theorem 11.5 and argue exactly the same way. Notice only that there, there is a small difference. We can no longer show that $c_1 \geq \alpha$, but merely that $c_{m+1} \geq \alpha$. □

11.3 A Superlinear Problem with Odd Nonlinearity

The symmetric MPT is an efficient tool for obtaining multiplicity results in a semilinear Dirichlet problem with odd nonlinearities. Indeed, consider the problem

$$(\mathcal{P}) \quad \begin{cases} -\Delta u(x) = f(x, u(x)) & \text{in } \Omega \\ u(x) = 0 & \text{on } \partial\Omega, \end{cases}$$

where Ω is a bounded domain of $\mathbb{R}^N$ with smooth boundary, where $N \geq 3$, and $f: \mathbb{R} \to \mathbb{R}$ is supposed to be a Carathéodory function with potential $F(x, t) = \int_0^t f(x, s)\, ds$. Then, we have the result.

Theorem 11.8. *Suppose that f satisfies the following:*

  f1. f is odd in u, i.e., $f(x, -u) = -f(x, u)$.
  f2. There exists $p \leq 2^* = \frac{2N}{N-2}$ such that f satisfies the growth condition

$$|f(x, s)| \leq C(1 + |s|^{p-1})$$

  almost everywhere.
  f3. There are constants $\mu > 2$ and $r > 0$ such that for almost every $x \in \Omega$ and all $|s| \geq r$

$$0 < \mu F(x, s) \leq s.f(x, s).$$

Then $(\mathcal{P})$ admits an unbounded sequence of solutions.

Proof. The problem $(\mathcal{P})$ corresponds to the Euler-Lagrange equation of the functional

$$\Phi(u) = \frac{1}{2} \int_\Omega |\nabla u(x)|^2 \, dx - \int_\Omega F(x, u(x)) \, dx$$

on $H_0^1(\Omega)$.

Notice that because f is odd, Φ is even and $\Phi(0) = 0$. That is why no condition on the nonlinearity controlling the behavior of Φ near 0 was required.

As seen before in application of the (classical) MPT to the superlinear problem, the growth condition $f2$ implies that Φ is of class C^1 and that its critical points are weak solutions of $(\mathcal{P})$.

The functional Φ satisfies (PS) also. Indeed, by $f2$, the map $u \mapsto f(., u)$ takes bounded sets of $L^p(\Omega)$ into bounded sets of $L^{p/(p-1)}(\Omega) \subset H^{-1}(\Omega)$. But by Rellich's theorem, the embedding $H_0^1(\Omega) \overset{\hookrightarrow}{\to} L^2(\Omega)$ is compact. Therefore, the map $K: H_0^1(\Omega) \to H^{-1}(\Omega)$ where $K(u) = f(., u)$ is compact. And since $\Phi'(u) = -\Delta u - f(., u)$, it suffices, by Proposition 2.2, to show that any (PS) sequence $(u_n)_n$ is bounded in $H_0^1(\Omega)$. Using $f2$ and $f3$, this fact follows exactly as in the application of the MPT to the superlinear problem.

Denote by $0 < \lambda_1 \le \lambda_2 \le \lambda_3 \le \cdots$ the eigenvalues of $-\Delta$ on $H_0^1(\Omega)$ and by φ_j the eigenvalue corresponding to λ_j.

Claim 11.2. For k_0 sufficiently large, there exist $\rho > 0$ and $\alpha > 0$ such that for any $u \in X^+ = \text{span}\{\varphi_k; \, k \ge k_0\}$ with $\|u\|_{H_0^1(\Omega)} = \rho$,

$$\Phi(u) \ge \alpha.$$

Indeed, by $f2$, the embedding $H_0^1(\Omega) \hookrightarrow L^{2^*}(\Omega)$, and Hölder's inequality, we have for $u \in V^+$,

$$\Phi(u) \ge \frac{1}{2} \int_\Omega |\nabla u(x)|^2 \, dx - C \int_\Omega |u(x)|^p \, dx - C$$

$$\ge \frac{1}{2} \|u\|_{H_0^1(\Omega)}^2 - C\|u\|_{L^2}^r \|u\|_{L^{2^*}}^{p-r} - C$$

$$\ge \left(\frac{1}{2} - C_1 \lambda_{k_0}^{-r/2} \|u\|_{H_0^1(\Omega)}^{p-2} \right) \|u\|_{H_0^1(\Omega)}^2 - C_2,$$

where $r/2 + (p-r)/2^* = 1$. In particular, $r = N(1 - p/2^*) > 0$ and we may take $\rho = 2\sqrt{(C_2 + 1)}$ and choose $k_0 \in \mathbb{N}$ such that

$$C_1 \lambda_{k_0}^{-r/2} \rho^{p-2} \le \frac{1}{4}$$

to achieve the relation

$$\Phi(u) \ge 1 = \alpha \qquad \text{for all } u \in V^+ \text{ with } \|u\|_{H_0^1(\Omega)} = \rho.$$

Now, take $X^- = \text{span}\{\varphi_j; \, j < k_0\} = (X^+)^\perp$.

As in the proof of Theorem 7.3, condition $f3$ yields that on any finite dimensional subspace $W \subset H_0^1(\Omega)$, there is a $C_i = C_i(W) > 0$ such that

$$\sup_{u \in \partial B_R(0;W)} \Phi(u) \le C_1 R^2 - C_2 R^\mu + C_3 \to -\infty \text{ as } R \to \infty.$$

Apply then the symmetric MPT to get an unbounded sequence of critical values of Φ. $\square$

11.4 Inductive Symmetric MPTs

We present now some variants and generalizations of the symmetric MPT.

In [737, 786], Ruf showed that the critical values obtained via index theories, in some semilinear problems, may also be obtained by using the MPT in an inductive procedure.

Indeed, consider the variational principle for finding the eigenvalues of the Laplacian by the Courant principle (seen earlier). Let $\Omega \subset \mathbb{R}^N$ be a bounded domain and consider

$$\begin{cases} -\Delta u = \lambda u & \text{in } \Omega, \\ u = 0 & \text{on } \partial\Omega. \end{cases} \tag{11.14}$$

The associated functional is $\Phi(u) = \int_\Omega |\nabla u|^2 \, dx$ on $H_0^1(\Omega)$ and the kth eigenvalue is given by

$$\lambda_k = \inf_{E_k \in \mathcal{E}_k} \sup_{u \in E_k \cap S_{L^2}} \Phi(u),$$

where S_{L^2} denotes the unit L^2-sphere in $H_0^1(\Omega)$ and $\mathcal{E}_k$ is the set of all k-dimensional linear subspaces E_k of $H_0^1(\Omega)$.

Consider now the nonlinear eigenvalue problem

$$\begin{cases} -\Delta u + g(x, u) = \lambda u & \text{in } \Omega, \\ u = 0 & \text{on } \partial\Omega, \end{cases} \tag{11.15}$$

where $g: \overline{\Omega} \times \mathbb{R} \to \mathbb{R}$ is continuous, odd, and satisfies a suitable growth condition. Setting, for $\rho > 0$,

$$\mathcal{B}_k(\rho) = \{B \subset \rho S_{L^2} \subset H_0^1(\Omega); \gamma(B) \ge k\},$$

where γ is the genus, one finds the eigenvalues of (11.15) by

$$\mu_k(\rho) = \inf_{B \in \mathcal{B}_k} \sup_{u \in B} \Phi(u),$$

where $\Phi(u) = \frac{1}{2} \int_\Omega |\nabla u|^2 + \int_\Omega G(x, u)$.

Remark 11.3. Similar constructions can be done for other compact groups provided a Borsuk-Ulam–type theorem is available. For example, an S^1-action developed in [118] is used in [786] following this procedure to study the Fučik spectrum associated to the operator $-\dfrac{d^2}{dt^2}$.

Generalized MPTs [737] are obtained as follows. Let

$$\Sigma_k = \big\{\sigma : D^k \to E \text{ continuous}; \ \sigma|_{\partial D^k} = I_d\big\},$$

where D^k denotes a k-dimensional closed ball in E. Then,

$$c = \inf_{\sigma \in \Sigma} \ \max_{u \in \sigma(D^k)} \Phi(u)$$

is a critical value provided that for every $\sigma \in \Sigma_k$ one has

$$\max_{\sigma(D^k)} \geq \max_{\partial D^k} \Phi(u) + \delta \qquad \text{for some } \delta > 0.$$

The way Ruf inductively used MPTs to get the values that were obtained earlier using the index theory is the following. Consider, for example, the nonlinear eigenvalue problem (11.15).

1. First, we have $\lambda_1(\rho) = \inf_{\rho S_{L^2}} \Phi(u)$. Denote by $\pm v_1 \in \rho S_{L^2}$ the corresponding eigenfunctions. Then, define

$$\Gamma_2 = \big\{\gamma_2 : [-1, 1] \to \rho S_{L^2} \subset H_0^1(\Omega) \text{ continuous}; \ \gamma_2(\pm 1) = \pm v\big\},$$

and set

$$c_2(\rho) = \inf_{\Gamma_2} \ \sup_{\gamma_2 \in \Gamma_2} \Phi(u).$$

The mountain pass procedure can be applied to Γ_2 provided that, for some $\delta > 0$,

$$\sup_{\gamma_2} \Phi(u) \geq \lambda_1 + \delta \qquad \text{for every} \quad \gamma_2 \in \Gamma_2,$$

in which case one obtains that $c_2(\rho) = \lambda_2(\rho)$, the second (variational) eigenvalue of (11.15), is a critical value.

2. To proceed to the next step, choose for a given $\varepsilon > 0$ a path $\gamma_{2,\varepsilon} \in \Gamma_2$ such that $\sup_{\gamma_{2,\varepsilon}} \Phi(u) \leq c_2 + \varepsilon$. Think of this path as defined on the semi-circle $S_+^1 = \{(x_1, x_2) \in \mathbb{R}^2; \ x_1^2 + x_2^2 = 1 \text{ and } x_2 \geq 0\}$ and extend it to the full circle $S^1 \subset \mathbb{R}^2$ by oddness, that is, set $\tilde{\gamma}_{2,\varepsilon} : S^1 \to S_{L^2}, \tilde{\gamma}_{2,\varepsilon} = -\gamma_{2,\varepsilon}$ on $S_-^1 = -S_+^1$. Then, set

$$\Gamma_3 = \{\gamma_3 : D^2 \subset \mathbb{R}^2 \to S_{L^2} \text{ continuous}; \ \gamma_3|_{\partial D^2} = \tilde{\gamma}_{2,\varepsilon}\},$$

where D^2 is the unit disk in $\mathbb{R}^2$. This family depends on ε. If now

$$c_3(\rho, \varepsilon) = \inf_{\Gamma_3} \ \sup_{\gamma_3 \in \Gamma_3} \Phi(u) > c_2(\rho),$$

then $c_3(\rho, \varepsilon)$ is a critical value that does not depend on ε; that is, $c_3(\rho, \epsilon) = \lambda_3(\rho)$.

3. To continue, choose again $\gamma_{3,\varepsilon} \in \Gamma$ such that $\sup_{\gamma_{3,\varepsilon}} \Phi \leq c_3 + \varepsilon$. Then, thinking of this mapping as defined on the hemisphere $S_+^2 = \{(x_1, x_2, x_3) \in \mathbb{R}^3; \ \sum_{i=1}^3 x_i^2 = 1 \text{ and } x_3 \geq 0\}$, extend it again by oddness to $\tilde{\gamma}_{3,\epsilon} : S^2 \to \rho S_{L^2}$. Defining the family

$$\Gamma_4 = \big\{\gamma_4 : D^3 \subset \mathbb{R}^3 \to \rho S_{L^2} \text{ continuous}; \ \gamma_4|_{\partial D^3} = \tilde{\gamma}_{3,\varepsilon}\big\},$$

one sets

$$c_4(\rho, \varepsilon) = \inf_{\Gamma_4} \sup_{\gamma_4 \in \Gamma_4} \Phi(u),$$

and so on.

4. In general,

$$c_k(\rho, \varepsilon) = \inf_{\Gamma_k} \sup_{\gamma_k \in \Gamma_k} \Phi(u). \tag{11.16}$$

The same approach works also for the S^1-symmetry and the Fučik spectrum problem as evoked earlier.

An advantage of the inductive symmetric mountain pass theorem is that it also works for functionals with *perturbed symmetry*. Indeed, consider the problem

$$\begin{cases} -\Delta u + g(x, u) = \lambda u + f & \text{in } \Omega, \\ u = 0 & \text{on } \partial \Omega, \end{cases} \tag{11.17}$$

where the forcing term $f \in L^2(\Omega)$ and $g \colon \overline{\Omega} \times \mathbb{R} \to \mathbb{R}$ is continuous, odd, and its associated potential satisfies $\frac{1}{t^2} G(x, t) \to \frac{1}{2}\beta(x)$ as $|t| \to +\infty$ where $\beta \in C(\overline{\Omega})$.

The corresponding action Φ_f is clearly not symmetric, but is considered a perturbation of the eigenvalue problem (11.15). Keeping the notation $c_k(\rho, \varepsilon)$ for the values given by (11.16), set

$$a_k(\varepsilon) = \lim_{\rho \to \infty} \frac{1}{\rho^2} c_k(\rho, \varepsilon)$$

and

$$b_{k+1} = \lim_{\rho \to \infty} \frac{1}{\rho^2} c_{k+1}(\rho, \varepsilon).$$

Then, we get the following result.

Theorem 11.9 (Ruf). *Assume that for some $k \in \mathbb{N}$ and $\varepsilon > 0$, we have $a_k(\varepsilon) < \lambda < b_{k+1}(\varepsilon)$ and assume that the action Φ_f associated to the perturbed problem satisfies* (PS). *Then* (11.17) *has a solution for every $f \in L^2(\Omega)$ obtained via an inf sup argument.*

11.5 The Fountain Theorem

We shall now see a generalization of the symmetric MPT due Bartsch [91] known as the *fountain theorem*. It is presented nicely by Willem in [957] using his quantitative deformation lemma. Bartsch and Willem are the authors of a *dual version* of the fountain theorem (see [957]).

First, we state an equivariant version of the quantitative deformation lemma of Willem (Lemma 4.2).

Lemma 11.10. *Assume that a compact group G acts isometrically on the Banach space X. Let $\Phi \in C(X; \mathbb{R})$ be equivariant and $S \subset X$ be invariant. Assume that $c \in \mathbb{R}$,*

$\varepsilon, \delta > 0$ *satisfy* (4.2). *Then, there exists* $\eta \in \mathcal{C}([0, 1] \times X; X)$ *satisfying properties*
i.–vi. of Lemma 4.2 and
v. $\eta(t, .)$ *is equivariant for every* $t \in [0, 1]$.

The fountain theorem depends on the notion of *admissible action*, which is a form of Borsuk-Ulam–type condition.

Definition 11.3. Assume that the compact group G acts *diagonally* on V^k where V is a finite dimensional space:

$$g(v_1, \ldots, v_k) = (gv_1, \ldots, gv_k).$$

Then, the action of G is admissible if every continuous equivariant map $\partial U \to V^{k-1}$ has a zero, where U is an open bounded invariant neighborhood of 0 in V^k, $k \geq 2$.

In other words, there does not exist an equivariant map $\partial U \to V^{k-1} \setminus \{0\}$. It is then clear that an admissible action has a trivial fixed point set reduced to $\{0\}$ and that the antipodal action of $G = \mathbb{Z}/2$ on $V = \mathbb{R}$ is admissible by the Borsuk-Ulam theorem. In fact, it suffices to restrict $\mathcal{U}$ to the unit ball in V^k (cf. Bartsch [93]).

Consider the following situation:
A_1. The compact group G acts isometrically on the Banach space $X = \overline{\bigoplus_{j \in \mathbb{N}} X_j}$, the spaces X_j are invariant, and there exists a finite dimensional space V such that for every $j \in \mathbb{N}$, $X_j \equiv V$ and the action of G on V is admissible.

Denote

$$Y_k = \bigoplus_{j=0}^{k} X_j,$$

$$Z_k = \overline{\bigoplus_{j=k}^{\infty} X_j},$$

$$B_k = \{u \in Y_k; \|u\| \leq \rho_k\},$$

$$N_k = \{u \in Z_k; \|u\| = r_k\},$$

where $\rho_k > r > 0$.

Then the following intersection (linking) property holds.

Proposition 11.11. *Under condition* A_1, *if* $\gamma \in \mathcal{C}(B_k; X)$ *is equivariant and if* $\gamma|_{\partial B_k} = Id$, *we have the intersection result* $\gamma(B_k) \cap N_k \neq \varnothing$.

Theorem 11.12. *Under assumption* A_1, *let* $\Phi \in \mathcal{C}^1(X; \mathbb{R})$ *be an invariant functional. Define, for* $k \geq 2$,

$$c_k = \inf_{h \in \Gamma_k} \max_{u \in B_k} \Phi(h(u)),$$
$$\Gamma_k = \{\gamma \in \mathcal{C}(B_k; X); \ \gamma \text{ is equivariant and } \gamma|_{\partial b_k} = Id\}.$$

If

$$b_k = \inf_{\substack{u \in Z_k \\ \|u\|=r_k}} \Phi(u) > a_k = \max_{\substack{u \in Y_k \\ \|u\|=\rho_k}} \Phi(u),$$

then $c_k \geq b_k$, and for every $\varepsilon \in]0, (c_k - a_k)/2[$, $\delta > 0$ and $\gamma \in \gamma_k$ such that

$$\max_{B_k} \Phi \circ h \leq c_k + \varepsilon, \tag{11.18}$$

there exists $u \in X$ such that

   a. $c_k - 2\varepsilon \leq \Phi(u) \leq c_k + 2\varepsilon$,
   b. $\text{dist}(u, \gamma(B_k)) \leq 2\delta$,
   c. $\|\Phi'(u)\| \leq 8\varepsilon/\delta$.

Proof. By the aforementioned intersection property, $c_k \geq b_k$. Supposing the thesis was false, and applying Lemma 11.10 with $S = h(B_k)$, the hypotheses imply that

$$c_k - 2\varepsilon > a_k \tag{11.19}$$

$$h(B_k) \subset \Phi^{c_k + \varepsilon}. \tag{11.20}$$

We define $\beta(u) = \eta(1, h(u))$ where η is given by Lemma 11.10. For every $u \in \partial B_k$, we obtain from (11.19) that

$$\beta(u) = \eta(1, \gamma(u)) = \eta(1, u) = u.$$

Since by condition v, β is equivariant, it follows that $\beta \in \Gamma_k$. We obtain from (11.20)

$$\max_{u \in B_k} \Phi(\beta(u)) = \max_{u \in B_k} \Phi(\eta(1, \gamma(u))) \leq c_k - \varepsilon,$$

contradicting the definition of c_k. $\square$

Theorem 11.13 (The Fountain Theorem, Bartsch). *Let $\Phi \in C^1(X; \mathbb{R})$ be an invariant functional and suppose A_1. If for every $k \in \mathbb{N}$ there exists $\rho_k > r_k > 0$ such that*

$A_2.$ $a_k = \max_{\substack{u \in Y_k, \\ \|u\| = \rho_k}} \Phi(u) \leq 0,$
$A_3.$ $b_k = \inf_{\substack{u \in Z_k \\ \|u\| = r_k}} \Phi(u) \to \infty$ *as* $k \to \infty$, *and*
$A_4.$ Φ *satisfies* $(PS)_c$ *for every* $c > 0$,

then Φ has an unbounded sequence of critical values.

Proof. For k large enough, $b_k > 0$. The former quantitative theorem implies the existence of a sequence $(u_n)_n \subset X$ satisfying

$$\Phi(u_n) \to c, \qquad \text{and} \qquad \Phi'(u_n) \to 0.$$

It follows from A_4 that c_k is a critical value of Φ, since $c_k \geq b_k$ and $b_k \to \infty$ as $k \to \infty$.
   $\square$

In this case $V = \mathbb{R}$ with the antipodal action of $G = \mathbb{Z}/2$, which is admissible by the classical Borsuk-Ulam theorem, the fountain theorem contains the symmetric MPT.

Comments and Additional Notes

The Ljusternik-Schnirelman theorem is a subject widely covered in the literature. Here will make some complementary remarks.

◇ 11.I Dual Ljusternik-Schnirelman Theorem

Coming back to Theorem 11.1, there is another way to obtain critical points of $\Phi|_{S^{N-1}}$.
Define

$$b_k = \sup_{A \in \Gamma_k} \min_{u \in A} \Phi(u), \qquad 1 \le k \le N.$$

Clearly $b_1 \ge b_2 \ge \cdots \ge b_N$ and we can show, as for Theorem 11.1, that the b_k's are critical values of $\Phi|_{S^{N-1}}$.

Notice that $c_1 = \min_{S^{N-1}}$ because if $x \in S^{N-1}$, then $\{x\} \cup \{-x\} \in \Gamma_1$. Moreover, $c_1 = b_N$. To see this, it suffices to remark that $\Gamma_N = \{S^{N-1}\}$. Otherwise, we can get a contradiction.

Indeed, if not, there is some $A \ne S^{N-1} \in \Gamma_N$, then there would exist $y \in S^{N-1} \setminus A$. Without loss of generality we can suppose that $y = (\underbrace{0, \ldots, 0}_{N-1}, 1)$.

The projection $P(u) = (u_1, \ldots, u_{N-1}, 0) \in \mathcal{C}(A; \mathbb{R}^{N-1} \setminus \{0\})$ and is odd. Therefore, $\gamma(A) \le N - 1$, a contradiction. This is reported in [748].

Similarly $c_N = b_1 = \max_{S^{N-1}} \Phi$. And always, as noticed by Rabinowitz, it is not known whether $c_j = b_{N-j+1}$ if $j \notin \{1, N\}$. However, by using the cohomological index of Fadell and Rabinowitz [385] instead of the genus, the corresponding minimax values c_j^* and b_j^* verify $c_j^* = b_{N-j+1}^*$.

◇ 11.II Relation between Ljusternik-Schnirelman Category and Krasnoselskii Genus

Rabinowitz [733] always proved the following *relation between Ljusternik-Schnirelman category and Krasnoselskii genus*.

Proposition. *Suppose $A \subset \mathbb{R}^N \setminus \{0\}$ is compact and symmetric and let $\tilde{A} = A/\mathbb{Z}_2$ with antipodal points identified. Then, $\gamma(A) = \mathrm{cat}_{\mathbb{R}^N \setminus \{0\}/\mathbb{Z}_2}(\tilde{A})$.*

Krasnoselskii already remarks [534, footnote on p. 358] that *"the genus of a set on a sphere coincides with the category of the image of the set in the projective space obtained by identifying points of a sphere which are symmetric with respect to the center."*

◇ 11.III Instability under Perturbation in the Ljusternik-Schnirelman Theorem

In Theorem 11.2, if we add a noneven perturbation (arbitrarily small) to the original even function, then we will not necessarily get infinitely many critical points as shown in the following example by Krasnoselskii [534, p. 379].

Consider the weakly continuous quadratic functional

$$\Phi(x) = \sum_{n=1}^{\infty} \frac{1}{n^2}(x, e_n)^2$$

on the unit ball of a separable Hilbert space H spanned by an orthonormal system $(e_i)_i$. The numbers $1/n^2$ are critical values of the functional Φ. For any $\delta > 0$, there is a $k > 0$ such that the functional

$$\Psi(x) = - \sum_{n=k+1}^{\infty} \frac{1}{n^2}(x, e_n)^2$$

has norm less than δ. The perturbed functional

$$\Phi(x) + \Psi(x) = \sum_{n=1}^{k} \frac{1}{n^2}(x, e_n)^2$$

is a degenerate quadratic functional and has only a finite number of critical points.

◇ 11.IV An Equivariant Ljusternik-Schnirelman Theory for Noneven Functionals

In [361], Ekeland and Ghoussoub develop an equivariant Ljusternik-Schnirelman theory for noneven functionals. They observe that the equivariant Ljusternik-Schnirelman min-max levels for the original functional Φ and for the even functional $\psi(x) = \max\{\Phi(x), \Phi(-x)\}$ are the same. This is used to show that the equivariant min-max procedure, if applied to a nonsymmetric functional Φ, gives either the usual critical points or points x such that $\Phi(x) = \Phi(-x)$ and $\Phi'(x) = \lambda\Phi'(-x)$, for some $\lambda > 0$.

For a detailed discussion, see the notes of Chapter 23. Now we give some specific notes on the symmetric MPT.

◇ 11.V A Ljusternik-Schnirelmann Theorem without the Palais-Smale condition

The theorem of Lusternik and Schnirelmann, as reformulated by Palais [694] or Schwartz [824], gives a lower bound, in terms of the Ljusternik-Schnirelmann category, for the number of critical points of C^1 real-valued functions on a complete Riemannian manifold. The functions are required to be bounded below (or above) and to incorporate the (PS) condition. Without the (PS) condition, for example, for the exponential function of a single real variable, the result is no longer true. Indeed the exponential has no critical points, although it is bounded from below and the real line has category one.

In [503], James introduces a variant of the Ljusternik-Schnirelmann theorem that is valid without the (PS) condition. It reduces to the classical theorem when the manifold is compact and the condition is satisfied. In the noncompact case, when the condition is not satisfied it provides new information.

◇ 11.VI Further Developments of the Symmetric MPT

Most of the following papers were developed with the though in mind of allowing the use of a general Lie group instead of $\mathbb{Z}_p$, S^1, passing by the torus and the p-torus. They are in majority due to Bartsch, Clapp, and Puppe.

1. In [95] (critical point theory for indefinite functionals with symmetries), Bartsch and Clapp consider strongly indefinite functionals Φ that are invariant under a compact Lie group action. A linking theorem and two generalizations of the symmetric MPT are presented.

2. Let X be a Hilbert space and G a compact Lie group acting orthogonally on X. Let $\Phi \in C^1(X, \mathbb{R})$ be a strongly indefinite functional, invariant with respect to the action of G. In [95], Bartsch and Clapp introduce an equivariant version of the limit relative category of Fournier et al. [408] to find critical points of Φ. This category is employed in order to obtain generalizations of the symmetric MPT and the symmetric linking theorem.

3. Let G be a compact Lie group acting linearly on a real Banach space E of infinite dimension. Let $\Phi \colon E \to \mathbb{R}$ be a G-invariant function. Bartsch et al. prove in [100] a generalized version of the MPT for actions of any compact Lie group, extending the symmetric MPT and a recent result of Clapp and Puppe [240] concerning torus and p-torus actions.

4. In [240], Clapp and Puppe extend the classical critical point theory of Ljusternik and Schnirelman to actions of a compact Lie group in such a way that also cases of nonfree group actions can be theated successfully. They use a relative version of the Ljusternik-Schnirelman category.

5. In [241], Clapp and Puppe develop a version of equivariant critical point theory particularly adapted to finding closed geodesics by variational methods and use it to improve the known lower bounds for the number of "short" closed geodesics on some closed Riemannian manifolds.

6. In [240], Clapp and Puppe develop a method to study the critical set of a functional $f \colon M \to \mathbb{R}$ possessing some symmetry where M is a G-equivariant absolute neighborhood retract. The method works for any compact Lie group.

7. In [100], Bartsch et al. prove a generalized version of the MPT, namely for actions of any compact Lie group. This result extends classical ones [50, 386] concerning $\mathbb{Z}/2$ and S^1 actions, and a recent result [240] concerning torus and p-torus actions.

8. In [94], the authors prove a generalized MPT and a bifurcation theorem for symmetric potential operators for a compact Lie group G acting orthogonally on a Hilbert space E.

Concerning the fountain theorem:

1. Admissible representations can be classified completely using an algebraic criterion [93, Theorem 3.7]. From [93], the admissible representations are precisely those that have the "dimension property" used by Benci [105].

2. The MPT version of Bartsch et al. [100] allows the summands X_j to be different representations of G, depending on j. (They are not necessarily isomorphic to V.) On the other hand, their version does not allow all admissible representations. Examples are given in [93, Chapter 3].

Some Nonsmooth Symmetric MPT Versions. A series of nonsmooth extensions of the Ljusternik-Schnirelman theory and the MPT exist in the literature. The first one in the case of locally Lipschitz functionals was given by Chang [202]. In [894, Section 6], Szulkin gave an extension of the Ljusternik-Schnirelman theory to functions of the form $I = \Phi + \psi$, where $\Phi \in C^1(X, \mathbb{R})$ and $\psi : X \to (-\infty, +\infty]$ is convex and lower semicontinuous (see Chapter 14) and applied it to variational inequalities and to boundary value problems, whereas he considered the symmetric MPT in [891]. The case of continuous functionals has also been treated in some papers beginning with Degiovanni and Marzocchi [310]. See also [306, 308, 640].

Morse Indices at Critical Points Related to the Symmetric MPT. In [898] (Morse indices at critical points related to the symmetric mountain pass theorem and applications), Tanaka considers an even functional I and its min-max critical levels b_n obtained using the symmetric MPT. He proved the existence, for each $n > 0$, of a critical point u_n at level b_n whose Morse index is less than or equal to n and of a critical point v_n at a level smaller than or equal to b_n whose Morse index plus nullity is greater than or equal to n.

These results are analogous to those known for nonsymmetric functionals already obtained by Hofer [481] and by Lazer and Solimini [555].

◇ *11.VII Some Applications*

For some applications of the symmetric MPT and related results to study multiplicity of nonlinear problems, the reader may consult [110, 197, 269, 461, 464, 485, 639, 760, 963, 964].

◇ *11.VIII The Problem of Stability under Perturbation*

A question that was treated from the beginning by Krasnoselskii when studying the effects of symmetry is the problem of *stability under perturbation.*

> In the present section, we shall develop methods, employing ideas contained in some of L.A. Ljusternik's work, which allow us to establish the existence of a denumerable number of stable critical values of an even functional – they do not disappear under small perturbations by odd functionals.
>
>   Krasnosel'skii [534].

Indeed, it is important to try to know whether this is really the symmetry that permitted obtaining these multiplicity results. In this spirit, it is interesting to study the stability under *small* perturbations. This question, in connection with the symmetric MPT, was treated by Rabinowitz in [745]. This will be detailed in Chapter 23, where some of the contributions made by Bahri and Berestycki [83], Struwe [868] and more recently by Ekeland et al. [371], Bolle [134], and Bolle et al. [135] will be given.

The assumption restricting the growth of the potential G associated to the nonlinearity in the study of the perturbed problem, in the result of Ruf, has been removed in the case of periodic solutions of Sturm-Liouville equations in [301].

◇ *11.IX Reviews and Survey Papers*

Many such papers have been written by Rabinowitz. We cite, for example, [749], where he provides an excellent review of minimax methods and their use in variational problems. He focuses especially on the case of symmetric functionals, concentrating on Z_2 and S^1-symmetries. As an example of an index theory the notion of genus is introduced and its main properties are proved. Then it is shown how, using these index theories and some variants of the deformation theorem, one obtains multiplicity results for the solutions of symmetric variational problems. Some applications to elliptic problems and to Hamiltonian systems with a superquadratic Hamiltonian are given. And the problem of perturbation from symmetry is also discussed.

We also cite the very interesting monograph [93] by Bartsch where he extends some classical topological methods in critical point theory (category, cup length, and the Conley Index) to obtain multiplicity results. He gives a detailed account of many of his contributions on the subject.

12

The Structure of the Critical Set in the MPT

Our principal result shows that the "typical" critical point obtained from the Mountain Pass Theorem is a saddle point of mountain-pass type as one would expect from the nature of the construction.

P. Pucci and J. Serrin, The structure of the critical set in the mountain pass theorem. *Trans. Am. Math. Soc.*, **91**, no. 1 (1987)

In this chapter, we are concerned with the *structure of the critical set* $\mathbb{K}_c$ in the situation of the MPT. Research on this topic was first motivated by the investigation of the question: "*does $\mathbb{K}_c$ really contain a saddle point as it would be expected from its construction?*" without requiring nondegeneracy conditions.

We will see first a local description of the behavior of a functional near a critical point given by the MPT. Then, we will review in detail some results concerning the structure and the nature of the points in this critical set.

The main references on the subject are the papers by Hofer [481,483], Pucci and Serrin [726–728], Ambrosetti [40], Ghoussoub and Preiss [427], the book by Ghoussoub [423], and Fang [387,388].

We saw in Chapter 7 the interpretation of the geometric conditions in the MPT, to which it owes its name. From this geometry and the minimax characterization of the critical value, it is expected that the critical points obtained are *saddle points* (according to Definition 12.1 to follow). Such conjecture seems to be natural. Indeed, in [733], Rabinowitz wondered whether $\mathbb{K}_c$ must be necessarily be made of saddle points, even though he noticed that this need not be the case when X is finite dimensional if the "mountain ridge" surrounding 0 everywhere has the same height, because a critical point would be a local maximum rather than a saddle point. Nevertheless, we will see among other results that indeed the critical set contains always a saddle point when the underlying space is infinite dimensional.

Besides the interest in the structure of the critical set in the MPT for its own, authors are nowadays exploiting the *nature* of the critical points in applications rather than their mere *existence*. We cite for example, a famous paper by Ekeland and Hofer [363]. The nature of this critical set makes the MPT the best choice when looking for unstable solutions.

12.1 Definitions and Terminology

We begin by fixing the notations and terminology we will use in the sequel. As always, by Φ we mean a real C^1-functional defined on a Banach space X. We suppose also that Φ satisfies $(PS)_c$ for all real values c unless the contrary is explicitly expressed.

Definition 12.1. A point $z \in X$ is said to be a *local minimum* if there is a neighborhood N of z such that

$$\Phi(z) \leq \Phi(x), \qquad \text{for all } x \in N.$$

The set of local minima corresponding to the level c will be denoted by $\mathcal{M}_c$.

A *local maximum* for Φ is a local minimum for $-\Phi$. We will not adopt any particular notation for the set of local maxima of Φ.

A point z is a *saddle point* if it is a critical point that is neither a local maximum nor a local minimum. That is, for each neighborhood N of z, there exist $x, y \in N$ such that

$$\Phi(x) < \Phi(z) < \Phi(y).$$

The set of saddle points of level c is denoted $\mathcal{S}_c$.

The point z is a *proper local maximum* if it is a local maximum and

$$z \in \overline{\Phi^c}, \text{ the adherence of } \Phi^c.$$

The set of such points will be denoted by $\mathcal{P}_c$.

We have the following immediate properties on the critical set $\mathbb{K}_c$.

1. The case of a local minimum and local maximum are not mutually exclusive, for it suffices to consider a functional Φ that is constant in a neighborhood of z.
2. The case of a local minimum and a proper local maximum are mutually exclusive. This explains the terminology of *proper* local maximum.
   Indeed, suppose by contradiction that $z \in \mathcal{M}_c \cap \mathcal{P}_c$. Then, there would exist two neighborhoods N_1 and N_2 of z such that

$$\Phi(x) \geq \Phi(z) \qquad \text{for all } x \in N_1, \tag{12.1}$$

$$\Phi(x) \leq \Phi(z) \qquad \text{for all } x \in N_2, \tag{12.2}$$

and

$$z \in \overline{\Phi^c}. \tag{12.3}$$

Hence, by (12.1) and (12.2), $\Phi \equiv \Phi(z) = c$ on the neighborhood $N_1 \cap N_2$ of z. But, by the relation (12.3), there is a sequence $(z_n)_n \subset \Phi^c$ such that $z_n \to z$. Therefore, $z_n \in N_1 \cap N_2$ for n greater than some n_0, which is a contradiction.

$\square$

3. So, any point in $\mathbb{K}_c$ is either
    - a local minimum,
    - a proper local maximum, or
    - a saddle point
    according to the preceding definition.

So, we have

$$\mathbb{K}_c = \mathcal{M}_c \cup \mathcal{P}_c \cup \mathcal{S}_c,$$

where these three sets are *mutually disjoint*. Indeed by property 3, a point $z \in \mathbb{K}_c$ is either in $\mathcal{S}_c$, in $\mathcal{M}_c$, or z is a local maximum. Suppose that $z \notin \mathcal{S}_c \cup \mathcal{M}_c$ and let us show that z is indeed in $\mathcal{P}_c$. By contradiction, if $z \notin \overline{\Phi^c}$, there exists a neighborhood N of z such that $N \cap \Phi^c = \varnothing$. Hence, the restriction of Φ to N is greater than or equal to c and then z would be a local minimum. $\square$

We already know that by (PS)$_c$, the critical set $\mathbb{K}_c$ in the MPT is compact. So, $\mathcal{S}_c$ and $\mathcal{P}_c \cup \mathcal{S}_c$ are also *compact*. Moreover, we have that the intersection

$$\mathcal{P}_c \cap \mathcal{S}_c \subset \overline{\Phi^c}.$$

Mountain Pass Points

The notion of a saddle point as defined here is not necessarily the same as what is known in general for convex-concave functions of the type $\Phi(x, y) = x^2 - y^2$ on the real plane. Simple examples show that the critical set $\mathbb{K}_c$ in the MPT need not include any point having this form. Nevertheless, there is a kind of "specific" points in $\mathbb{K}_c$, in the particular situation when it is made of isolated critical points, for example, as shown by Hofer [481].

Definition 12.2. Let $z \in X$ with $\Phi(z) = c$. It is said to be of *mountain pass type (Hofer)* if any of its neighborhoods N satisfies

$$N \cap \Phi^c \text{ is nonempty and not path-connected.}$$

Heuristically, a mountain pass point in the *landscape of the MPT* is in general a point that is crossed going from one set of lower points to another and at the moment of the crossing one may still see higher points.

Consider the following definitions.

Definition 12.3. Let A and B be two subsets of X. We say that A is a *subcomponent* of B if A is a subset of some component of B.

Remark 12.1. Suppose that x and y are two points in a subcomponent A of an open set B. Then, there exists a continuous path $\gamma : [0, 1] \to B$ joining x and y (i.e., $\gamma(0) = x$ and $\gamma(1) = y$). A slightly modified version of the definition of a mountain pass point due to Pucci and Serrin is the following.

Definition 12.4. Consider a point $z \in X$ such that $\Phi(z) = c$. Then, z is of *mountain pass type (Pucci-Serrin)* if, for any neighborhood N of z, the set

$$N \cap \Phi^c \text{ is not a subcomponent of } \Phi^c.$$

Notice that

a mountain pass point (MPP) in the sense of Pucci and Serrin is also an MPP according to the definition of Hofer. Hence, if z is not an MPP according to Hofer, then for all *suitably small* neighborhoods N of x, the set $N \cap \Phi^c$ *must* be a subcomponent of Φ^c.

When $X = \mathbb{R}$, it is obvious that an *isolated critical point u_0 of mountain pass type* is a *strict maximum*, whereas when $\dim X \geq 2$, it is necessarily a *saddle point*. Indeed, suppose by contradiction that u_0 is not a saddle point. Then, there exists some neighborhood N of u_0 such that

$$\Phi(u) \leq \Phi(u_0) = c, \qquad \text{for all } u \in N.$$

Since u_0 is also an isolated critical point, it follows that there is an open ball B centered at u_0 such that

$$\Phi(u) \leq c, \qquad \Phi'(u) \neq 0, \qquad \text{for all } u \in B \setminus \{u_0\}.$$

This implies that $B \setminus \{u_0\} \subset \Phi^c$ (any point $u \in B \setminus \{u_0\}$ such that $\Phi(u) = c$ would have been a local maximum and hence a critical point), so that

$$B \cap \Phi^c = B \setminus \{u_0\}$$

is a nonempty path-connected set. A contradiction with the fact that any neighborhood of u_0 is not path-connected. $\square$

The preceding proof is due to Pucci and Serrin, but as we can see in the upper right figure of the examples (Figure 12.1) of MPPs (for $\Phi(x, y) = -(|x| - |y|)^2$), an MPP that is not isolated needs not be a saddle point even when $\dim X \geq 2$. In fact, in the situation of the MPT, when $\dim X \geq 2$ we have a very interesting result concerning the structure of the critical set (cf. Corollary 12.13 by Fang).

12.2 On the Nature of $\mathbb{K}_c$

We will begin our discussion of the nature of the critical set $\mathbb{K}_c$ with some results by Pucci and Serrin [726, 728] and by Hofer [483]. They can be derived from the general principle of Ghoussoub and Preiss and also constitute particular cases of the subsequent work of Fang on the subject. The approach of Ghoussoub and Preiss and that of Fang will be described here. So, their proofs are omitted. The results of Fang [387] are also presented in Ghoussoub's book [423], so only the proof of his main result will be given there for the convenience of the reader. For the detailed proofs of Fang's results, the reader is referred to [423].

Without supposing the geometry of the MPT on Φ, but still requiring (PS), the following result holds.

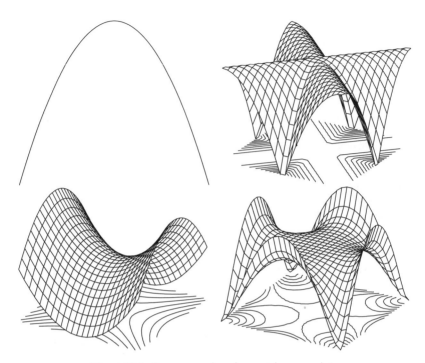

Figure 12.1. Some examples of mountain pass points.

Theorem 12.1 (Pucci-Serrin [726]). *Let* 0 *be a local minimum of* Φ*. Then, the following alternative holds:*

- *Either there exists a second critical point* *that is not a local minimum,*
- *or* 0 *is an* absolute minimum *and the set of absolute minima is* connected

when we are in the specific situation of the MPT; that is, Φ satisfies $(PS)_c$ and there are $e \in X$, $a \in \mathbb{R}$, and $R \in \mathbb{R}^+$ such that

$$\Phi|_{S(0,R)} \geq a > \max\left\{\Phi(0),\, \Phi(e)\right\}, \qquad \text{where } ||e|| > R.$$

Also, let c be the critical value defined by minimaxing the class of all continuous paths joining 0 and e. What happens is the following. The first result (chronologically speaking) in this direction is due to Hofer.

Theorem 12.2. *The critical* $\mathbb{K}_c$ *contains*

- *either a mountain pass point (in the sense of Hofer)*
- *or a local minimum.*

and if $\mathbb{K}_c$ *is made of isolated points, it contains a mountain pass point.*

Remark 12.2.
    1. When $X = \mathbb{R}$, an MPP is a local maximum. This remark will be more suggestive later.

2. The two situations of Theorem 12.2 indeed occur as we can see on simple examples in the real plane.

Pucci and Serrin [726] proved the following results the same year.

Theorem 12.3. *Let $\mathbb{K}_c$ be the set of critical points in the MPT. Then,*

    a. *If $\mathbb{K}_c$ does not separate 0 and e, then $\mathbb{K}_c$ contains a saddle point.*
    b. *If $\partial \mathbb{K}_c$ is connected and contains a point that is not a local maximum, then $\mathbb{K}_c$ contains a saddle point.*

As a consequence of condition a, we get the second part of the result of Hofer (Theorem 12.2). Moreover:

1. If there exists a path $\gamma_0 \in \Gamma$ such that $\gamma_0([0, 1])$ contains no critical point of Φ or $\max_{t \in [0,1]} \Phi(\gamma_0(t)) < c$, then $\mathbb{K}_c$ must possess a saddle point.
2. Suppose that X is an infinite dimensional Banach space. Since, by $(PS)_c$, the critical set $\mathbb{K}_c$ is compact, and in infinite dimensional spaces a compact set cannot separate two points in its complement, $\mathbb{K}_c$ contains at least one saddle point. This point shows that the conjecture of Rabinowitz is indeed true.

The results of [726] are proved using a very interesting result to be compared to two other results by Taubes (Proposition 7.2) and Shafrir (Theorem 7.10).

Theorem 12.4. *Let $\mathbb{K}_c$ be the set of critical points in the MPT and let $\mathcal{U}$ be an arbitrary neighborhood of $\mathbb{K}_c$. Then, there exists $\varepsilon = \varepsilon(\mathcal{U}) > 0$ such that*

$$\max_{t \in [0,1]} \Phi(\gamma(t)) < c + \varepsilon \qquad \textit{implies that} \qquad \gamma([0, 1]) \cap \mathcal{U} \neq \varnothing.$$

Then, in the next paper of Pucci and Serrin [728], it became very clear that whether the dimension of the space X is finite or not plays a crucial role in the nature of the critical set $\mathbb{K}_c$. The two situations differ because the set $\mathbb{K}_c$ is compact by $(PS)_c$, and a compact set cannot separate two points in an infinite dimensional space, whereas this is possible in finite dimension.

Roughly speaking,

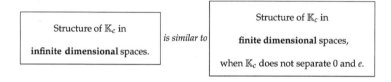

The Infinite Dimensional Case

The main result in the *infinite dimensional case* is the following.

Theorem 12.5. *Let X be an infinite dimensional Banach space. Then,*

- *either $\mathbb{K}_c$ contains a saddle point of mountain pass type*
- *or $\overline{\mathcal{M}_c}$ intersects at least two components of $\mathcal{S}_c$.*

The Finite Dimensional Case

While in the *finite dimensional case*, we have to distinguish between two cases.

A. $\mathbb{K}_c$ Possibly Separate 0 and e. This case is more general and more subtle.

Theorem 12.6. *Let X be a finite dimensional Banach space. Then,*

- *either $\mathbb{K}_c$ contains a saddle point of mountain pass type,*
- *or $\mathcal{M}_c \neq \varnothing$ and $\overline{\mathcal{M}_c}$ intersects at least two components of $(\mathcal{S}_c \cup \mathcal{P}_c)$.*

In the finite dimensional case, an MPP need not be a saddle point, as we remarked earlier.

B. $\mathbb{K}_c$ Does Not Separate 0 and e. We have then exactly the statement of the situation of the infinite dimensional case.

Theorem 12.7. *Let X be a finite dimensional Banach space and assume moreover that $\mathbb{K}_c$ does not separate 0 and e. Then,*

- *either $\mathbb{K}_c$ contains a saddle point of mountain pass type,*
- *or $\overline{\mathcal{M}_c}$ intersects at least two components of $\mathcal{S}_c$.*

12.2.1 Fang's Refinements

With the general mountain pass principle of Ghoussoub and Preiss (Theorem 9.6), the study of $\mathbb{K}_c$ becomes easier. Indeed, Ghoussoub and Preiss [427] proved the following.

Theorem 12.8 (Ghoussoub-Preiss, [427, Theorem (1.ter) (a)]). *Let $\Phi \in C^1(X; \mathbb{R})$. Consider the* inf max *value,*

$$c = \inf_{\gamma \in \Gamma} \max_{t \in [0,1]} \Phi(\gamma(t)),$$

where Γ is the set of all continuous paths joining two points u and v in X. Suppose that F is a closed subset of X such that

$$H = F \cap \{u \in X; \ \Phi(u) \geq c\}$$

separates u and v. Assume that Φ satisfies $(PS)_{F,c}$.
If $F \cap \mathcal{P}_c$ contains no compact set that separates u and v, then the following alternative holds:

a. *either $F \cap \mathcal{M}_c \neq \varnothing$,*
b. *or $F \cap \mathbb{K}_c$ contains a saddle point.*

This is always the case in infinite dimensional Banach spaces. We will need in the proof the following technical lemma (see Kuratowski [543, Sections 57 and 49]).

Lemma 12.9. *Assume F is a closed subset of X separating two points u and v. Then, there exists a closed connected subset $\tilde{F}$ of F separating u and v such that $\tilde{F} = \partial \mathcal{U} = \partial V$ where $\mathcal{U}$, V are components of $X \setminus \tilde{F}$ containing u and v, respectively.*

Proof of Theorem 12.8. Suppose that $F \cap \mathcal{M}_c = \varnothing = F \cap \mathcal{S}_c$. Let $\tilde{F} = F \cap \{\Phi \geq c\}$. By Lemma 12.9, there is a closed connected subset $\hat{F} \subset \tilde{F}$ that separates u and v, too. Note that $\hat{F} \cap \mathbb{K}_c = \hat{F} \cap \mathcal{P}_c$. Since $\hat{F} \cap \mathcal{P}_c$ is open in $\hat{F}$, $\hat{F} \cap \mathcal{P}_c$ is a compact set closed and open in $\hat{F}$. And since $\hat{F}$ is connected, either $\hat{F} \cap \mathcal{P}_c = \varnothing$ or $\hat{F} \cap \mathcal{P}_c = \hat{F}$. The first case is impossible since $\hat{F} \cap \mathcal{P}_c = \hat{F} \cap \mathbb{K}_c \neq \varnothing$ according to an earlier result by Ghoussoub and Preiss [427, Theorem (1.ter) (a)]. Hence, $\hat{F} \subset \mathcal{P}_c$ and the corollary is proved. So, $F \cap \mathcal{P}_c$ will contain the compact set $F \cap \mathcal{P}$ which will contain the compact set $\hat{F} \cap \mathcal{P}_c = \hat{F}$ which separates u and v. $\square$

An easy consequence that will be useful to prove Fang's result on the structure of the critical set is the following.

Corollary 12.10. *Take Φ, X, $\{u, v\}$, c, and F as above. Assume that Φ satisfies $(PS)_{\mathcal{N}_\varepsilon(F \cup \mathbb{K}_c), c}$ and that u, $v \notin \overline{\mathcal{M}_c}$. If $\mathcal{P}_c$ does not contain a compact subset that separates u and v, then $\mathcal{S}_c \neq \varnothing$.*

Proof. By $(PS)_{\mathcal{N}_\varepsilon(F \cup \mathbb{K}_c), c}$, we know that $\mathbb{K}_c$ is compact. Suppose by contradiction that that $\mathcal{S}_c$ is empty. For each $x \in \mathcal{M}_c$, there exists a ball $B(x, \varepsilon_x)$ such that $B(x, \varepsilon_x) \cap \Phi^c = \varnothing$. Set

$$N = \bigcup_{x \in \mathcal{M}_c} B(x, \varepsilon_x).$$

Then, $\mathcal{M}_c \subset N \subset X \setminus \Phi^c$. Since u, $v \notin \overline{\mathcal{M}_c}$ and $\overline{\mathcal{M}_c}$ is compact, we may assume that u, $v \notin \overline{N}$. Set

$$F_0 = (F \setminus N) \cup \partial N.$$

Then, $\inf_{x \in F_0} \eta(x) \geq c$ and F_0 separates u, v. Moreover, $F_0 \cap (\mathcal{M}_c \cup \mathcal{P}_c) = \varnothing$. By Theorem 12.8, $\mathcal{P}_c \cap F_0$ and $\mathcal{P}_c$ must contain a compact subset that separates u and v – a contradiction. $\square$

We continue with an important result due to Fang [387]. This result exploits the MPP (Theorem 9.6) by Ghoussoub and Preiss to study the structure of the critical set in the MPT without requiring any nondegeneracy condition. It is possible to use it to get the earlier results already obtained by Hofer [481] and Pucci and Serrin [726–728].

The interesting approach of Fang [387, 388, 390] is also described in Ghoussoub's book [423].

Recall the following notions of connectedness (see [543]).

Definition 12.5. Consider two disjoint subsets A and B of X and a nonempty subset C of X. We say that A and B are *connected through* C (or C *connects* A and B) if there is no $F \subset C \cup A \cup B$ both relatively open and closed such that

$$A \subset F \quad \text{and} \quad F \cap B = \varnothing.$$

Lemma 12.11. *Let* (X, dist) *be a locally connected metric space and let* F_0 *be a closed subset of* X *such that it separates two distinct points* u *and* v. *Let* $(Z_i)_{i=1}^n$ *be* n *mutually disjoint open subsets of* X *such that* $u, v \notin \cup_{i=1}^n \overline{Z_i}$.

Let G *be an open subset of* $X \setminus F_0$ *and denote* $Y_i = Z_i \setminus G$. *Then, the set*

$$F_1 = \left[F_0 \setminus \left(\bigcup_{i=1}^n Z_i \right) \right] \cup \bigcup_{i=1}^n (\partial Y_i \setminus T_i)$$

separates u *and* v.

Moreover, if $(A_i)_{i=1}^n$ *are* n *nonempty connected components of* G *and* $(T_i)_{i=1}^n$ *are relatively open subsets of* $Z_i \cap \partial A_i$ *such that* $T_i \cap \partial L = \varnothing$ *for any connected* L *of* G *with* $L \neq A_i$, *then*

$$F_2 = \left[F_0 \setminus \left(\bigcup_{i=1}^n Y_i \right) \right] \cup \bigcup_{i=1}^n \partial Y_i$$

also separates u *and* v.

Proof. See Fang [387,390] or Ghoussoub [425].   □

Theorem 12.12 (Structure of the Critical Set in the MPT, Fang). *Suppose that we are in the situation of the general mountain pass principle of Ghoussoub and Preiss, that* $u, v \notin \mathcal{S}_c \cup \mathcal{M}_c$ *and* Φ *satisfies* (PS)$_{\mathcal{N}_\varepsilon(F \cup \mathbb{K}_c),c}$. *Then, one of the following three alternatives holds:*

1. $\mathcal{P}_c$ *contains a compact subset that separates* u *and* v.
2. $\mathbb{K}_c$ *contains a saddle point of mountain pass type.*
3. *There are finitely many components of* Φ^c, *say* $(C_i)_{i=1}^n$, *are such that for any* $1 \leq i, j \leq n$, *and* $i \neq j$,

$$\mathcal{S}_c = \bigcup_{i=1}^n \mathcal{S}_c^i, \qquad \mathcal{S}_c^i \cap \mathcal{S}_c^j = \varnothing,$$

where $\mathcal{S}_c^i = \mathcal{S}_c \cap \overline{C_i}$. *Moreover, at least two of them* $\mathcal{S}_c^{i_1}, \mathcal{S}_c^{i_2}$ $(1 \leq i_1, i_2 \leq n, \ i_1 \neq i_2)$ *are such that the sets* $\overline{\mathcal{M}_c} \cap \mathcal{S}_c^{i_1}, \overline{\mathcal{M}_c} \cap \mathcal{S}_c^{i_2}$ *are nonempty and connected through* $\mathcal{M}_c$.

The proof of this result that appears in Fang's thesis [387] is also in Ghoussoub's book [425]. But since it is very important, we recall it for the convenience of the reader.

Proof. Suppose assertions 2 and 3 are not true. We will show that assertion 1 holds true.

By the (PS) condition, $\mathcal{S}_c$ is compact. If $\mathcal{S}_c$ was nonempty, we would conclude using Corollary 12.10. So, suppose by contradiction that it is empty.

Claim 12.1. There exist finitely many components $(C_i)_{i=1}^n$ of Φ^c and $\delta_1 > 0$ such that

$$\Phi^c \cap N_{\delta_1} \mathcal{S}_c \subset \bigcup_{i=1}^n C_i. \tag{12.4}$$

Indeed, if not, there would be a sequence $(x_n)_n$ in $\mathcal{S}_c$ and a sequence $(C_n)_n$ of different components of Φ^c such that $\mathrm{dist}(x_n, C_n) \to 0$. But then any limit point of $(x_n)_n$ would be a saddle point of mountain pass type for Φ, a contradiction with our assumption that assertion 2 is false.

For each $i = 1, 2, \ldots, n$, set

$$\mathcal{S}_c^i = \mathcal{S}_c \cap \overline{C_i}.$$

They are all compact and mutually disjoint. Also, we have that

$$\mathcal{S}_c = \bigcup_{i=1}^n \mathcal{S}_c^i. \tag{12.5}$$

Claim 12.2. There are n mutually disjoint open sets $(N^i)_{i=1}^n$ such that $u, v \notin \cup_{i=1}^n \overline{N_i}$ and

$$\mathcal{S}_c \cup \mathcal{M}_c \subset \bigcup_{i=1}^n N^i \quad \text{and} \quad \mathcal{S}_c^i \subset N^i \quad \text{for all} \quad i = 1, 2, \ldots, n. \tag{12.6}$$

We have to consider two cases:

Case 1. $\mathcal{M}_c$ is Empty. Since by assumption $u, v \notin \mathbb{K}_c$, for each $i \in \{1, 2, \ldots, n\}$ there exists an open neighborhood N^i of $\mathcal{S}_c^i$ such that $u, v \notin N^i$. Since the sets $\mathcal{S}_c^i$ are mutually disjoint compact sets, the sets N^i may also be taken to be mutually disjoint.[1]

Case 2. $\mathcal{M}_c$ is Not Empty. So, we have n mutually disjoint compact sets $\mathcal{S}_c^i$ and a nonempty set $\mathcal{M}_c$. Moreover, all the pairs

$$\mathcal{S}_c^i \cap \overline{\mathcal{M}_c}, \quad \mathcal{S}_c^j \cap \overline{\mathcal{M}_c}, \quad i, j = 1, 2, \ldots, n, \ i \neq j$$

are not connected through $\mathcal{M}_c$ since assumption 3 was assumed false. Using Lemma 12.11, we can find n mutually disjoint open sets N^i such that (12.6) is satisfied. Since $u, v \notin \mathbb{K}_c$, we may assume that $u, v \notin N^i$.

This proves Claim 12.2.

Claim 12.3. There exists a closed set $\hat{F}$ such that $\hat{F}$ separates u and v and

$$\inf_{x \in \hat{F}} \Phi(x) \geq c \quad \text{and} \quad \hat{F} \cap (\mathcal{S}_c \cup \mathcal{M}_c) = \varnothing.$$

Indeed, set $Y_i^c = N^i \setminus \Phi^c$ for each $i \in \{1, 2, \ldots, n\}$. Then for each $i \in \{1, 2, \ldots, n\}$ and any $x \in \mathcal{S}_c^i$, there is a ball $B(x, \varepsilon_x)$ such that for each connected component $\mathcal{U}$ of

---

[1] A metric space is a regular topological space.

Φ^c with $C^i \neq \mathcal{U}$, $B(x, \varepsilon_x) \cap \mathcal{U} = \varnothing$. Otherwise, x would be a saddle point of mountain pass type, a contradiction with the fact that assumption 2 was assumed false. Set

$$T_i^c = \bigcup_{x \in \mathcal{S}_c^i} B(x, \varepsilon_x/2) \cap \partial C_i \cap N^i \cap N_{\delta_1}(\mathcal{S}_c^i). \tag{12.7}$$

We have that

$$\mathcal{S}_c^i \subset T_i^c, \qquad T_i^c \subset N^i \cap \partial C_i, \tag{12.8}$$

and T_i^c is open relative to $N^i \cap \partial C_i$. And for any component $\mathcal{U}$ of Φ^c such that $\mathcal{U} \neq C_i$, $T_i^c \cap \partial \mathcal{U} = \varnothing$. Now, let

$$\hat{F} = \left[(F \cap (X \setminus \Phi^c)) \setminus \left(\bigcup_{i=1}^{n} N^i \right) \right] \cup \left(\bigcup_{i=1}^{n} \partial Y_i^c \setminus T_i^c \right).$$

Then, $\inf_{x \in \hat{F}} \Phi(x) \geq c$. Since $F \cap (X \setminus \Phi^c)$ separates u and v and by Claims 12.1 and 12.2, (12.8) and the fact that $Y_i^c = N^i \setminus \Phi^c$, we can apply Lemma 12.11 with $A_i = C_i$, $G = \Phi^c$, $Z_i = N^i$, $Y_i = Y_i^c$, $T_i = T_i^c$ for $i = 1, 2, \ldots, n$ to conclude that $\hat{F}$ separates u and v.

On the other hand, since $\mathcal{M}_c \cap (\overline{\Phi^c} \setminus \Phi^c) = \varnothing$, we have by (12.6) and since $Y_i^c = N^i \setminus \Phi^c$ that $\partial Y_i^c \cap \mathcal{M}_c = \varnothing$. Therefore, by (12.5) and (12.7), we have that $\cup_{i=1}^n (\partial Y_i^c \setminus T_i^c) \cap (\mathcal{S}_c \cup \mathcal{M}_c) = \varnothing$. Hence, $\hat{F} \cap (\mathcal{M}_c \cup \mathcal{S}_c) = \varnothing$. This proves Claim 12.3.

So, by Theorem 12.8, $\hat{F} \cap \mathcal{P}_c$ and hence $\mathcal{P}_c$ must contain a compact subset that separates u and v. This implies assertion 1 and finishes the proof of the theorem. $\square$

As corollaries, we get all the aforementioned results.

The following interesting result on the cardinal of the critical set in the MPT is also proved by Fang in [387].

Corollary 12.13. *Suppose that* $\dim X \geq 2$. *Under the hypotheses of Theorem 12.12, the following alternative holds.*

    a. $\mathbb{K}_c$ *contains a saddle point of mountain pass type.*
    b. The cardinal of $\mathcal{P}_c$ *is at least the same as the continuum.*
    c. The cardinal of $\mathcal{M}_c$ *is at least the same as the continuum.*

This last result implies the following result to be compared to Corollary 9.7.

Corollary 12.14. *Suppose that* Φ *has a local maximum and a local minimum on a Banach space* X. *If* Φ *satisfies* (PS) *and* $\dim X \geq 2$, *then* Φ *has necessarily a third critical point.*

Proof. Suppose that u_1 is a local minimum and u_2 is a local maximum. If Φ is not bounded from below, then we have a mountain pass situation with u_1 as an initial point and Corollary 12.13 yields either an infinite number of critical points or a saddle point of mountain pass type which is necessarily distinct from both u_1 and u_2. On the other hand, if Φ is bounded from below, since it satisfies (PS), we know that (see Chapter 10) Φ is coercive. Hence, we have a mountain pass situation for $-\Phi$ with u_2 as initial point. Again, Corollary 12.13 applies and this proves our result. $\square$

Comments and Additional Notes

In this chapter, the distinction between the finite and the infinite dimension of X is the most visible. And the more interesting case to study is not when dim $X = \infty$.

◇ 12.1 Morse Index at Critical Points in the MPT

Consider a C^2-functional Φ and u is a critical point of Φ. We recall that the (generalized) Morse index, $m(u)$, of u is the dimension of $E^0 \oplus E^-$ where E^0 is the kernel of Φ'' and E^+ (resp. E^-) is the subspace where Φ'' is positive (resp. negative) definite. And u is nondegenerate if $E^0 = \{0\}$.

An interesting result concerning the Morse index at critical points in the MPT is the following.

Theorem 12.15. *Suppose that Φ satisfies the assumptions of the MPT and suppose that $\mathbb{K}_c$ is discrete. Then, there exists $u^* \in \mathbb{K}_c$ such that $m(u^*) \leq 1$. Moreover, if u^* is nondegenerate, then $m(u^*) = 1$.*

It has been found independently by Ambrosetti [40] in the nondegenerate case and by Hofer [482] in the general case (of possibly degenerate critical points).

A Sketch of the Proof of $m(u^*) = 1$ When u^* Is Nondegenerate (cf. [43]). Without loss of generality, we can take $u^* = 0$. By contradiction, suppose that $m(u^*) \geq 2$. Then, $E = E^- \oplus E^+$, with dim $E^- \geq 2$, and each u can be written $u = u^- + u^+$, where $u^\pm \in E^\pm$.

If $u^* = 0$ is nondegenerate, by the Morse lemma, up to a regular change of coordinates we have

$$\Phi(u) = c - \|u^-\|^2 + \|u^+\|^2 + R(u),$$

where $R(0) = R'(0) = 0$. Consider the neighborhood $\mathcal{U}$ of $u^* = 0$,

$$\mathcal{U} = \left\{ u = u^- + u^+; \ \|u^-\| < \alpha, \ \|u^+\| < \beta, \right\}, \qquad \beta > \alpha > 0. \tag{12.9}$$

For all $u \in \overline{\mathcal{U}}$, such that $\|u^+\| = \beta$,

$$\Phi(u) \geq c - \alpha^2 + \beta^2 + o(\alpha^2 + \beta^2)$$

and, hence, taking $\beta > \alpha > 0$ small enough, we get that

$$\inf\left\{ \Phi(u); \ u \in \overline{\mathcal{U}}, \ \|u^+\| = \beta \right\} \geq d > c. \tag{12.10}$$

Let $\delta > 0$ be such that $\delta < d - c$. By the definition of c there exists $\gamma \in \Gamma$ such that $\Phi(\gamma(t)) \leq c + \delta$, for all $t \in [0, 1]$. For the same $\delta > 0$ and to $\mathcal{U}$ given by (12.9) we find $\eta \in C(E; E)$ such that

$$\eta(\Phi^{c+\delta} \setminus \mathcal{U}) \subset \Phi^{c-\delta}. \tag{12.11}$$

Then, $\eta \circ \gamma \in \Gamma$. If γ does not intersect $\mathcal{U}$, (12.11) yields a contradiction. Let $t_0, t_1 \in]0, 1[$ be such that $\gamma(t_0) = z_0$ and $\gamma(t_1) = z_1 \in \partial\mathcal{U}$ while $\gamma(t) \notin \overline{\mathcal{U}}$, for all $t < t_0$ and $t > t_1$. Since $\Phi(z_i) \leq c + \delta < d$, (12.10) implies $\|z_i^-\| = \alpha$, $\|z_i^+\| < \beta$ for $i = 0, 1$. Let σ_i be the segment joining z_i and z_i^-. Then the restriction $\Phi|_{\sigma_i} \leq c + \delta$.

Last, if dim $E^- \geq 0$, we can connect z_0^- and z_1^- by an arc τ contained in $\partial U \cap e^-$. In particular, $\Phi|_\tau < c$. Let $\overline{\gamma}$ be the path which coincides with γ for $t \in [0, t_0] \cup [t_1, 1]$ and with $\{\sigma_0\} \cup \{\sigma_1\} \cap \{\tau\}$ elsewhere. Then $\overline{\gamma} \in \Gamma$, $\Gamma|_{\overline{\gamma}} \leq c + \delta$ and $\{\overline{\gamma}\} \cap U = \varnothing$. Then, by (12.11), $\Phi|_{\eta \circ \overline{\gamma}} \leq c - \delta$, which is a contradiction. So, $m(u^*) \leq 1$. If $m(u^*) = 0$, we could take a neighborhood U of u^* in such a way that $\Phi|_{\partial U} \geq d > c$ and the conclusion follows as before.   □

◇ 12.II Topological Degree at a Mountain Pass Point

And now, we will see a local description of the behavior of a functional near a critical point given by the MPT. In [482], Hofer proved that under some conditions on Φ, the Leray-Schauder *topological degree* at a critical point of mountain pass type is -1. This is visible on C^2-functionals Φ on a Hilbert space whose gradient has the form of a compact perturbation of the identity. Indeed, its linearization at a nondegenerate critical point given by the MPT must have exactly one negative eigenvalue. Consequently, its local degree is -1.

Let $\Phi \colon U \subset H \to \mathbb{R}$ be a C^2-functional where U is some nonempty open subset of a Hilbert space H. Assume that the gradient of Φ has the form $\Phi' = I_d + K$ where K is compact. Assume also that for any $u_0 \in \mathbb{K}$, the *smallest* eigenvalue λ_1 of the linearization $\Phi''(u_0) \in \mathcal{L}(H)$ at u_0, referred to in the sequel as the first eigenvalue, is simple.

Theorem 12.16 (Topological Degree at an MPP, Hofer). *Suppose that $u_0 \in U$ is an isolated critical point of Φ, of mountain pass type. Then the local topological degree at u_0 is -1.*

Without the particular form required on the functional in this statement, the result is not true. Indeed, consider in the real plane the function

$$\Phi(x, y) = x^4 + y^4 - 8x^2 y^2,$$

and the two points $e_0 = (1, -1)$ and $e_1 = (1, -1)$. Then, they satisfy

$$-6 = \max\{\Phi(e_0), \Phi(e_1)\} < \inf_{\gamma \in \Gamma} \max_{t \in [0, 1]} \Phi(\gamma(t)) = 0,$$

where Γ is the usual set of all continuous paths connecting e_0 and e_1 (Figure 12.2). It satisfies the *strong form* of the MPT of Hofer (see the following notes) and $\mathbb{K}_0$ is reduced to the singleton $\{(0, 0)\}$, necessarily of mountain pass type, whose local degree is -3 and not -1. This example is due to Hofer [482] where the proof of the aforementioned result may be found.

The assumptions required on Φ are not restrictive in the sense that they are verified in second-order elliptic problems.

Indeed, consider the differential equation

$$\begin{cases} -\Delta u = f(x, u) & \text{in } \Omega, \\ \quad u = 0 & \text{on } \partial\Omega, \end{cases} \tag{12.12}$$

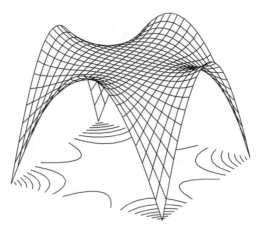

Figure 12.2. A mountain pass point with a topological degree different from -1.

where Ω is a bounded domain with smooth boundary of $\mathbb{R}^N$, $N \geq 3$. Assume that $f \in C^1(\overline{\Omega} \times \mathbb{R}; \mathbb{R})$ and that the following growth condition holds:

$$\left| \frac{\partial f(x,s)}{\partial s} \right| \leq C(1 + |s|^{\sigma-1}), \qquad \forall (x,s) \in \overline{\Omega} \times \mathbb{R},$$

and

$$1 \leq \sigma \leq (N+2)/(N-2).$$

Then the associated functional Φ is of class C^2 on $H_0^1(\Omega)$. Denote by λ_1 the first eigenvalue of $\Phi''(u_0)$ at a critical point u_0 of Φ and assume $\lambda_1 = 0$. Denote by $b(x) = f'(x, u_0(x))$, $b \in C(\Omega)$ and consider the eigenvalue problem

$$\begin{cases} -\Delta u = \hat{\lambda} bu & \text{in } \Omega, \\ u = 0 & \text{in } \partial\Omega. \end{cases}$$

It possesses a smallest positive eigenvalue $\hat{\lambda} = 1$ with a corresponding eigenfunction u_1 that does not change sign and spans a one-dimensional space. By the fact that $\hat{\lambda} = 1$, the functional Φ has the form required in Hofer's result.

◇ 12.III A Counterexample

Pucci and Serrin reported in [728] that when $\mathbb{K}_c$ consists of isolated critical points or reduces to a single point, one may think that it should contain not only a mountain pass point z but also z must have the property that for any of its neighborhood N, the sets $N \cap \Phi_c$ and $N \cap \Phi^c$ are nonempty and not path-connected. However, this is untrue, as shown by the following example by Pucci and Serrin [728].

Example 12.1. Consider the function

$$\Phi(x, y, z) = x^2 + y^2 + z^2(3 - z).$$

Then the MPT applies with $e = (0, 0, 3)$ and $a = R = 1$. The critical value c equals 4 and $\mathbb{K}_c = \mathbb{K}_4 = \{(0, 0, 2)\}$, while $z^* = (0, 0, 2)$ is a saddle point of MP type for which the sets

$$\{(x, y, z) \in \mathbb{R}^3; \ (x, y, z) \in B(z^*, r) \text{ and } \Phi(x, y, z) > 4\}, \ r > 0$$

are nonempty and path-connected.

◇ 12.IV Stability Properties of Critical Values Obtained by Minimax Procedures

According to [362], critical values obtained as minimax values enjoy good *stability* properties. When a critical point is degenerate, even a C^1-perturbation can destroy the corresponding critical value. But for a class of compact subsets Γ in X, the functional

$$\Phi \mapsto c(\Phi, c) = \inf_{\gamma \in \Gamma} \max_{\gamma} \Phi$$

is continuous in the C^0-topology; that is,

$$\max_{X} \| \Phi(x) - \Psi(y) \| \to 0 \qquad \text{implies that} \qquad |c(\Phi, \Gamma) - c(\Phi, \Psi)| \to 0.$$

In [115], Benkert studied C^1-functionals $\Phi^* = \Phi + \Psi : M \to \mathbb{R}$ that satisfy the (PS) condition; M is either a Banach space X or a bounded hypersurface in X. If c is a minimax critical value of Φ then there exists a corresponding minimax critical value c_* of Φ_*, which is close to c, provided Ψ is *small*. Applications to semilinear Dirichlet problems are given.

◇ 12.V Classification of Critical Points for Nonsmooth Functionals

In [390], Fang extends the smooth theory of classification and localization seen earlier to the case of continuous functionals in the spirit of what has been done in Chapter 16. Moreover, as in his thesis [387], he also obtains similar results on the structure of the critical set generated by min-max principles. He considers homotopy, cohomotopy, and homology classes.

◇ 12.VI A Strong Form of the MPT

Consider a triplet $(e_0, e_1, c) \in E \times \mathbb{R} \times \mathbb{R}$ such that

$$c = \inf_{\gamma \in \Gamma} \max_{t \in [0,1]} \Phi(\gamma(t)) > \max\{\Phi(e_0, e_1)\},$$

where Γ is, as usual, the set of all continuous paths joining e_0 and e_1. Such a triplet is called a *mountain pass characterization* by Hofer when Φ satisfies $(PS)_c$.

> In the statement of the MPT, if we assume only that (e_0, e_1, c) is a mountain pass characterization, without requiring that e_0, or e_1 is a minimum, then $\mathbb{K}_c \neq \emptyset$.

13

Weighted Palais-Smale Conditions

Plus ça change, plus c'est la même chose.
[The more that changes, the more it's the same.]
> Alphonse Kaar, reported in *Dictionnaire des citations de langue française* by
> P. Ripert, Bookking International, Paris, 1995

When nonlinear problems have a variational structure but do not present enough compactness to use the MPT in its classical form, we have to study carefully the behavior of the Palais-Smale sequence for the inf max level in the statement of the MPT. In this chapter, we will review some of the results that refined the MPT to treat this situation. We will discuss some variants of the MPT where "weighted" Palais-Smale conditions, weaker than the classical one, appear. Then, we will see a very interesting procedure, attributable to Corvellec [257], for deducing new critical point theorems with weighted Palais-Smale conditions from older ones with the standard Palais-Smale condition just by performing a change of the metric of the underlying space X.

In many nonlinear problems with a variational structure, the standard methods of calculus of variations, including the MPT, do not apply in a direct way because of *lack of compactness*.

Technically, the lack of compactness (failure of (PS)) arises when dealing, for example, with semilinear elliptic problems on the whole space $\Omega = \mathbb{R}^N$, or even on a bounded domain Ω but with $p = 2^* = 2N/(N-2)$ for $N \geq 3$, the embedding of $H^1(\Omega)$ in $L^p(\Omega)$ being no more compact. As noticed, for example, by Brézis [147] and Willem [957], this difficulty appears in some problems because of their invariance under some transformations (*noncompact groups* like dilatations (the case of the Sobolev exponent) and translations (the case of a domain like $\mathbb{R}^N$)).

Nevertheless, a careful study of the behavior of the Palais-Smale sequences may provide useful information that leads sometimes to a solution.

In this chapter, we will not discuss one of the different ways used to overcome this lack of compactness. The reader is referred to the many good books available in the literature, especially [957] by Willem which focuses particularly on this subject. We will deal with this situation *in relation* to the MPT. More precisely, we will review some results giving some additional (second-order) information on the Palais-Smale sequences in the situation of the MPT and we will also see some variants of the MPT that *allow the competing paths in* Γ *to roam freely* but require (PS) on bounded regions and control the growth of $\|\Phi'(u)\|^{-1}$ near infinity. This has the same effect as (PS); it allows one to deal with unbounded regions in a uniform way.

Another way is to restrict the competing paths to a bounded region. This can be done only if one can be assured that the paths will not leave the region as they approach the optimal one. This second approach is discussed in Chapter 21.

13.1   "Weighted" Palais-Smale Conditions

We will discuss some versions of the Palais-Smale condition with *weights*.

As expressed by Schechter [793], Cerami [189] and Bartolo, Benci, and Fortunato [114] require the (usual) Palais-Smale condition (PS) on bounded regions, while they control the growth $\|\Phi'(u)\|^{-1}$ near infinity, and this has the *same effect* as (PS). This allows one to deal with unbounded regions in a uniform way. In the next section, we will see an ingenious fashion, attributed to Corvellec [257], to get new critical point theorems with "weighted" Palais-Smale conditions from the old ones.

But beforehand, we will get an idea about the way people used to deal with these newer and weaker forms of (PS). Beginning with [114], it was easily seen that the Palais-Smale condition at the level c, $(PS)_c$, was equivalent to the combination of the following two assumptions.

   i. Every bounded sequence $(u_n)_n$ such that

$$(\Phi(u_n))_n \to_{n\to\infty} 0 \qquad \text{and} \qquad \Phi'(u_n) \to_{n\to\infty} 0$$

      possesses a convergent subsequence.

  ii. $\big((\Phi(u_n))_n$ bounded and $\|u_n\| \to_{n\to\infty} \infty \big)$ implies that $\big(\|\Phi'(u_n)\| \geq \alpha > 0$ for n sufficiently large$\big)$.

Assumption ii may be also written as follows:
There are $R > 0, \sigma > 0$, and $\alpha > 0$ such that

$$\left(u \in \Phi_{c-\sigma}^{c+\sigma} \text{ and } \|u\| \geq R \right) \qquad \text{implies that} \qquad \|\Phi'(u)\| \geq \alpha.$$

We know that the MPT holds true for this condition; this is just the classical MPT.

A condition introduced by Cerami [189] supposed that

$$\|\Phi'(u)\|(1 + \|u\|) \geq \alpha$$

in assumption ii, while in [114], Benci, Bartolo, and Fortunato considered

$$\|\Phi'(u)\|\|u\| \geq \alpha.$$

A natural question that arises then is whether we can go even further. Would we still be able to derive a similar critical point theorem if we suppose in assumption ii that the functional $\alpha(u)$ *tends quickly* to 0 when we approach 0. We will see that this is indeed the case. Our approach will be based on a yet unpublished work of El Amrouss and Moussaoui (personal communication, 1998).

Consider the following weighted (PS).

Definition 13.1. A real-valued C^1-functional Φ defined on a Banach space X is said to satisfy the $(C)_c^{\alpha(\cdot)}$ condition if

i. every bounded sequence $(u_n)_n$ such that

$$\begin{cases} (\Phi(u_n))_n \text{ is bounded, and} \\ \Phi'(u_n) \to_{n \to \infty} 0 \end{cases}$$

  possesses a convergent subsequence.
ii. There are constants $R > 0$, and $\sigma > 0$ such that

$$\left(u \in \Phi_{c-\sigma}^{c+\sigma} \text{ and } \|u\| \geq R \right) \qquad \text{implies that} \qquad \|\Phi'(u)\| \geq \alpha(\|u\|),$$

  where $\alpha : \,]0, \infty[\to \,]0, \infty[$ is continuous and satisfies
  1. $\dfrac{1}{\alpha(.)}$ is locally Lipschitz continuous on $]0, \infty[$, and
  2. the real-valued function $\rho \mapsto t^+\rho$ defined from $]0, \infty[$ to $[0, \infty]$ is continuous, where $[0, t^+\rho[$ is the half interval on which the maximal solution of the Cauchy problem

$$\begin{cases} \dfrac{du(t)}{dt} = \dfrac{1}{\alpha(u(t))} \\ u(0) = \rho > 0 \end{cases}$$

  is defined.

The functional Φ is said to satisfy $(C)^{\alpha(\cdot)}$ if it satisfies $(C)_c^{\alpha(\cdot)}$ for any $c \in \mathbb{R}$.

Example 13.1.
  1. Let $\alpha : \,]0, \infty[\to \,]0, \infty[$ be continuous such that $\dfrac{1}{\alpha(.)}$ is locally Lipschitz continuous on $]0, \infty[$ and $\int_1^\infty \alpha(s)\,ds = +\infty$. Then, we may check easily that $t_\rho^+ = +\infty$ for any $\rho > 0$.
  2. For $\alpha(s) = \dfrac{1}{s^q}$, where $q > 1$, we get $t_\rho^+ = \dfrac{1}{q-1}(1/\rho)^{q-1}$ for any $\rho > 0$.
  3. For $\alpha(s) = \dfrac{s^q}{\exp(s^{q+1})}$, where $q > 1$, we get $t_\rho^+ = \dfrac{1}{q+1}\exp(-\rho^{q+1})$ for any $\rho > 0$.

Notice that when $\alpha(s) = $ constant (we are in the first case, $\int_1^\infty \alpha(s)\,ds = +\infty$), $(C)^{\alpha(\cdot)}$ reduces to the usual (PS) condition, while when $\alpha(s) = A/s$ where $A = $ constant > 0 or $\alpha(s) = A/(1+s)$ (we are still in the first case) we get, respectively, the forms of (PS) from [114] and [189].

With $(C)_c^{\alpha(\cdot)}$, it is possible to prove a deformation lemma that can be used to get critical point theorems as usual.

Theorem 13.1 (Deformation Lemma, El Amrouss and Moussaoui). *Let X be a Banach space and $\Phi\colon X \to \mathbb{R}$ a C^1-functional that satisfies $(C)_c^{\alpha(\cdot)}$ for some $c \in \mathbb{R}$. For $\bar\varepsilon \in]0, \sigma[$ (where σ is given in assumption ii of the definition of $(C)_c^{\alpha(\cdot)}$) and $\delta > 0$, there exist $\varepsilon \in]0, \bar\varepsilon[$ and $\eta \in C(X \times \mathbb{R}; \mathbb{R})$ such that*

    i. $\eta(0, x) = x$, *for any $x \in X$,*
    ii. $\eta(t, .)$ *is a homeomorphism on X for any $t \in [0, 1]$,*
    iii. $\eta(0, x) = x$, *for any $t \in [0, 1]$ if $x \in \Phi_{c-\bar\varepsilon}^{c+\bar\varepsilon}$,*
    iv. $\Phi(\eta(., x))$ *is nondecreasing in $[0, 1]$, for any $x \in X$,*
    v. $\eta(1, \Phi_{c-\varepsilon} \setminus (\mathbb{K}_c)_\delta) \subset \Phi_{c+\varepsilon}$,
    vi.   *if $\mathbb{K}_c = \varnothing$, then $\eta(1, \Phi_{c-\varepsilon}) \subset \Phi_{c+\varepsilon}$,*
    vii. $\sup_{\substack{t\in[0,1] \\ x\in A}} \|\eta(t, x)\| + \|\eta^{-1}(t, x)\| < \infty$, *for any bounded subset $A \subset X$,*
    viii.   *when Φ is even, $\eta(t, .)$ is odd for any $t \in [0, 1]$.*

The proof of this deformation lemma is reported through a series of lemmata.

Lemma 13.2. *Let W be locally Lipschitz on X. For each $x \in X$, the Cauchy problem*

$$
\begin{cases}
\dfrac{d\eta(t, x)}{dt} = W(\eta(t, x)) \\[2mm]
\eta(t, x) = 2x > 0
\end{cases}
$$

has a unique maximal solution defined on an open interval $]w_-(x), w_+(x)[$ containing 0. The set $D = \big\{(t, x);\ w_-(x) < t < w_+(x)\big\}$ is open in $\mathbb{R} \times X$ and $\eta\colon D \to X$, $(t, x) \mapsto \eta(t, x)$ is continuous. Moreover, $x \mapsto w_-(x)$, $x \mapsto w_+(x)$ are l.s.c. from X to $]0, +\infty[$.

See Palais [697].

The next lemma is a direct consequence of a theorem on differential inequalities (cf. [546]).

Lemma 13.3 (Differential Inequalities Theorem). *Let h be a locally Lipschitz function defined on $]0, \infty[$ and let u be the unique solution of the Cauchy problem,*

$$
u'(t) = h(u(t)), \qquad u(0) = a,
$$

defined on $[0, b[$, $b > 0$.
Let v be a functional with values in X. If v is a solution of the differential inequality

$$
|D_r\|v(t)\|\,| \le h(\|v(t)\|) \text{ in }]-b, b[, \qquad and \ \|v(t)\| \le u(0),
$$

that is, v is continuous and its right derivative D_r satisfies the above inequality, then

$$
\|v(t)\| \le u(t) \quad and \quad \|v(-t)\| \le u(t), \text{ for all } t \in [0, b[.
$$

We can also easily prove the following result.

Lemma 13.4. *Let* $u : [a, b] \to E$ *be of class* C^1*; then* $D_r \|u(t)\|$ *exists on* $a \leq t < b$ *and*

$$|D_r \|u(t)\|\,| \leq \left\|\frac{d}{dt} u(t)\right\| \qquad in \ ]a, b[.$$

Now, we can proceed to the proof of the deformation lemma. The solution is obtained as a solution of a differential inequality.

Proof. By $(C)_c^{\alpha(\cdot)}$ assumption ii,

$$\|\Phi'(u)\| \geq \alpha(\|u\|), \ \text{for every} \ u \in \Phi_{c-\bar{\varepsilon}}^{c+\bar{\varepsilon}} \setminus B_R(0).$$

Consider $R' > R$ such that $(\mathbb{K}_c)_\delta \subset B_{R'}$.
Then, by $(C)_c^{\alpha(\cdot)}$ assumption i, there is $0 < \hat{\varepsilon} < \bar{\varepsilon}$ and $b > 0$ such that

$$\|\Phi'(u)\| \geq b, \ \text{for every} \ u \in (\Phi_{c-\hat{\varepsilon}}^{c+\hat{\varepsilon}} \setminus (\mathbb{K}_c)_{\delta/4}) \cap B'_R(0).$$

By the pseudo-gradient lemma, there exists a locally Lipschitz mapping $V : \tilde{X} \to X$ that may be chosen odd if Φ is even, such that

$$\|V(x)\| < 2 \qquad \text{and} \qquad \langle V(x), \Phi'(x)\rangle \geq \|\Phi'(x)\|.$$

Consider $\varepsilon_1 \in]0, \hat{\varepsilon}[$ and set $A = \Phi_{c-\hat{\varepsilon}}^{c+\hat{\varepsilon}}$ and $B = \Phi_{c-\varepsilon_1}^{c+\varepsilon_1}$. Define

$$g(x) = \frac{\text{dist}(x, X \setminus A)}{\text{dist}(x, B) + \text{dist}(x, X \setminus A)},$$

$$\ell(x) = \frac{\text{dist}(x, X \setminus A)}{\text{dist}(x, (\mathbb{K}_c)_{\delta/4}) + \text{dist}(x, (\mathbb{K}_c)_{\delta/2})},$$

and

$$h(s) = \begin{cases} (2\alpha(R'))^{-1} & \text{if } s \leq R', \\ (2\alpha(s))^{-1} & \text{if } s \geq R'. \end{cases}$$

Set

$$W(x) = \begin{cases} g(x)\ell(x)h(\|x\|)V(x) & \text{if } x \in \tilde{X}, \\ 0 & \text{otherwise.} \end{cases}$$

Then, W is well defined on X, locally Lipschitz, and it is odd if Φ is even. By Lemma 13.2, the Cauchy problem,

$$\begin{cases} \dfrac{d\eta(t, x)}{dt} = W(\eta(t, x)), \\ \eta(t, x) \, x, \end{cases}$$

has a unique maximal solution $\hat{\eta}$ defined on an open interval $]w_-(x), w_+(x)[$ containing 0.

Meanwhile, by Lemma 13.4, we have that $D_r||\hat{\eta}(t, x)|| \, | \leq \left\| \frac{d\hat{\eta}(t,x)}{dt} \right\|$. Since $\left\| \frac{d\hat{\eta}(t,x)}{dt} \right\| = ||W(\hat{\eta}(t, x))|| \leq 2h(\hat{\eta}(t, x))$, we get

$$D_r||\hat{\eta}(t, x)|| \, | \leq 2h(\hat{\eta}(t, x)).$$

By the definition of h, the problem

$$\begin{cases} u'(t) = \dfrac{1}{\alpha u(t)} \\ u(0) \, ||x|| \geq R' \end{cases}$$

admits a nondecreasing solution, which is also a solution of

$$\begin{cases} u'(t) = 2h(u(t)) \\ u(0) \, ||x|| \geq R', \end{cases}$$

and conversely. Using Lemma 13.3, we can easily check that

$$\begin{cases} w_+(x) \geq t^+_{||x||} & \text{if } ||x|| \geq R \\ w_-(x) \leq -t^+_{||x||} & \text{if } ||x|| \geq R \end{cases}$$

and

$$\begin{cases} w_+(x) \geq t^+_{R'} & \text{if } ||x|| \leq R \\ w_-(x) \leq -t^+_{R'} & \text{if } ||x|| \leq R. \end{cases}$$

Let q be the functional defined by

$$\begin{cases} q(u) = 1 & \text{if } u \in B_{R'+\delta/2} \\ q(u) = 0 & \text{if } u \in X \setminus B_{R'+\delta} \\ 0 \leq q(u) \leq 1 & \text{otherwise.} \end{cases}$$

Then q is Lipschitz in X. Set $\chi(x) = q(x) = W(x)$, then χ is locally Lipschitz and bounded. (W is bounded on bounded sets.) Consider $k > 0$ such that $\delta/(2k) \leq 1$ and $||\chi(x)|| \leq k$ for all $x \in X$. Then, the solution $\xi(t, x)$ of the problem

$$\frac{d\xi}{dt} = \chi(\xi), \qquad \xi(0, x) = x,$$

is well defined and continuous on $\mathbb{R} \times X$ and satisfies

$$||\xi(t, x)|| \leq ||x|| + k|t|, \qquad \text{for every } t \in \mathbb{R}, \ x \in X.$$

By Lemma 13.2,

$$\hat{\eta}(t, x) = \eta(t, x) \qquad \text{for every } t \in \left[\frac{-\delta}{2k}, \frac{\delta}{2k} \right] \text{ and } x \in \overline{B_{R'}}.$$

Our deformation η is defined by

$$
\begin{cases}
\xi(t\, t_{R'}^+(\delta/2k), x) & \text{if } t \in [-1,1],\, t_{R'}^+ < 1 \text{ and } x \in \overline{B_{R'}}, \\
\xi(t(\delta/2k), x) & \text{if } t \in [-1,1],\, t_{R'}^+ \ge 1 \text{ and } x \in \overline{B_{R'}}, \\
\xi(t\, t_{\|x\|}^+(\delta/2k), x) & \text{if } t \in [-1,1],\, t_{\|x\|}^+ < 1 \text{ and } x \in X \setminus \overline{B_{R'}}, \\
\xi(t(\delta/2k), x) & \text{if } t \in [-1,1],\, t_{\|x\|}^+ \ge 1 \text{ and } x \in X \setminus \overline{B_{R'}}.
\end{cases}
$$

It is continuous on $[-1,1] \times X$ and satisfies the assumption i–viii. $\quad\square$

Theorem 13.5 (Linking Theorem, El Amrouss and Moussaoui). *Suppose that* $\Phi \in C^1(X;\mathbb{R})$ *and that there exists a nonempty bounded and closed set* Q *satisfying the following:*

H_1. *There exists* $\beta \in \mathbb{R}$ *such that for any* $\gamma \in \Gamma = \{\gamma\, homeomorphism\ on\ X;$
$\gamma|_{\partial Q} = Id\}$,

$$
\begin{cases}
\alpha = \sup_{\partial Q} \Phi < \beta & and \\
\gamma(Q) \cap \Phi^\delta \ne \varnothing & for\ any\ \delta < \beta,
\end{cases}
$$

H_2. $\sup_{u \in Q} \Phi(u) < \infty$.

Suppose that Φ *satisfies* $(C)_c^{\alpha(.)}$ *where*

$$
c = \inf_{\gamma \in \Gamma} \sup_{u \in Q} \Phi(\gamma(u)).
$$

Then c is a critical value of Φ such that $c \ge \beta$.

Proof. By contradiction, suppose that c is regular, that is, $\mathbb{K}_c = \varnothing$. Let $\bar{\varepsilon} < \min(\sigma, \beta - \alpha)$, then by the former deformation lemma, there is an isomorphism $\eta(1,.): X \to X$ that satisfies

$$
\eta(1,x) = x \qquad \text{when} \qquad x \notin \Phi_{c-\bar{\varepsilon}}^{c+\bar{\varepsilon}},
$$

$$
\eta(1, \Phi_{c-\varepsilon}) \subset \Phi_{c+\varepsilon}, \tag{13.1}
$$

$$
\sup_{u \in Q} \left[\|\eta(1, \gamma(x))\| + \|\eta^{-1}(1, \gamma(x))\| \right] < \infty, \qquad \forall \gamma \in \Gamma. \tag{13.2}
$$

By the second part in H_1, there is a sequence $(x_n)_n \subset Q$ such that

$$
\beta - \frac{1}{n} \le \Phi(\gamma(x_n)), \qquad \forall n \ge 1,\ \forall \gamma \in \Gamma.
$$

So

$$
\beta \le \sup_n \Phi(\gamma(x_n)), \qquad \forall \gamma \in \Gamma,
$$

and hence $\beta \leq c$. By H_2, it follows that $\beta \leq c \leq \infty$.
By the first part in H_1,

$$\Phi(x) \leq \alpha < \beta, \qquad \text{for all } x \text{ on } \partial Q.$$

Hence, $\Phi(x) < c - \varepsilon$ because $\bar{\varepsilon} < \beta - \alpha$ and $\beta \leq c$.
From assumption ii of the deformation lemma, it follows that

$$\eta(1, x) = x \qquad \text{on } \partial Q. \tag{13.3}$$

By (13.2) and (13.3), $\eta^{-1}(1, .) \circ \gamma$ lies to Γ and by definition of c, there is $\bar{x} \in Q$ such that

$$\Phi(\eta^{-1}(1, \gamma(\bar{x}))) \geq c - \varepsilon.$$

So that by (13.1), we get

$$c + \varepsilon \leq \Phi(\eta(1, [\eta^{-1}(1, \gamma(\bar{x}))])) = \Phi(\gamma(\bar{x})).$$

Since this is possible for all $\gamma \in \Gamma$,

$$c + \varepsilon \leq \inf_{\gamma \in \Gamma} \sup_{x \in Q} \Phi(\gamma(x)),$$

which is contradiction with the definition of c.   □

This result extends many linking theorems, including the MPT and the versions by Bartolo, Benci, and Fortunato and Cerami.

13.2 Changing the Metric

For the material of this section, the reader is supposed to be acquainted with the metric approach to critical point theory presented in Chapter 16.

Consider a metric space (X, dist) and a continuous function $\Phi \colon X \to \mathbb{R}$. We will now see a very interesting procedure, attributable to Corvellec [257], for deducing a new critical point theorem with (PS) with a *weight* from older ones with the standard (PS) just by changing the metric of the space X.

Theorem 13.6. *Let (X, dist) be a metric space, $\tilde{A}$ be a nonempty subset of X, and $\beta \colon [0, +\infty[\to]0, +\infty[$ be continuous. Then, there exists a metric $\widetilde{\text{dist}}$ on X that is topologically equivalent to* dist *and such that*

a. *For any subset B of X, it holds that*

$$\widetilde{\text{dist}}(B, \tilde{A}) \geq \int_0^{\text{dist}(B, \tilde{A})} \beta(t)\, dt, \tag{13.4}$$

while if $\int_0^{+\infty} \beta(t)\, dt = +\infty$, then $(X, \widetilde{\text{dist}})$ is complete if and only if (X, dist) is complete.

b. *If* $\Phi: X \to \mathbb{R}$ *is continuous and* $|\widetilde{d}\Phi|$ *denotes the weak slope of* Φ *with respect to the metric* $\widetilde{\mathrm{dist}}$, *then*

$$|\widetilde{d}\Phi|(u) = \frac{|d\Phi|(u)}{\beta(\mathrm{dist}\,(u, \tilde{A}))} \qquad \text{for every } u \in X.$$

Proof. See the paper by Corvellec [257, pp. 270–273]. □

Remark 13.1. When X has a rich(er) topology, namely, X is a Banach space, $\tilde{A}$ is the origin and $\beta(t) = 1/(1+t)$, the metric $\widetilde{\mathrm{dist}}$ is the *Cerami metric* (see [189] or [360, p. 138]) where it is used in the context of critical point theory for smooth functions. This metric is used in [67] in the context of the critical point theory for continuous functions defined on complete metric space.

Therefore, $(X, \widetilde{\mathrm{dist}})$ is complete by Theorem 13.6a and the (PS) condition with weight $1/\beta(\|u\|) = 1 + \|u\|$ is just the usual (PS) condition for Φ in the metric space $(X, \widetilde{\mathrm{dist}})$.

Theorem 13.7. *Let X be a complete metric space, $\Phi: X \to \mathbb{R}$ continuous, and A, B two subsets of X such that B links A. Set*

$$a = \inf_A \Phi, \qquad b_0 = \sup_B \Phi, \qquad b - \inf_{\psi \in \Phi_B} \sup_{B \times [0,1]} \Phi \circ \psi,$$

where Φ_B is the set of contractions of B in X, and assume that $a, b \in \mathbb{R}$ and $b_0 > a$.
If $0 < \rho < \mathrm{dist}\,(B \cap \Phi_a, A)$ (resp. $0 < \rho < \mathrm{dist}\,(B, \Phi^b \cap A)$) and $\beta: [0, +\infty[\to]0, +\infty[$ is a continuous function such that

$$\int_0^{+\infty} \beta(t)\,dt = +\infty \qquad \text{and} \qquad \int_0^{\rho} \beta(t)\,dt \geq b_0 - a,$$

then, for any $\delta > 0$, there exists $u \in X$ with

$$\begin{cases} a - \delta \leq \Phi(u) \leq b + \delta \\ |d\Phi| \leq \beta(\mathrm{dist}\,(u, A)) \quad (\text{resp. } |d\Phi| \leq \beta(\mathrm{dist}\,(u, B))). \end{cases}$$

Proof. Consider the case $0 < \rho < \mathrm{dist}\,(B \cap \Phi_a, A)$. Let $\widetilde{\mathrm{dist}}$ be the metric given by Theorem 13.6, with $\tilde{A} = A$. Then $(X, \widetilde{\mathrm{dist}})$ is complete and

$$\widetilde{\mathrm{dist}}\,(B \cap \Phi^a, A) \geq \int_0^{\mathrm{dist}\,(B \cap \Phi_a, A)} \beta(t)\,dt > \int_0^{\rho} \beta(t)\,dt \geq b_0 - a > 0,$$

while $|\widetilde{d}\Phi|(u) = \dfrac{|d\phi|(u)}{\beta(\mathrm{dist}\,(u, A))}$ for every $u \in X$.

Applying Theorem 20.7b in $(X, \widetilde{\mathrm{dist}})$ with $\alpha = \widetilde{\mathrm{dist}}\,(B \cap \Phi_a, A)$ and $\rho = 1$ yields the conclusion.

In the case $0 < \rho < \mathrm{dist}\,(B, A \cap \Phi^{b_0})$, consider the metric $\widetilde{\mathrm{dist}}$ associated with β and $\tilde{A} = B$. □

Remark 13.2. For $\beta(t) = \sigma$, a positive constant, we recover Theorem 20.7b. This result extends Theorems 2.1 and 2.2 of Schechter [809], where the space X is a Banach space,

Φ is of class C^1, and β is a positive nondecreasing function. (Note that if $\inf \beta > 0$, Theorem 13.7 holds without assuming $b_0 > a$ thanks to Theorem 20.7a).

Comments and Additional Notes

Always concerning weighted Palais-Smale conditions, in the first note of Chapter 5 on page 53, the following form of (PS) by Silva and Teixeira [846] is reported.

Let $f: E \to \mathbb{R}$ be a C^1-mapping on the real Banach space E. Denote by Λ the set of nonincreasing, locally Lipschitz continuous functions $\phi: (0, \infty) \to (0, \infty)$ such that $\int_0^\infty \phi(t) \, dt = \infty$. Then, f is said to satisfy the *generalized Palais-Smale condition with respect to* $\phi \in \Lambda$ at $c \in \mathbb{R}$ if every sequence $(u_m)_m$ of points in E satisfying $f(u_m) \to c$ and $\|f'(u_m)\|/\phi(\|u_m\|) \to 0$ $(m \to \infty)$ has a convergent subsequence. The value $c \in \mathbb{R}$ is an admissible level for f if c is a regular value of f or c is an isolated critical value with discrete critical points over $f^{-1}(c)$.

Silva and Teixeira prove that, for an admissible critical level c, $f^{-1}(c)$ is arcwise-connected or f has a critical point with different critical value from c. This result also implies the MPT.

◇ 13.1 A Half MPT with a Loose End

The Palais-Smale condition has the drawback that it requires sets of the form $|\Phi(u) - c| < \varepsilon$, $\|\Phi'(u)\| < \varepsilon$ to be bounded for $\Phi > 0$ sufficiently small. Substitutes for this condition have been proposed by several authors [114, 189] but they must deal with $\Phi(u)$ and $\Phi'(u)$ in unbounded regions. Schechter [800] introduced a method that requires only a knowledge of $\Phi(u)$ and $\Phi'(u)$ on a bounded region.

The context of the Hampwile theorem and the ideas it presents are behind many variants of the MPT obtained by Schechter and Tintarev. Some of them that are quite interesting are presented here.

The MPT supposes $\max\{\Phi(0), \Phi(e)\} < \rho \le \inf_{S(0,\delta)}$ where $\|e\| > \delta$, while in [793], the author uses only assumptions on $\Phi(u)$ and $\Phi'(u)$ in a ball of radius $\|e\|$ and a local compactness conditions. No boundary conditions are supposed even though they intervene in the proof. Moreover, the condition $\Phi(0) < \rho$ is removed. The result is described as a *half MPT with a loose end*.

In fact, the starting point is the idea that if we are not seeking critical points of Φ but only eigenvectors, $u \in X \setminus \{0\}$ and $\lambda \in \mathbb{R}$ such that

$$\Phi'(u) = \lambda u.$$

It is expected that we would be able to relax some of the requirements made in the MPT because we are asking for less, and in [793], it is indeed the case.

Let H be a Hilbert space, $\Phi \in C^1(B(0, R); \mathbb{R})$ such that there is $e \in B(0, R)$, and a $\delta > 0$ such that $\delta < \|e\|$ and

$$\Phi(e) < \rho = \inf_{\|u\|=\delta} \Phi(u).$$

Let $\Gamma = \{h\colon [0, 1] \to B(0, R) \text{ continuous}; \|h(0)\| \leq \delta \text{ and } h(1) = e\}$. Defining

$$c = \inf_{h \in \Gamma} \max_{0 \leq s \leq 1} \Phi(h(s)),$$

clearly, $c \geq \rho$.

Two extensions of the Hampwile theorem are given in [793].

Theorem 13.8 (First Extension of the Hampwile Theorem, Schechter). *Assume that*

$$|(\Phi'(u), u)| \leq M, \qquad u \in B(0, R). \tag{13.5}$$

Then there is a sequence $(u_n)_n \subset B(0, R)$ *such that*

$$\Phi(u_n) \to c, \tag{13.6}$$

$$\Phi'(u) - \frac{(\Phi'(u), u)u}{\|u\|^2} \to 0 \qquad \textit{strongly in } H, \tag{13.7}$$

$$\limsup |(\Phi'(u), u)| \leq 0, \tag{13.8}$$

$$\|u_n\| \geq \delta, \qquad \forall u, \tag{13.9}$$

$$u_n \to u \textit{ weakly in } H, \tag{13.10}$$

and

$$\Phi'(u_n) \to \lambda u \textit{ weakly for some } \lambda \leq 0.$$

Theorem 13.9 (Second Extension of the Hampwile Theorem, Schechter). *Assume that there are* $\varepsilon > 0$, $\Theta < 1$ *such that*

$$(\Phi'(u), u) + \Theta\|u\|\,\|\Phi'(u)\| \geq 0$$

for all $u \in B(0, R)$ *satisfying*

$$|\Phi(u) - c| < 2\varepsilon \qquad \textit{and} \qquad \delta < \|u\| \leq R.$$

Then there is a sequence $(u_n)_n \subset B(0, R)$ *satisfying* (13.6)–(13.10) *and*

$$\Phi'(u) \to 0 \textit{ strongly in } H.$$

The second form implies the first one. Indeed, to prove it, it suffices to note that because every sequence in $B(0, R)$ has a weakly convergent subsequence, we need only to prove the existence of a sequence satisfying (13.6)–(13.10).

Moreover, the quantities $(\Phi'(u_n), u_n)$ and $(\Phi'(u_n), u_n)/\|u\|^2$ are bounded by (13.5). Thus, the second sequence has a sequence converging to some $\lambda \leq 0$. Hence,

$$\Phi'(u_n) = \Phi'(u_n) - \frac{(\Phi'(u_n), u_n)u_n}{\|u\|^2} + \frac{(\Phi'(u_n), u_n)u_n}{\|u\|^2}$$

converges weakly to λu. It therefore suffices to prove the existence of a sequence satisfying (13.6)–(13.9).

See also [806] (The *Hampwile alternative*) by Schechter.

◇ *13.II Complementary Readings*

The very interesting book [957] (Minimax theorems) by Willem is concerned with critical point theory of minimax type and its applications to partial differential equations lacking compactness. He uses his quantitative deformation lemma, which proves to be a particularly efficient tool.

The reader may also consult, for example, the papers [30, 78, 183, 218, 445, 595, 797, 846, 991] and especially to [147] by Brézis.

IV

The Landscape Becoming Less Smooth

14

The Semismooth MPT

Soon after Baire had introduced in 1908 the concept of semicontinuity for real valued functions, Tonelli, in 1914, recognized semicontinuity as one of the relevant properties of the functionals of the calculus of variations.

L. Cesari, *Optimization. Theory and applications*, Springer-Verlag, 1983

A notion of critical points for continuous convex perturbations of C^1-functionals defined on a Banach space X is introduced, and an appropriate version of the MPT is proved. The same minimax argument serving to obtain a critical level in the classical version is used here too, to get a critical value corresponding to the new notion of critical point.

The main result of this chapter is taken from Szulkin [891]. It extends critical point theory to functionals of the form $\Phi + \Psi$ in a real Banach space X, where $\Phi : X \to \mathbb{R}$ is a C^1-functional, and $\Psi : X \to \mathbb{R} \cup \{+\infty\}$ is a convex and lower semicontinuous (l.s.c.) perturbation. Denote by $\mathcal{D}(\Psi) = \{u \in X; \ \Psi(x) < +\infty\}$ the *effective domain* of Ψ. The most important point here is to find an appropriate definition of nonregularity/regularity in this particular context and the rest of the program of projecting C^1-critical point theory for functionals of this form *should be a priori easy*. This semismooth MPT was inspired by the presentation of the MPT by Mawhin and Willem [627, Theorem 4.3], which constitutes an *approximate* version of the nonsmooth minimax theorem of Shi [835] (see Chapter 15).

The notion of differentiability suitable for handling functionals like Ψ is the *subdifferentiability*. The *subdifferential of Ψ at a point* $u \in \mathcal{D}(\Psi)$ is the set

$$\partial \Psi = \{u^* \in X^*; \ \Psi(v) - \Psi(u) \geq \langle u^*, v - u \rangle \text{ for all } v \in X\},$$

where X^* is the dual space of X and $\langle ., . \rangle$ is the duality pairing between X^* and X. A point $u \in \mathcal{D}(\Psi)$ is *critical* if $0 \in \partial \Psi(u)$.

So, for a functional $f = \Phi + \Psi$, where $\Phi \in C^1(X; \mathbb{R})$ and $\Psi : X \to \mathbb{R} \cup \{+\infty\}$ is convex and l.s.c., a way to define a notion of critical points is the following.

Definition 14.1. A point $u \in \mathcal{D}(\Psi)$ is said to be *critical* of f if

$$0 \in \Phi'(u) + \partial \Psi(u),$$

that is,

$$-\Phi'(u) \in \partial\Psi(u),$$

or equivalently, if u satisfies

$$\langle\Phi'(u), v - u\rangle + \Psi(v) - \Psi(u) \geq 0, \qquad \text{for all } v \in X. \qquad (14.1)$$

Remark 14.1. If $\Psi \equiv 0$, then $f = \Phi \in \mathcal{C}^1(X;\mathbb{R})$ and this definition coincides with the usual one.

14.1 Preliminaries

We will suppose in all the sequels that $\Phi \in \mathcal{C}^1(X;\mathbb{R})$ and $\Psi : X \to \mathbb{R}\{+\infty\}$ is convex, l.s.c., and proper (that is, $\Psi \not\equiv +\infty$). We shall also use the following forms of Palais-Smale condition.

First, the functional f satisfies $(PS)_{Sz}$ if any sequence $(u_n)_n$ such that $f(u_n) \to c \in \mathbb{R}$ and

$$\langle\Phi'(u_n), v - u_n\rangle + \Psi(v) - \Psi(u_n) \geq -\varepsilon_n\|v - u_n\|, \qquad \text{for all } v \in X, \qquad (14.2)$$

where $\varepsilon_n \to_{n\to+\infty}$, possesses a convergent subsequence.

And f satisfies $(PS)'_{Sz}$ if any sequence $(u_n)_n$ such that $f(u_n) \to c \in \mathbb{R}$ and

$$\langle\Phi'(u_n), v - u_n\rangle + \Psi(v) - \Psi(u_n) \geq \langle z_n, v - u_n\rangle, \qquad \text{for all } v \in X, \qquad (14.3)$$

where $z_n \to 0$ as $n \to \infty$, possesses a convergent subsequence.

Equation (14.3) means that

$$z_n \in \Phi'(u_n) + \partial\Psi(u_n),$$

or equivalently,

$$z_n - \Phi'(u_n) \in \partial\Psi(u_n),$$

which shows the similarity with the usual (PS) condition. The $(PS)_{Sz}$ and $(PS)'_{Sz}$ sequences are defined by analogy to $(PS)_c$ sequences.

Remark 14.2. If $\Psi \equiv 0$, then $f = \Phi \in \mathcal{C}^1(X;\mathbb{R})$ and this definition of the (PS) condition coincides with the usual one.

Proposition 14.1. *Conditions* $(PS)_{Sz}$ *and* $(PS)'_{Sz}$ *are equivalent.*

The proof of Proposition 14.1 follows immediately using the following geometric result that asserts that a l.s.c. convex function χ, such that $\chi(0) = 0$, is bounded from below by a linear functional with norm less or equal to 1.

Lemma 14.2. *Let* $\chi : X \to \mathbb{R} \cup \{+\infty\}$ *be a l.s.c. convex function such that* $\chi(0) = 0$. *If*

$$\chi(x) \geq -\|x\|, \qquad \text{for all } x \in X,$$

then there exists $z \in X^*$ such that $\|z\| \leq 1$ and

$$\chi(x) \geq \langle z, x \rangle, \quad \text{for all } x \in X.$$

For the proofs of the lemma and of Proposition 14.1, the reader is referred to [891].

Proposition 14.3. *Suppose $f = \Phi + \Psi$, as described earlier, and satisfies* (PS)$_{Sz}$. *Let $(u_n)_n$ be a* (PS)$_{Sz}$ *sequence and u an accumulation point of $(u_n)_n$; then $u \in \mathbb{K}_c$. In particular, the set $\mathbb{K}_c$ is compact.*

Proof. For some subsequence, still denoted by $(u_n)_n$, we have that $u_n \to u$. Passing to the limit in (14.2) and using the fact that $\Psi(u_n) \geq \Psi(u)$, we obtain that u is a critical point. But, the inequality (14.1) cannot be strict for $v = u$, so $\lim_{n \to \infty} \Psi(u_n) = \Psi(u)$. Therefore, $f(u_n) \to f(u) = c$, and $u \in \mathbb{K}_c$.

To prove that $\mathbb{K}_c$ is compact, take a sequence $(u_n)_n \subset \mathbb{K}_c$. Then, $f(u_n) = c$ and (14.2) is true with $\varepsilon_n = 0$ for all n. So, a subsequence of $(u_n)_n$ converges to some $u \in X$; but by the first part of the proposition, $u \in \mathbb{K}_c$. □

Remark 14.3. If $f = \Phi + \Psi$ is bounded from below and satisfies (PS)$_{Sz}$, then $c = \inf_{u \in X} f(x)$ is a critical value. When $\Psi \equiv 0$, this result is classical and, like the classical result, it can be easily proved. So, the proof is omitted.

14.1.1 An Appropriate Deformation Lemma

We will see now a weak form of the deformation lemma which will be combined to Ekeland's variational principle to prove the semismooth MPT.

This is *unusual* since, in general, the deformation lemma and Ekeland's principle are "competitive" results.

We call a *deformation* a function $\eta : X \times [0, /overlines] \to X$ such that $\eta(., 0) \equiv I_{dW}$, the identity on W. We adopt the notation $\eta_s(.) = \eta(., s)$, where $0 \leq s \leq \bar{s}$. First, we give some technical lemmata. Set $\mathcal{B}_{c,\varepsilon,\delta} = f_{c-\varepsilon}^{c+\varepsilon} \setminus (\mathbb{K}_c)_\delta$ where $(\mathbb{K}_c)_\delta$ is a δ-neighborhood of $\mathbb{K}_c$.

Lemma 14.4. *Suppose that $f = \Phi + \Psi$ satisfies* (PS)$_{Sz}$. *Then, for each $\bar{\varepsilon} > 0$, there exists $\varepsilon \in]0, \bar{\varepsilon}[$ such that, if $u_0 \in \mathcal{B}_{c,\varepsilon,\delta}$,*

$$\langle \Phi'(u_0), v_0 - u \rangle + \Psi(v_0) - \Psi(u_0) < -3\varepsilon.\|v_0 - u_0\| \tag{14.4}$$

for some $v_0 \in X$.

Proof. Suppose by contradiction that there exists a sequence $(u_n)_n \subset X \setminus (\mathbb{K})_\delta$ such that

$$\begin{cases} f(u_n) \to c, \\ \langle \Phi'(u_n), v - u_n \rangle + \Psi(v) - \Psi(u_n) \geq -\dfrac{1}{n}\|v_0 - u_0\|, & \text{for all } v \in X. \end{cases}$$

Then, by (PS)$_{Sz}$ and Proposition 14.3, a subsequence of $(u_n)_n$ converges to $u \in \mathbb{K}_c$. But, for all n we have $u_n \notin (\mathbb{K})_\delta$, a contradiction. □

Lemma 14.5. *If ε, u_0, v_0 are as in Lemma 14.4, then there exists $\mu_0 = \mu(u_0, v_0) > 0$ such that for each $u \in B(u_0, \mu)$ and $u \in B(u_0, \mu/2)$, we have*

$$\langle \Phi'(z), v_0 - z \rangle + \Psi(v_0) - \Psi(u) \le -3\varepsilon \|v_0 - u\|. \tag{14.5}$$

Proof. Suppose by contradiction that there exist two sequences $(z_n)_n$ and $(u_n)_n$ such that

$$z_n \rightarrow u_0, \qquad u_n \rightarrow u_0$$

and

$$\langle \Phi'(z_n), v_0 - z_n \rangle + \Psi(v_0) - \Psi(u_n) > -3\varepsilon \|v_0 - u_n\|.$$

Passing to the limit, we could get

$$\langle \Phi'(u_0, v_0 - u_0 \rangle + \Psi(v_0) - \Psi(u_0) \ge -3\varepsilon \|v_0 - u_0\|,$$

which contradicts Lemma 14.4. $\square$

Lemma 14.6 (Deformation Lemma). *Suppose that $f = \Phi + \Psi$ satisfies* (PS)$_{Sz}$*. Then for each $\overline{\varepsilon} > 0$, $\delta > 0$, there exists $0 < \varepsilon < \overline{\varepsilon}$ such that for each compact set $A \subset \mathcal{B}_{c,\varepsilon,\delta}$, we can find a closed set $W \supset A$, $\overline{s} > 0$ and a continuous deformation $\eta_s : X \to X$, $0 \le s \le \overline{s}$ such that*

   i. $\|\eta_s(u) - u\| \le s, \forall u \in X,$
   ii. $\eta_s(u) = u$ in $f^{c-\varepsilon}$,
   iii. $I(\eta_s(u)) \le I(u) - 2\varepsilon s, \forall u \in W,$
   iv. $I(\eta_s(u)) \le I(u), \forall u \in X,$
   v. $\sup_{u \in W} I(\eta_s(u)) - \sup_{u \in W} I(u) \le -2\varepsilon s.$

Proof. Consider ε as above and $A \subset \mathcal{B}_{\chi,\varepsilon,\delta}$ compact. For every $u \in A$, we can find $v \in X$, $\mu = \mu(u) > 0$ such that the conclusion of Lemmata 14.4 and 14.5 hold with u, v, μ instead of u_0, v_0, μ_0. The corresponding μ may be chosen small enough such that

$$\begin{cases} f|_{B(u_0,\mu)} > c - \varepsilon, \\ \mu < \|v - u\|. \end{cases}$$

We may also consider, by choosing μ small enough, that we have a Lipschitz constant K for Φ on $\cup B(u, \mu)$.

Take a finite covering with the sets $V_i = B(u_i, \mu_i/4)$, $i = 1, \ldots, n$ of A. Consider a partition of unity $\{\sigma_i\}_{i=1}^n$ subordinate to the covering $\{U_i\}$ such that

$$\begin{cases} \text{supp } \sigma_i \subset U_i, \\ \sum_i \sigma_i \le 1 \text{ in } X, \\ \sum_i \sigma_i = 1 \text{ on } \cup_{i=1}^n V_i, \\ 0 \le \sigma_i \le 1. \end{cases}$$

We define the deformation $\eta_s : X \to X$, $s \ge 0$ by

$$\eta_s(u) = u + s \sum_{i=1}^n \sigma_i(u) \frac{v_i - u}{\|v_i - u\|}. \tag{14.6}$$

We observe that since $v_i \notin B(u_i, \mu_i)$, η_s is well defined and continuous in s and u. Property i follows immediately from (14.6), while ii follows from $\eta_s(u) = u$ for $u \in X \setminus \cup U_i$.

We will write in the sequel, for short, $\eta_s(u) = u + sw$, $\|w\| \le 1$. Then, we have

$$\Phi(u + sw) - \Phi(u) = s\langle \Phi'(z), w\rangle, \tag{14.7}$$

where $z = u + \tau w$, $0 < \tau < s$. So, we obtain the following sequence of inequalities:

$$\Phi(u + sw) \le \Phi(u) + s\langle \Phi'(z), \sum_i \sigma_i(u)\frac{v_i - u}{\|v_i - u\|}\rangle \le$$

$$\le \Phi(u) + s\sum_i \frac{\sigma_i(u)}{\|v_i - u\|}\langle \Phi'(z), v_i - u\rangle \le$$

$$\le \Phi(u) + s\sum_i \frac{\sigma_i(u)}{\|v_i - u\|}\langle \Phi'(z), v_i - z\rangle + s\sum_i \frac{\sigma_i(u)}{\|v_i - u\|}\langle \Phi'(z), z - u\rangle \le$$

$$\le \Phi(u) + s\sum_i \frac{\sigma_i(u)}{\|v_i - u\|}\langle \Phi'(z), v_i - z\rangle + s\sum_i \frac{\sigma_i(u)}{\|v_i - u\|}K\|z - u\|.$$

And since $\|z - u\| = \|\tau w\| \le s$, we obtain that

$$\Phi(u + sw) \le \Phi(u) + s\sum_i \frac{\sigma_i(u)}{\|v_i - u\|}\langle \Phi'(z), v_i - z\rangle + s^2 \sum_i \frac{\sigma_i(u)}{\|v_i - u\|}K$$

$$\le \Phi(u) + s\sum_i \frac{\sigma_i(u)}{\|v_i - u\|}\langle \Phi'(z), v_i - z\rangle + s^2 \sum_i K\frac{\mathrm{sign}\sigma_i(u)}{\|v_i - u\|}\sigma_i(u) \quad (*)$$

$$\le \Phi(u) + s\sum_i \frac{\sigma_i(u)}{\|v_i - u\|}\langle \Phi'(z), v_i - z\rangle + Ms^2 \sum_i \sigma_i(u)$$

where M is a constant.

On the other hand, we have

$$\eta_s(u) = u + sw = \underbrace{\left(1 - s\sum_i \frac{\sigma_i(u)}{\|v_i - u\|}\right)}_{\beta} u + s\sum_i \frac{\sigma_i(u)}{\|v_i - u\|}v_i.$$

For $s \le \bar{s}$ sufficiently small, $\beta \le 1$, and by the convexity of Ψ, we get that

$$\Psi(\eta_s(u)) \le (1 - s\sum_i \frac{\sigma_i(u)}{\|v_i - u\|})\Psi(u) + s\sum_i \frac{\sigma_i(u)}{\|v_i - u\|}\Psi(v_i). \tag{14.8}$$

From $(*)$ and (14.8), we obtain

$$f(\eta_s(u)) \le f(u) + s\sum_i \frac{\sigma_i(u)}{\|v_i - u\|}[\langle \Phi'(z), v_i - z\rangle + \Psi(v_i) - \Psi(u)] + Ms^2 \sum_i \sigma_i(u).$$

We take $\bar{s}$ such that if $s \le \bar{s}$ then $z \in B(u_i, \mu_i)$ and we use Lemma 14.5 to conclude that

$$f(\eta_s(u)) \le f(u) + s\left(\sum_i \sigma_i(u)(Ms - 3\varepsilon)\right). \tag{14.9}$$

In particular, if $u \in W = \cup \overline{V}_i$, we have

$$f(\eta_s(u)) \leq f(u) + s(Ms - 3\varepsilon), \qquad (14.10)$$

and for $s \leq \varepsilon/M$, we obtain

$$f(\eta_s(u)) \leq f(u) - 2\varepsilon s \sum_i \sigma_i(u), \qquad \forall u \in X. \qquad (14.11)$$

From (14.11), we immediately get properties iii, iv, and v. $\square$

14.2 Semismooth MPT

We can state now and prove the MPT announced in the beginning.

Theorem 14.7 (Semismooth MPT, Szulkin). *Suppose that $f = \Phi + \Psi \colon X \to \mathbb{R} \cup \{+\infty\}$ satisfies* $(PS)_{Sz}$ *and is such that*

$$\inf_{\partial S(0,\rho)} > \theta = \max\{f(0), f(e)\}$$

for some $e \notin \overline{B(0, \rho)}$.
Then Φ has a critical value $c \geq \beta$ characterized by

$$c = \inf_{\gamma \in \Gamma} \sup_{t \in [0,1]} f(\gamma(t)),$$

where

$$\Gamma = \big\{\gamma \in \mathcal{C}([0, 1]; X); \ \gamma(0) = 0 \ and \ \gamma(1) = e\big\}.$$

Remark 14.4. As we see, this version is also a generalization of the MPT. Indeed, when $\Psi \equiv 0$, $f = \Phi$ is in $\mathcal{C}^1(X; \mathbb{R})$ and the notions of critical point and (PS) condition coincide with the usual ones.

Let Z be a topological space and X be a real space. It is well known that in $\mathcal{C}(X; Y)$, the set of bounded mappings [1] from Z to X, endowed by the distance of the uniform convergence,

$$\mathrm{dist}\,(f, g) = \sup_{z \in Z} \| f(z) - g(z) \|,$$

is a Banach space.

Lemma 14.8. *Suppose that $f \colon X \to \mathbb{R} \cup \{+\infty\}$ is l.s.c. Then, the function $\Pi \colon \mathcal{C}(Z; X) \to \mathbb{R} \cup \{+\infty\}$ defined by*

$$\Pi(\gamma) = \sup_{z \in Z} f(\gamma(z))$$

is also l.s.c.

---

[1] By a bounded mapping from Z to Y in $\mathcal{C}(X; Y)$, we mean a function f such that $f(Z)$ is bounded in X. This is the case, for example, for all functions in $\mathcal{C}(X; Y)$ if Z is compact.

Proof. Let $\gamma_n \to \gamma$. Since f is l.s.c.,

$$f(\gamma(z)) \leq f(\gamma_n(z)), \qquad \text{for all } z \in Z.$$

Then,

$$\Pi(\gamma) = \sup_{z \in Z} f(\gamma(z)) \leq \liminf_n \left(\sup_{z \in Z} f(\gamma_n(z)) \right) = \liminf_n \Pi(\gamma_n).$$

$\square$

Lemma 14.9. *$f \circ \gamma$ is continuous on $[0, 1]$ for every $\gamma \in \Gamma$ such that $\Pi(f) < \infty$.*

We will see in the next chapter that it is in fact locally Lipschitz continuous.

Proof. Since $[0, 1]$ is compact, Ψ is bounded on $f([0, 1])$. Moreover, Ψ is bounded and l.s.c. on the convex envelope $\text{co} f([0, 1])$ of $f([0, 1])$. Now it is sufficient to prove that the restriction of $\Psi|_{\text{co} f([0,1])}$ is continuous. Let $x_0 \in \text{co} f([0, 1])$ and let N be a neighborhood of x_0 such that

$$\Psi|_{N \cap \text{co} f([0,1])} \leq a < \infty.$$

We may suppose that $x_0 = 0$ and $\Psi(x_0) = 0$. So, if $v \in \tau N \cap \text{co} f([0, 1])$ with $0 \leq \tau \leq 1$, we obtain

$$\Psi(v) \leq (1 - \tau)\Psi(0) + \tau \Psi \left(\frac{v}{\tau} \right) \leq \tau a.$$

Therefore, for $v \in \text{co} f([0, 1])$, when $v \to 0$,

$$\limsup \Psi(v) \leq 0.$$

This and the lower semicontinuity of Ψ yield the continuity of Ψ on $\text{co} f([0, 1])$. $\square$

Proof of Theorem 14.7. Suppose by contradiction that c is not a critical value of f. Let $\bar{\varepsilon} = c - \eta$ and $\varepsilon < \bar{\varepsilon}$ be given by the deformation lemma for $\delta = 0$. We may use Ekeland's variational principle with Π on Γ to obtain $\gamma \in \Gamma$ with $\Pi(\gamma) \leq c + \varepsilon$ and

$$\Pi(\gamma') - \Pi(\gamma) \geq -\varepsilon \, \text{dist}\,(\gamma, \gamma'), \qquad \forall \gamma' \in \Gamma. \tag{14.12}$$

Consider the set $A = \{\gamma(t); \, t \in [0, 1], \, f(u) \in [c - \varepsilon, c + \varepsilon]\}$. Then neither 0 nor e belong to A because $\max\{f(0), f(e)\} = \theta < c - \varepsilon$. Moreover, A is compact by the continuity of $f \circ \gamma$ $A \subset \mathcal{B}_{c,\varepsilon,0}$.

In the proof of the deformation lemma, we can choose the covering $B_i(u_i, \mu_i)$ of A such that $B_i(u_i, \mu_i) \cap \{0, e\} = \varnothing$.

Set $g = \eta_s \circ f$, where η is the deformation given by the deformation lemma. Then it is clear by property ii that $g \in \Gamma$ because $\eta_s = I_d$ on $\{0, e\}$.

By property i, we have that

$$\text{dist}\,(\gamma, g) \leq s. \tag{14.13}$$

Now, $\Pi(g) = \sup_{t \in [0,1]} f(g(t)) > c - \varepsilon$. This means that

$$\Pi(g) = \sup_{t \in [0,1]} I(g(t)) = \sup_{t \in [0,1]} f(\eta_s \circ \gamma(t)) = \sup_{t \in A} f(\eta_s(t)).$$

We obtain from property v of the deformation lemma that

$$\Pi(g) - \Pi(f) = \sup_{t \in A} f(\eta_s(t)) - \sup_{t \in A} f(t) \le -2\varepsilon s.$$

With (14.13), we get a contradiction with (14.12), so c is a critical value of f. $\square$

14.2.1 Semismooth Parents of the MPT

We signal some critical point theorems from the linking family, proved also by Szulkin [891], for functionals of the type $f = \Phi + \Psi$: a multidimensional MPT for functionals of the type $\Phi + \Psi$ satisfying $(PS)_{Sz}$ (Chapter 8).

Theorem 14.10 (Semismooth Generalized MPT, Szulkin). *Suppose that* $f = \Phi + \Psi : X \to \mathbb{R} \cup \{+\infty\}$ *satisfies* $(PS)_{Sz}$, *where* $X = X_1 \oplus X_2$ *and* $\dim X_1 < \infty$. *Suppose also that*

    i. *there are constants* $\theta, \rho > 0$ *such that* $\rho|_{\partial B(0,\rho) \cap X_2} \ge \theta$,
    ii. *there is a constant* $R > \rho$ *and* $e \in X_2$ *with* $\|e\| = 1$ *such that*

$$f|_{\partial Q} \le 0,$$

where $Q = \left(\overline{B(0,R)} \cap X_1 \right) \oplus \{re;\ 0 \le r \le R\}$.

Then f *has a critical value* $c \ge \theta$ *characterized by*

$$c = \inf_{\gamma \in \Gamma} \sup_{x \in Q} f(\gamma(x)),$$

where

$$\Gamma = \left\{ \gamma \in \mathcal{C}(Q; X);\ \gamma|_{\partial Q} = Id_{\partial Q} \right\}.$$

The proof of the MPT given here applies with no change to the general situation of a closed set S and a compact Q such that S and ∂Q *link* (see Chapter 19). In particular, it applies to the case of the generalized MPT.

The semismooth results seen in this chapter have been used to solve certain inequalities with single- and multivalued operators that appear in elliptic boundary value problems.

Comments and Additional Notes

◇ 14.I The Original Proof of the Semismooth MPT

The proof of the deformation lemma given here is an adaptation of a proof due to Lefter [557] for functionals $f = \Phi + \Psi$ where Φ is locally Lipschitz and Ψ is convex and l.s.c. (See also the last note.) The original proof by Szulkin follows.

Proof of Theorem 14.7. First of all, since for all $\gamma \in \Gamma$, the intersection $\gamma([0, 1]) \cap \partial B(0, \rho) \ne \varnothing$, we have that $c \ge \eta$.

By contradiction, supposing that c is not a critical value of f, we get that $N = \varnothing$ is a neighborhood of $\mathbb{K}_c$. Therefore, we may use the deformation lemma with $N = \varnothing$ and $\bar{\varepsilon} = c$ to obtain an $\varepsilon \in]0, \bar{\varepsilon}[$ satisfying the properties of the lemma.

By definition of c, the level set $f^{c-\varepsilon/4}$ is not path-connected, and 0 and e lie in different path components W_0 and W_e. Because $\eta_s \circ \gamma$ may not be in Γ when $\gamma \in \Gamma$, the set Γ is not suitable for our purposes. So, we will use an auxiliary family of mappings defined from $[0, 1]$ to X. Let

$$\Gamma_1 = \left\{ \gamma \in C^1([0, 1]; X); \ \gamma(0) \in W_0 \cap f^{c-\varepsilon/2} \text{ and } \gamma(1) \in W_e \cap f^{c-\varepsilon/2} \right\}$$

and

$$c_1 = \inf_{\gamma \in \Gamma_1} \sup_{t \in [0,1]} f(\gamma(t)).$$

Notice that $c_1 = c$.

Indeed, since $\Gamma \subset \Gamma_1$ we have $c \leq c_1$. And by contradiction, if $c_1 < c$, there would be paths lying in $f^{c-\varepsilon/4}$ (recall that $\gamma(0) \in W_0$ and $\gamma(1) \in W_e$ that are two path components of $f^{c-\varepsilon/4}$); there is $\overline{\gamma} \in \Gamma$ such that

$$\sup_{t \in [0,1]} f(\overline{\gamma}(t)) < c.$$

Claim 14.1. The set Γ_1 is a closed subset of $C([0, 1]; X)$ and then it is a complete metric space.

Indeed, let $(\gamma_n)_n$ be a sequence in Γ_1 such that $\gamma_n \to \gamma$. Denoting $\gamma(0) = u$ and $\gamma_n(0) = u_n$, by the l.s.c. of f, we have that

$$f(u) \leq \liminf_n f(u_n) \leq c - \frac{\varepsilon}{2}.$$

And by the convexity of Ψ and the continuity of Φ, we get, respectively, that

$$\Psi(tu_n + (1 - t)u) \leq t\Psi(u_n) + (1 - t)\Psi(u)$$

and

$$\Phi(tu_n + (1 - t)u) \leq t\Psi(u) + \delta_n \leq t\Phi(u_n) + (1 - t)\Phi(u) + 2\delta_n,$$

where $\delta_n \to 0$ and $t \in [0, 1]$.

Therefore, for n large enough,

$$f(tu_n + (1 - t)u) \leq tf(u_n) + (1 - t)f(u) + 2\delta_n$$

$$\leq c - \frac{\varepsilon}{2} + 2\delta_n \qquad\qquad (14.14)$$

$$\leq c - \frac{\varepsilon}{4}.$$

So, the segment joining u_n to u lies in W_0. In particular, $u \in W_0$. Since also $f(u) \leq c - \varepsilon/2$, we get that $u \in W_0 \cap f^{c-\varepsilon/2}$. Likewise, we prove that $f(1) \in W_e \cap f^{c-\varepsilon/2}$, that is, $\gamma \in \Gamma_1$. Now that we have proved that Γ_1 is complete, we can use Lemma 14.8 with $Z = \Gamma_1$.

Applying Ekeland's variational principle on Γ_1 for the functional Π and ε, we get that for $\gamma \in \Gamma_1$ such that $\Pi(\gamma) \leq c + \varepsilon$,

$$\Pi(\gamma_1) - \Pi(\gamma) \geq -\varepsilon \operatorname{dist}(\gamma_1, \gamma) \qquad \text{for all } \gamma \in \Gamma_1. \tag{14.15}$$

Let $A = \gamma([0, 1])$ and η_s be the deformation given in the deformation lemma. Set $\tilde{\gamma} = \eta_s \circ \gamma$. For s small enough, $\eta_s \circ \gamma \in \Gamma_1$. Indeed,

- if $f(\gamma(0)) \in]c - \varepsilon, c - \varepsilon/2[$, then by property v we get

$$f(\eta_s \circ \gamma(0)) \leq f(\gamma(0)) \leq c - \frac{\varepsilon}{2}.$$

- And if $f(\gamma(0)) \leq c - \varepsilon$, by property iii we have that

$$f(\eta_s \circ \gamma(0)) \leq f(\gamma(0)) + 2s \leq c - 2\varepsilon.$$

Hence, $\eta_s \circ \gamma(0) \in W_0 \cap f^{c-\varepsilon/2}$. Likewise, $\eta_s \circ \gamma(1) \in W_e \cap f^{c-\varepsilon/2}$. So, the function $\eta_s \circ \gamma$ is in Γ_1. And since, by property i, $\operatorname{dist}(\gamma, \tilde{\gamma}) \leq s$, it follows from properties v and iv that

$$-2\varepsilon s \geq \Pi(\tilde{\gamma}) - \Pi(\gamma) \geq -\varepsilon \operatorname{dist}(\gamma, \tilde{\varepsilon}) \geq -\varepsilon s,$$

which is impossible. This shows that $\mathbb{K}_c \neq \varnothing$ and then completes the proof. □

◇ 14.II A Regularization Procedure for l.s.c. Functions f Such That $f(u) + c\|u\|^2$ Is Convex for Some $c \geq 0$

For a l.s.c. function $f: V \to \mathbb{R}\{+\infty\}$ such that $f(u) + c\|u\|^2$ is convex for some $c \geq 0$, there is a regularization procedure due to Lasry [369, Lemma 7] that associates to f a family of functions $(f_\varepsilon)_{0 < \varepsilon < 1/c}$ such that

$$\begin{cases} f_\varepsilon \in C^1(X; \mathbb{R}), \\ f_\varepsilon(u) \to f(u), & \text{for all } u \in X. \end{cases}$$

Furthermore, $f_\varepsilon(u) \leq f(u)$ for all $u \in X$ and f and f_ε have the same critical points. And f_ε satisfies the (PS) condition whenever f satisfies (PS)$_{\text{Sz}}$.

◇ 14.III Some Applications of Semismooth MPT

In addition to the paper by Szulkin, the semismooth MPT was used recently in some other papers.

In [976], Yang considers the existence of multiple solutions of elliptic obstruction problems.

In [329], Dinca and Pasca study periodic solutions of superlinear convex autonomous Hamiltonian systems.

In [466], Halidias and Papageorgiou study a quasilinear elliptic problem with multivalued terms.

◇ *14.IV Another Semismooth Critical Point Theory*

A critical point theory (existence results and multiplicity) for functionals of the form $I = \Phi + \Psi$, with $\Phi \in \mathcal{C}^1(X; \mathbb{R})$ and Ψ a proper, convex, and l.s.c. functional, possibly invariant under the action of a compact Lie group of linear isometries, is developed in Chapters 2 and 3 of the book by Motreanu and Panagiotopoulos [654], including the MPT, the saddle point theorem, and the main results for even functionals. The corresponding definition of a critical point u is

$$\Phi^\circ(u, v - u) + \Psi(v) - \Psi(u) \geq 0 \qquad \text{for all } v \in X,$$

where Φ° is the generalized directional derivative of Φ (see the next chapter). This is also the definition adopted by Lefter [557]. This theory is used in subsequent chapters of [654] (4, 5, 6, 7, and 8) to the study hemivariational inequalities.

When $\Phi \in \mathcal{C}^1(X; \mathbb{R})$ and Ψ proper, convex, and l.s.c., $f = \Phi + \Psi$ is locally Lipschitz. A critical point theory for Lipschitz functionals exists and is well established. It will be described in the next chapter.

15

The Nonsmooth MPT

Just as "nonlinear" is understood in mathematics to mean "non necessarily linear,"
we intend the term "nonsmooth" to refer to certain situation in which smoothness
(differentiability) of the data is not necessarily postulated.

F. Clarke, *Optimization and nonsmooth analysis*, Wiley, 1983.

In the previous chapter we treated continuous convex perturbations of C^1-functionals. In this one, we will assume a weakened smoothness assumption on functionals. Namely, we will suppose that our functional is only *locally Lipschitz*. And we will prove an appropriate MPT for this kind of functionals.

The first and the most famous version of the MPT for locally Lipschitz continuous functionals is due to Chang [202] who proved it by *adapting* the deformation lemma to this particular situation.

His approach is exactly the one we saw for C^1-functionals "minus" the fact that there was no result about the existence of a pseudo-gradient vector field for locally Lipschitz functionals. He succeeds in establishing such a result using the Hahn-Banach theorem and the basic properties of the subdifferential calculus and generalized gradients in the sense of F.H. Clarke. The rest of his procedure then becomes "standard." The paper [202] of Chang proved to be an *invaluable* reference when dealing with locally Lipschitz critical point theory.

Nevertheless, we adopted an other nice approach due to Shi [835] where we will touch the on beauty and strength of Ekeland's principle. The content of this chapter is based on a nice course given by Shi at the first school on nonlinear functional analysis and applications to differential equations held at Trieste in 1996.

15.1 Notions of Nonsmooth Analysis

Let X be a Banach space and consider a function $F \colon X \to \mathbb{R} \cup \{+\infty\}$ with a nonempty *effective domain*. Recall that the effective domain is the set $\mathcal{D}(F) = \{x \in X; \ F(x) <$

$+\infty\}$. The functional

$$F^\circ(x;h) = \limsup_{\substack{t\to 0^+ \\ y\to x}} \frac{F(y+th) - F(y)}{t} \qquad (15.1)$$

is called the *Clarke directional derivative* of F at x with respect to the direction h. The following properties of the Clarke directional derivative that may be proved easily (cf. [243]).

1. $F^\circ(x;h) = (-F)^\circ(x;-h)$.
2. Let $F: X \to \mathbb{R} \cup \{+\infty\}$ be locally Lipschitz on int $\mathcal{D}(F)$, the interior of the domain of F. Then, for any $x \in$ int $\mathcal{D}(F)$, the functional $h \mapsto F^\circ(x;h)$ is *sublinear and continuous* on X and $(x, h) \mapsto F^\circ(x;h)$ is *upper semi-continuous*.
3. If F is locally Lipschitz at x with Lipschitz constant C_x, then $h \mapsto F^\circ(x;h)$ is Lipschitz continuous with the same Lipschitz constant.
4. When F is convex on X and continuous at $x \in X$, it is easy to verify that

$$F^\circ(x;h) = F'(x;h) \qquad \text{for all } h \in X.$$

In 1975, Clarke introduced the concept of *generalized gradient* for locally Lipschitz functions to generalize the notion of a *subdifferential* for a continuous convex function. Using the subdifferential of F at x, the generalized gradient of F at x is defined by

$$\partial F(x) = \{x^* \in X^*; \langle x^*, h \rangle \le F^\circ(x;h), \ \forall h \in X\}. \qquad (15.2)$$

Proposition 15.1. *Let F be locally Lipschitz at $x \in X$ with Lipschitz constant C_x. Then,*

i. there exists $\delta > 0$ such that

$$\|x^*\| \le C_x, \qquad \text{for all } y \in [x + \delta B] \text{ and all } x^* \in \partial F(y).$$

ii. $\partial F(x)$ is a w^*-compact convex set whose support function $h \mapsto F^\circ(x;h)$.
iii. For each $h \in X$, there exists $x^* \in \partial F(x)$ such that

$$\langle x^*, h \rangle = F^\circ(x;h).$$

iv. If $x_n \to x$, $x_n^* \in \partial F(x_n)$ and x^* is a w^*-cluster point of $(x_n^*)_n$, then $x^* \in \partial F(x)$.

Proof. The proof is not difficult and may be found, for example, in [243]. $\square$

Remark 15.1. When a functional F is continuously differentiable at some point x, then $\{F'(x)\} = \partial F(x)$; but when F is only Gâteaux-differentiable at x, the sets $\{F'(x)\}$ and $\partial F(x)$ may be different.

Proposition 15.2. *Let $(F_i)_{i=1}^m$ be locally Lipschitz at $x \in X$ and $F = \max_i F_i$. Then,*

$$\partial \left(\sum_{i=1}^m F_i \right)(x) \subset \sum_{i=1}^m \partial F_i(x) \qquad (15.3)$$

and

$$\partial F(x) \subset \mathrm{co}\left\{\partial F_i(x); \ i \in I_F(x)\right\}, \tag{15.4}$$

where

$$I_F(x) = \{i; \ F_i(x) = F(x)\}.$$

Proof. By Definition 15.1, we have that

$$\left(\sum_{i=1}^m F_i\right)^\circ (x;h) \le \sum_{i=1}^m F_i^\circ(x;h) \qquad \text{for all } h \in X.$$

Hence, to prove (15.3) it is enough to show that the support function of the set $K = \sum_{i=1}^m \partial F_i(x)$ is $h \mapsto \sum_{i=1}^m F_i^\circ(x;h)$. It obvious that

$$\sigma_K(h) \le \sum_{i=1}^m F_i^\circ(x;h) \quad \text{for all } h \in X.$$

And since $\partial F_i(x)$ is w^*-compact for $i \in \{1, \ldots, m\}$, the set $\sum_{i=1}^m \partial F_i(x)$ is also w^*-compact. Therefore,

$$\sigma_K(h) = \max\left\{\langle x^*, h\rangle; \ x^* \in \sum_{i=1}^m \partial F_i(x)\right\},$$
$$\ge \sum_{i=1}^m \max\left\{\langle x^*, h\rangle; \ x^* \in \partial F_i(x)\right\},$$
$$= \sum_{i=1}^m F_i^\circ(x;h).$$

To prove (15.4), we note that

$$F^\circ(x;h) \le \max_{i \in I_F(x)} F_i^\circ(x;h)$$

and that for $K_1 = \mathrm{co}\left\{\partial F_i(x); \ i \in I_F(x)\right\}$, we have that for any $h \in X$,

$$\sigma_{K_1}(h) = \sup_{x^* \in K_1} \langle x^*, h\rangle,$$
$$\ge \max_{i \in I_F(x)} \max\left\{\langle x^*, h\rangle; \ x^* \in \partial F_i(x)\right\},$$
$$= \max_{i \in I_F(x)} F_i^\circ(x;h).$$

Hence, (15.4) holds. $\square$

The role of generalized gradients in variational methods may be well understood by the study of minimization problems. We recall some well-known results in nonsmooth analysis that will be needed in the sequel. The following one is obvious.

Proposition 15.3. *Let $F\colon X \to \mathbb{R}\cup\{+\infty\}$ be proper. If F has a local minimizer or maximizer $x \in X$, then $0 \in \partial F(x)$.*

This proposition has the following important consequence.

Theorem 15.4 (Mean-Value Theorem). *Let* $x, y \in X$ *and* F *be a locally Lipschitz function on an open set containing the line segment* $[x, y]$*. Then, there exists* $\theta \in]0, 1[$ *such that*

$$F(y) - F(x) \in \langle \partial F(x + \theta(y - x)), y - x \rangle.$$

Proof. Define $\phi : [0, 1] \to \mathbb{R}$ by

$$\phi(t) = F(x + t(y - x)) + t[F(y) - F(x)].$$

Then $\phi(0) = \phi(1) = F(x)$ and there exists $\theta \in (0, 1)$, a minimum or a maximum of ϕ. Hence, $0 \in \partial\phi(\theta)$. The theorem is obtained by verifying that

$$\partial\phi(\theta) \subset F(x) - F(y) + \langle \partial F(x + \theta(y - x)), y - x \rangle.$$

□

Using generalized gradients, the following *form* of Ekeland's variational principle holds.

Proposition 15.5. *Let* X *be a Banach space and* $F: X \to \mathbb{R}$ *be locally Lipschitz and bounded from below. Then, there exists a sequence* $(x_n)_n \subset X$ *such that*

i. $F(x_n) \to \inf F$,
ii. $\inf_{x^* \in \partial F(x_n)} \|x^*\| \to 0$.

The (PS) condition for a locally Lipschitz function $F: X \to \mathbb{R}$ is expressed as

$$(PS)_{Lip} \quad \begin{cases} \text{Any sequence } (x_n)_n \subset X \text{ such that} \\ F(x_n) \text{ is bounded and } \inf_{x^* \in \partial F(x_n)} \|x^*\| \to 0, \\ \text{admits a convergent subsequence of } (x_n)_n. \end{cases}$$

And a sequence that satisfies the relation in (PS)*Lip* is called a (PS)*Lip* sequence.

Corollary 15.6. *Let* X *be a Banach space and* $F: X \to \mathbb{R}$ *be locally Lipschitz and bounded from below. If* F *satisfies* (PS)*Lip, then* F *has a minimum.*

The bounded character of F and (PS)*Lip* plays the role of *coercivity*. Recall that the coercivity is a sufficient condition for the existence of minima for a l.s.c. convex function F on a reflexive Banach space. Indeed, we have also the following result.

Proposition 15.7. *Let* X *be a Banach space and* $F: X \to \mathbb{R}$ *be locally Lipschitz and bounded from below. If* F *satisfies* (PS)*Lip, then* F *is coercive.*

Proof. Setting

$$C = \sup_{b \in \mathbb{R}} \left(\{x \in X; \; F(x) \le b\} \text{ is bounded} \right),$$

we shall prove that $C = +\infty$. This should imply the coercivity of F. In fact, if $C < +\infty$, then $F^{C-2^{-n}}$ is bounded, but $F^{C+2^{-n}}$ is not. So, there exists a real number $R_n \ge n$ such that

$$F(x) > C - 2^{-n}, \qquad \text{for all } x \text{ such that } \|x\| \ge R_n.$$

Since $F^{C+2^{-n}}$ is unbounded, there is $x_n \in F^{C+2^{-n}}$ such that

$$\|x_n\| \geq R_n + 2 \qquad \text{and} \qquad F(x_n) \leq C + 2^{-n}. \qquad (15.5)$$

Applying Ekeland's principle to the metric space

$$X_n = \{x \in X; \ \|x\| \geq R_n\},$$

we obtain the existence of $x'_n \in X_n$ such that

$$C - 2^{-n} < F(x'_n) \leq F(x_n) \leq c + 2^{-n}, \qquad (15.6)$$

$$\|x'_n - x_n\| \leq 1, \qquad (15.7)$$

and for all $x \neq x'_n$, $x \in X_n$:

$$F(x) > F(x'_n) - 2^{-(n-1)}\|x'_n - x_n\|. \qquad (15.8)$$

The inequalities (15.6), (15.7), and (15.8) imply that

$$0 \in \partial F(x'_n) + 2^{(1-n)}(\overline{B})^*.$$

Therefore, by (PS)$_{Lip}$, the sequence $(x'_n)_n$ admits a convergent subsequence. However, from (15.5) and (15.7) we have

$$\|x'_n\| \geq \|x_n\| - \|x_n - x'_n\| > R_n + 1 \geq n + 1 \to +\infty,$$

which is impossible.   □

15.2 Nonsmooth MPT

This section is devoted to the study of an MPT for locally Lipschitz functions on a Banach space. The version we will present uses Ekeland's principle and is attributable to Shi [835]. As mentioned earlier, an analogous result, earlier than that of Shi, has been proved by Chang in [202] using a specific deformation lemma adapted to Lipschitz functionals.

Theorem 15.8 (Nonsmooth MPT, Shi). *Let* $f: X \to \mathbb{R}$ *be locally Lipschitz contin-uous. Consider an open neighborhood* Ω *of* 0 *and a point* $e \notin \overline{\Omega}$ *such that*

$$\max\{f(0), f(e)\} < c_0 \leq \inf_{\partial\Omega} f. \qquad (15.9)$$

Then, setting

$$c = \inf_{\gamma \in \Gamma} \max_{t \in [0,1]} f(\gamma(t)) \geq c_0, \qquad (15.10)$$

where

$$\Gamma = \{\gamma \in \mathcal{C}([0, 1]; X); \ \gamma(0) = 0 \text{ and } \gamma(1) = e\}, \qquad (15.11)$$

there exists a sequence $(x_n)_n \subset X$ *such that*

$$f(x_n) \to c \qquad \text{and} \qquad \inf_{x^* \in \partial f(x_n)} \|x^*\| \to 0. \qquad (15.12)$$

If, in addition, f satisfies (PS)$_{Lip}$, then c is a critical value of f; that is, there exists $\bar{x} \in X$ such that $f(\bar{x}) = c$ and $0 \in \partial f(\bar{x})$.

The idea of the proof is the following. Consider

$$F(\gamma) = \max_{t \in [0,1]} f(\gamma(t))$$

as a function defined on the closed linear subspace Γ of $C([0, 1]; X)$. It is easy to verify that F is locally Lipschitz on Γ. Then, by Ekeland's variational principle, F has almost minimizers satisfying some particular relations. Using this form, we shall then establish the existence of a (PS)$_{Lip}$ sequence.

The proof is decomposed into several propositions. They will be proved for a general case, in which $[0, 1]$ is replaced by a compact metric space K. So, it possible, using this result, to get some extensions and variants of the MPT.

Proposition 15.9. *Let $f: X \to \mathbb{R}$ be a locally Lipschitz mapping and K a compact metric space. Then, the functional $F: C(K; X) \to \mathbb{R}$, defined by*

$$F(\gamma) = \max_{t \in K} f(\gamma(t)) \qquad \text{for all } \gamma \in C(K; X)$$

is locally Lipschitz on $C(K; X)$.

Proof. Consider $\gamma \in C(K; X)$; the set $\gamma(K)$ is compact as a continuous image of the compact K. Since f is locally Lipschitz, for any $t \in K$, there are $\delta_t > 0$ and $C_t > 0$ such that

$$|f(x_1) - f(x_2)| \le C_t \|x_1 - x_2\| \qquad \text{for all } x_1, x_2 \in \big[\gamma(t) + \delta B\big]. \qquad (15.13)$$

Then, $\big(\gamma(t) + \delta_t B\big)_{t \in K}$ constitutes an open covering of $\gamma(K)$ and there are $t_1, \dots, t_k \in K$ such that

$$\gamma(K) \subset \bigcup_{i=1}^{k} \big[\gamma(t_i) + \delta_{t_i} B\big]. \qquad (15.14)$$

On the other hand, by the Lebesgue lemma there exists a Lebesgue number $\delta > 0$ depending on $\gamma(K)$ such that for any $x \in \gamma(K)$,

$$[x + \delta B] \subset \big[g(t_i) + \delta_{t_i} B\big], \qquad (15.15)$$

for some i. Set $C_\gamma = \max_{1 \le i \le k} C_{t_i}$. By (15.13)–(15.15), we have that

$$|f(x_1) - f(x_2)| \le C_\gamma \|x_1 - x_2\| \qquad \text{for all } t \in K,\ x_1, x_2 \in \gamma(t) + \delta B.$$

Thus, when $h_1, h_2 \in C(K; X)$ satisfy

$$\|h_i - \gamma\|_{C(K;X)} = \max_{t \in K} \|h_i(t) - \gamma(t)\| < \delta \qquad \text{for } i = 1, 2,$$

we have that

$$\begin{aligned}
|F(h_1) - F(h_2)| &= \big| \max_{x \in K} f(h_1(t)) - \max_{t \in K} f(h_2(t)) \big|, \\
&\le \max_{t \in K} |f(h_1(t)) - f(h_2(t))|, \\
&\le C_\gamma \max_{t \in K} \|h_1(t) - h_2(t)\| = C_\gamma \|h_1 - h_2\|.
\end{aligned}$$

$\square$

Proposition 15.10. *Let f, F, X, and K be as in* Proposition 15.9 *and set*

$$M(\gamma) = \{s \in K; \; f(\gamma(s)) = F(\gamma) = \max_{t \in K} f(\gamma(t))\}.$$

Then,

$$F^{\circ}(\gamma; h) \le \max_{s \in M(\gamma)} f^{\circ}(\gamma(s); h(s)) \qquad \text{for all } h \in \mathcal{C}(K; X).$$

Proof. We choose two sequences $(u_i)_i \subset \mathcal{C}(K; X)$ and $(\lambda_i)_i \subset \mathbb{R}^+$ such that $\|u_i - \gamma\| = \max_{t \in K} |u_i(t) - \gamma(t)| \to 0$, $\lambda_i \to 0^+$ as $i \to \infty$ and

$$F^{\circ}(\gamma; h) = \lim_{i \to \infty} \frac{F(u_i + \lambda_i h) - F(u_i)}{\lambda_i}. \qquad (15.16)$$

For any $s_i \in M(u_i + \lambda_i h)$, $i = 1, 2, \ldots$, it follows that

$$\frac{F(u_i + \lambda_i h) - F(u_i)}{\lambda_i} \le \frac{f(u_i(s) + \lambda_i h(s_i)) - f(u_i(s_i))}{\lambda_i}. \qquad (15.17)$$

By the mean-value theorem, there exist $\theta_i \in]0, 1[$ and $x_i^* \in \partial f(u_i(s_i) + \theta_i \lambda_i h(s_i))$ such that

$$\frac{f(u_i(s_i) + \lambda_i h(s_i)) - f(u_i(s_i))}{\lambda_i} = \langle x_i^*, h(s_i) \rangle \qquad \text{for } i = 1, 2, \ldots. \qquad (15.18)$$

Since K is a compact metric space, $(s_i)_i$ has a convergent subsequence denoted also by $(s_i)_i$, such that $s_i \to s \in K$. Then,

$$u_i(s_i) + \theta_i \lambda_i h(s_i) \to \gamma(s).$$

By Proposition 15.1, the sequence $(x_i^*)_i$ has a w^*-cluster point $x^* \in \partial f(\gamma(s))$. We may suppose that $\langle x_i^*, h(s) \rangle \to \langle x^*, h(s) \rangle$ and then, by the relations (15.16)–(15.17), we have that

$$F^{\circ}(\gamma; h) \le \lim_{i \to \infty} \langle x_i^*, h(s_i) \rangle \le \lim_{i \to \infty} \langle x_i^*, h(s_i) - h(s) \rangle + \lim_{i \to \infty} \langle x_i^*, h(s) \rangle.$$

Finally, we have to check that $s \in M(\gamma)$. Since $s_i \in M(u_i + \lambda_i h)$, we get that

$$f(u_i(s_i) + \lambda_i h(s_i)) \ge f(u_i(t) + \lambda_i h(t)) \qquad \text{for all } t \in K.$$

Taking i to infinity, we conclude that

$$f(\gamma(s)) \ge f(\gamma(t)) \qquad \text{for all } t \in K.$$

$\square$

Proposition 15.11. *Let f, F, X, and K be as before and $K_0 \subset K$ be a closed subset of K. If for some $\gamma \in \mathcal{C}(K; X)$*

$$M(\gamma) \subset K \setminus K_0, \qquad (15.19)$$

and there exists $\varepsilon > 0$ such that for any $h \in \mathcal{C}_0(K; X)$, we have

$$F^{\circ}(\gamma; h) \ge -\varepsilon \|h\|, \qquad (15.20)$$

where

$$\mathcal{C}_0(K; X) = \{h \in C(K; X); \text{ for all } t \in K_0, \ h(t) = 0\}.$$

Then, there exists $s \in M(\gamma)$ such that

$$f^\circ(\gamma(s); v) \geq -\varepsilon \|v\| \qquad \text{for all } v \in X. \tag{15.21}$$

Proof. By contradiction, if such an s does not exist, then for any $t \in M(\gamma)$ there exists $v_t \in X$ with $\|v_t\| = 1$ such that

$$f^\circ(\gamma(t); v_t) < -\varepsilon.$$

Since γ is continuous and f° is upper semicontinuous, we have that, for any $t \in M(\gamma)$, there exists $v_t \in X$ with $\|v_t\| = 1$ and $\delta_t > 0$ such that

$$f^\circ(\gamma(s); v_t) < -\varepsilon \qquad \text{for all } s \in \mathfrak{B}(t, \delta) = \{s \in K; \ \text{dist}(s, t) < \delta_t\}. \tag{15.22}$$

The family $\{\mathfrak{B}(t, \delta_t)\}_{t \in M(\gamma)}$ forms an open covering of $M(\gamma)$. And from the compactness of $M(\gamma)$ and the relation (15.19), we may suppose that

$$K_0 \cap \mathfrak{B}(t, \delta_t) = \varnothing \qquad \text{for all } t \in M(\gamma). \tag{15.23}$$

So, there exist $t_1, \ldots, t_k \in M(\gamma)$ such that

$$M(\gamma) \subset \bigcup_{i=1}^{k} \mathfrak{B}(t_i, \delta_{t_i}). \tag{15.24}$$

For any $t \in K$, we define

$$\rho_0(t) = \min_{s \in M(\gamma)} \ \text{dist}(t, s), \tag{15.25}$$

$$\rho_i(t) = \min_{s \in K \setminus \mathfrak{B}(t_i, \delta_{t_i})} \ \text{dist}(t, s) \qquad \text{for } i = 1, 2, \ldots, k. \tag{15.26}$$

From (15.24), we have that

$$K = \left\{ \bigcup_{i=1}^{k} \mathfrak{B}(t_i, \delta_{t_i}) \right\} \cup \{K \setminus M(\gamma)\},$$

and it follows that

$$\sum_{i=0}^{k} \rho_i(t) > 0 \qquad \text{for all } t \in K. \tag{15.27}$$

Set

$$h(t) = \frac{\sum_{i=1}^{k} v_{t_i} \rho_i(t)}{\sum_{i=0}^{k} \rho_i(t)}.$$

Then, by (15.22)–(15.27), we get that

$$h \in \mathcal{C}_0(K; X) \qquad \text{and} \qquad \|h\| \leq 1.$$

And since $v \mapsto f^\circ(x; v)$ is sublinear,

$$f^\circ(\gamma(t); h(t)) \le \left(\sum_{i=1}^{k} \rho_i(t) f^\circ(\gamma(t); v_{t_i}) \right) / \sum_{i=0}^{k} \rho_i(t).$$

By (15.22), (15.25), and (15.26), for any $t \in M(\gamma)$, we have

$$\rho_0(t) = 0 \quad \text{and} \quad (\rho_i(t) > 0) \quad \text{implies that} \quad (f^\circ(\gamma(t); v_{t_i}) < -\varepsilon).$$

Then, from Proposition 15.10, it follows that

$$F^\circ(\gamma; h) \le \max_{s \in M(\gamma)} f^\circ(\gamma(s), h(s)) < -\varepsilon \|h\|,$$

which contradicts (15.20). Hence, (15.21) is proved.   □

Proof of Theorem 15.8. Set $K = [0, 1]$ and $K_0 = \{0, 1\}$. Since Ω separates 0 and e, we have, for any $\gamma \in \Gamma$,

$$\gamma(K) \cap \partial \Omega = \gamma([0, 1]) \cap \partial \Omega \ne \varnothing. \tag{15.28}$$

And by (15.9) and (15.11),

$$\max_{t \in [0,1]} f(\gamma(t)) \ge \inf_{\partial \Omega} f \ge c_0 > \max \{ f(\gamma(0)), f(\gamma(1)) \}. \tag{15.29}$$

Therefore, for all $\gamma \in \Gamma$,

$$M(\gamma) = \left\{ s \in [0, 1]; \ f(g(s)) = \max_{t \in [0,1]} f(\gamma(t)) \right\} \subset]0, 1[= K \setminus K_0.$$

The set Γ is a closed linear manifold of $\mathcal{C}([0, 1]; X)$, which is a complete metric space for the distance of the uniform convergence. We define $F : \Gamma \to \mathbb{R}$ by

$$F(\gamma) = \max_{t \in [0,1]} f(\gamma(t)) \qquad \text{for all } \gamma \in \Gamma.$$

Then, by Proposition 15.9, the functional F is locally Lipschitz on Γ and from (15.29) and (15.10), it is bounded from below. According to Ekeland's variational principle, for any positive sequence $(\varepsilon_n)_n$, where $\varepsilon_n \searrow 0^+$, there exists a sequence $(\gamma_n)_n \subset \Gamma$ such that

$$c \le F(\gamma_n) \le c + \varepsilon_n$$

and

$$F(u) > F(\gamma_n) - \varepsilon_n \|u - \gamma_n\|$$

for $n = 1, 2, \ldots$ and all $u \ne \gamma_n$.

Thus, for any $h \in \mathcal{C}_0([0, 1]; X)$, we have that

$$F^\circ(\gamma_n; h) \ge \limsup_{\lambda \searrow 0^+} \frac{F(\gamma_n + \lambda h) - F(\gamma_n)}{\lambda} \ge -\varepsilon_n \|h\| \qquad \text{for } n = 1, 2, \ldots .$$

By Proposition 15.11, there exists $s_n \in M(\gamma_n)$ such that

$$f(\gamma_n(s_n)) = F(\gamma_n),$$

$$f^\circ(\gamma_n(s_n); v) \geq -\varepsilon_n \|v\| \qquad \text{for all } v \in X \quad \text{and} \quad n = 1, 2, \ldots.$$

Then, setting $x_n = \gamma_n(s_n)$ for $n = 1, 2, \ldots$, we have that

$$f(x_n) \to c \quad \text{and} \quad 0 \in \partial f(x_n) + \varepsilon_n(\overline{B})^*.$$

□

Many applications to partial differential equations have been given (see Chang [202] for the earliest ones). A typical example is the Dirichlet problem:

$$\begin{cases} \Delta u = f(u), & \text{in } \Omega, \\ u = 0 & \text{on } \partial\Omega, \end{cases}$$

where f is only a locally bounded measurable function, and the corresponding functional,

$$\Phi(u) = \frac{1}{2} \int (\nabla u)^2 dx + \int_\Omega F(u(x)) \, dx,$$

is locally Lipschitz.

Comments and Additional Notes

Shi combined nonsmooth analysis with Ekeland's principle in a fantastic way to generate one of the earliest linking (nonsmooth) results. The nonsmooth MPT is behind the (smooth) version to be found in [623, 628] according to their respective authors, and it is this latter form that inspired Szulkin to establish his "semismooth" MPT, as he reported in [891].

◇ 15.I Some Variants of Nonsmooth Critical Point Theory

We describe briefly *some* papers that presented some forms of the nonsmooth MPT.

In [530, 531], Kourogenis and Papageorgiou combine the nonsmooth critical point theory for locally Lipschitz functionals with a nonsmooth counterpart of the Cerami condition (see Chapter 13) to obtain the following nonsmooth extension of the (C) condition and of (PS)$_{Lip}$:

for $\Phi \colon X \to \mathbb{R}$ locally Lipschitz, $c \in \mathbb{R}$, every sequence $(x_n)_{n\geq1} \subset X$ such that $\Phi(x_n) \to c \in \mathbb{R}$ as $n \to \infty$ and $(1 + \|x_n\|)m(x_n) \to 0$ as $n \to \infty$ has a strongly convergent subsequence, where $m(x) = \inf\{\|x^*\| : x^* \in \partial\Phi(x)\}$.

The authors then prove a general minimax principle that includes, of course, as a particular case, an MPT, and also the limiting case (see Chapter 9). They rely on a variant of the deformation lemma that uses a locally Lipschitz vector field instead of the pseudo-gradient vector field of the smooth case.

They use the generalized MPT to obtain a nontrivial solution of a nonlinear elliptic problem at resonance where they do not suppose the well-known Ambrosetti-Rabinowitz condition $0 < \mu F(t, x) \leq x f(t, x)$.

In [668] Motreanu and Varga extend the classical deformation lemma to functionals that are only locally Lipschitz continuous defined on reflexive Banach spaces and use it to establish some results on the existence of critical points for locally Lipschitz continuous functionals including an MPT.

In [648], Motreanu proves (Theorem 1) a critical point theorem that contains the MPT, the nonsmooth MPT, the multidimensional MPT, and the semismooth MPT.

In [233] (A general mountain pass principle for nondifferentiable functionals), Choulli et al. prove a mountain pass principle for locally Lipschitz functionals. It generalizes Theorem 9.6 of Ghoussoub and Preiss [427]. The proof is based on Ekeland's variational principle. The paper contains also the following interesting form of the pseudo-gradient lemma.

Lemma 15.12 (Choulli et al. [233]). *Let M be a compact metric space and let $\varphi \colon M \to 2^{X^*}$ be a set-valued mapping which is u.s.c. and $\varphi(x)$ is weak-$*$ compact and convex. Let*

$$\gamma = \inf\big\{\|x^*\|; \ x\varphi(t), \ t \in M\big\}.$$

Then, given $\varepsilon > 0$, there exists a continuous function $v \colon M \to X$ such that for all $t \in M$ and $x^* \in \varphi(t)$,

$$\|v(t)\| \leq 1 \qquad and \qquad \langle x^*, v(t) \rangle \geq \gamma - \varepsilon.$$

In [756] Radulescu proves a nonsmooth version of the MPT similar to that of Shi (with K, K^* satisfying assumptions similar to those in Brézis and Nirenberg's paper [153]). His proof relies on Ekeland's principle and on Lemma 15.12.

The paper [765] by Ribarska et al. contains an extension of the general mountain pass principle of Ghoussoub and Preiss (Theorem 9.6) to locally Lipschitz functions. Their main tool is a generalization to locally Lipschitz functions of the quantitative deformation lemma of Willem. This allows them to get a simple proof of the general mountain pass principle.

In [640], Mironescu and Radulescu use the Ekeland variational principle to obtain a result of Ljusternik-Schnirelman type for locally Lipschitz functionals. They use it to prove that a periodic multivalued problem of the forced pendulum admits at least two geometrically distinct solutions.

In [441], Goeleven studies semicoercive variational-hemivariational inequalities and also proves an MPT for locally Lipschitz functionals.

◇ *15.II A Ljusternik–Schnirelman–Type Theorem for Locally Lipschitz Functionals*

In [758] (A Lyusternik-Schnirelman type theorem for locally Lipschitz functionals with applications to multivalued periodic problems), Radulescu restudies the Ljusternik-Schnirelman theorem for locally Lipschitz functionals. This has already been done in Chang's paper [202, Theorem 3.2], which also contains a nonsmooth version of the Rabinowitz saddle point theorem.

◇ *15.III A Nonsmooth Mountain Impasse Situation*

In [920], Tintarev generalizes the results [817, 917, 918] to Banach spaces and nonsmooth functionals. For the details, see Chapter 21.

◇ *15.IV A MPT on C^1-Finsler Manifolds*

The paper [767] by Ribarska et al. contains a deformation lemma for locally Lipschitz continuous real-valued functions f on a complete Finsler manifold M of class C^1 without boundary. It is used to prove some corresponding versions of minimax theorems, including a theorem of Ljusternik-Schnirelman–type and an MPT.

◇ *15.V A Critical Point Theory for Locally Lipschitz Functionals on Locally Convex Closed Subsets of Banach Spaces*

In [942, 944], Wang establishes a minimax principle, including the MPT of course, for a locally Lipschitz functional Φ on *locally convex closed subsets S* of a Banach space. (See Chapter 25 for some details.)

16

The Metric MPT

Mathematical discoveries, small or great are never born of spontaneous genera-
tion. They always presuppose a soil seeded with preliminary knowledge and well
prepared by labour, both conscious and subconscious.

Bertrand Russel

Going further in nonsmoothness, we present now a variant of the MPT for continuous functionals
on metric spaces. Appropriate notions of critical point and Palais-Smale condition are defined
to handle this more general situation that still contains as particular cases the previous results
stated when more smoothness and regularity on the functional were supposed.

The metric MPT was discovered independently by Degiovanni and Marzocchi [310]
and Katriel [516]. They both use Ekeland's variational principle but in two different
ways. The method of Degiovanni and Marzocchi has become widely known these days.
Nevertheless, Katriel's approach will be more familiar to those who have read the
previous chapter devoted to the nonsmooth MPT.

16.1 Preliminaries

In both papers [310, 516], the notions of critical point and Palais-Smale condition are
defined in a very similar way although they use different terminology.

16.1.1 Critical Points of Continuous Functions in Metric Spaces

The definition of a critical point for a continuous function defined on a metric space
reduces to the usual one known in the smooth case when the functional is smooth. This
guarantees that the new theory extends the classical one.

Definition 16.1. Let (X, dist) be a metric space, $x \in X$, f a real function defined in
a neighborhood of x, and $\delta > 0$ a given positive real number. The point x is said to
be δ-*regular* if there are a neighborhood U of x, a constant $\alpha > 0$, and a continuous
mapping $\eta: U \times [0, \alpha] \to X$ such that for all $(u, t) \in U \times [0, \alpha]$:

1. dist $(\eta(u, t), u) \le t$,
2. $f(u) - f(\eta(u, t)) \ge \delta t$.

The functional η is then called a δ-*regularity mapping* for f at x, and x is called *regular* if it is δ-regular for some $\delta > 0$; otherwise it is a *critical point* of f.

We call the δ-*regularity constant* of f at a regular point x the value

$$\delta(f, x) = \sup \{\delta; \ f \text{ is } \delta\text{-regular at } x\}.$$

If x is a critical point of f, we set $\delta(f, x) = 0$.

The preceding notions and terminology are those of Katriel [516].

In [310], Degiovanni and Marzocchi independently adopted a similar definition. It is the same as the aforementioned one with the set $U = B(x, \alpha)$ as a neighborhood of x. They denote $\delta(f, x)$ by $|df|(x)$ and call it *weak slope* of f at x. Their statement is the following.

Definition 16.2. Let $f : X \to \mathbb{R}$ be a continuous function and let $x \in X$. We denote by $|df|(x)$ the supremum of the $\sigma \in [0, +\infty[$ such that there exist $\delta > 0$ and a continuous map $\eta : B(x, \delta) \times [0, \delta[\to X$ such that, for all $(u, t) \subset B(u, \delta) \times [0, 1]$,

  i. dist $(\eta(u, t), u) \le t$, and
  ii. $\Phi(\eta(u, t)) \le \Phi(u) - \sigma t$.

Note that $|df|(x)$ is an extended real number, that is, it can be infinite. The two definitions are obviously equivalent and the two notions coincide.

In the sequel we combine the two approaches of Katriel and Degiovanni and Marzocchi and each time use the one that seems more appropriate. We will also present the two proofs of the metric MPT. The proof of Katriel relies on Ekeland's principle, while the proof of Degiovanni and Marzocchi uses a specific deformation lemma.

Lemma 16.1. *If X is a Banach space and $f : X \to \mathbb{R}$ is of class C^1, then*

$$\delta(f, x) = \|f'(x)\|.$$

In particular, the notion of critical point presented in Definition 16.2 coincides with the usual one.

Proof. When x is a critical point of f, the relation is obviously true. Otherwise, consider a real number δ such that $0 < \delta < \|f'(x)\|$. Then, we can choose $v \in X$ with $\|v\| < 1$ and $f'(x).v > \delta$. Let $\eta(u, t) = u - tv$. It satisfies condition 1 in the definition of a regularity mapping and also the relation

$$\limsup_{\substack{u \to x \\ t \to 0^+}} \frac{f(u) - f(\eta(u, t))}{t} > \delta.$$

So condition 2 also holds; that is, f is δ-regular at x.

To finish the proof, it suffices to show that f is not δ-regular at x when $\delta > \|f'(x)\|$.

By contradiction, if f was δ-regular at x for some δ, we would have a δ-regularity mapping $\eta: U \times [0, \alpha] \to X$ on some neighborhood U of x. Then, combining conditions 1 and 2 in the former definition we get

$$f(u) - f(\eta(u, t)) \geq \delta \, \mathrm{dist}\,(u, \eta(u, t)) = \delta \|u - \eta(u, t)\|. \tag{16.1}$$

But f is differentiable; hence, there is a neighborhood V of x such that, for $u, v \in V$,

$$|f(u) - f(v)| < \delta \|u - v\|,$$

which is a contradiction with (16.1) when (u, t) is close enough to $(x, 0)$. □

Since a δ-regularity mapping for f at x is a δ-regularity mapping for f at any point in a neighborhood of x, the following result becomes intuitive.

Lemma 16.2. *The functional* $x \mapsto \delta(f, x) = |df|(x)$ *is l.s.c.*

Proof. Consider a sequence $(x_n)_n$ such that $x_n \to x$ as n goes to infinity. Let us show that $\liminf_n |df|(x_n) \leq |df|(x)$. Suppose that $\sigma = |df|(x) > 0$. So, for all $\sigma' \in (0, \sigma)$, there is $\delta > 0$ and $\mathcal{H}: B_\delta \times [0, \delta] \to X$ such that

$$\mathrm{dist}\,(\mathcal{H}(u, t), u) \leq t \qquad \text{and} \qquad f(\mathcal{H}(u, t)) \leq f(y) - \sigma' t.$$

There exists $N > 0$ such that $x_n \in B_{\delta/2}(x)$ for $n > N$. For $\delta/2 > 0$, consider $\mathcal{H}_1 = \mathcal{H}|_{B_{\delta/2}(x_n) \times [0, \delta/2]}$, so

$$|df|(x_n) \geq \sigma' \qquad \text{for} \qquad n > N.$$

Therefore, $\sigma \leq |df|(x)$. □

16.1.2 Palais-Smale Condition for Continuous Functions

Consider now the following definition of the Palais-Smale condition for continuous functions defined on metric spaces.

Definition 16.3. Let X and f be as before. We say that f satisfies the *Palais-Smale condition* (PS) if

> any sequence $(u_n)_n \subset X$ such that $f(x_n)_n$ is bounded and $\delta(f, x_n) = |df|(x_n) \to 0$ has a convergent subsequence.

Similarly, we say that f satisfies the *local Palais-Smale condition* (PS)$_c$, if

> any sequence $(u_n)_n \subset X$ such that $(f(x_n))_n \to c$ and $\delta(f, x_n) = |df|(x_n) \to 0$ has a convergent subsequence.

Remark 16.1. By the l.s.c. of the δ-regularity (weak slope), the limit of the convergent subsequence is necessarily a critical point.

16.2 Metric MPT

The MPT for continuous functions on metric spaces is based on these new notions of critical point and (PS) condition. It has been stated and proved independently by Degiovanni and Marzocchi [310] and Katriel [516]. Katriel's proof will not seem strange to us because it is an adaptation of the proof of the standard MPT given in [628], which was taken from the proof of the nonsmooth MPT of Shi we have just seen in the preceding chapter.

Theorem 16.3 (Metric MPT, Katriel Formulation). *Let f be a continuous function on a path-connected complete metric space X. Suppose that $x_1, x_2 \in X$, Γ is the set of continuous paths joining x_1 and x_2,*

$$\Gamma = \big\{\gamma : [0, 1] \to X;\ \gamma(0) = x_1, \gamma(1) = x_2 \text{ and } \gamma \text{ continuous}\big\},$$

and let $\chi : \Gamma \to \mathbb{R}$ be the function defined by

$$\chi(\gamma) = \max_{t \in [0,1]} f(\gamma(t)).$$

Let

$$c = \inf_{\gamma \in \Gamma} \chi(\gamma) > c_1 = \max\big\{f(x_1), f(x_2)\big\}. \tag{16.2}$$

For every $\varepsilon > 0$, there is a point $v \in X$ such that

$$c - \varepsilon \le f(v) \le c + \varepsilon$$

and

$$\delta(f, v) = |df|(v) \le \sqrt{\varepsilon}.$$

If f satisfies $(PS)_c$, then there is a critical point $\overline{x}$ of f such that $f(\overline{x}) = c$.

Proof. Of course, Γ is a complete metric space for the uniform distance

$$\rho(\gamma_1, \gamma_2) = \max_{t \in [0,1]} \text{dist}\,(\gamma_1(t), \gamma_2(t))$$

on which χ is continuous, as noticed many times in earlier chapters.

For $0 < \varepsilon < c - c_1$, choose $\gamma_1 \in \Gamma$ such that $\chi(\gamma_1) \le c + \varepsilon$. By Ekeland's principle, there is $\gamma_2 \in \Gamma$ such that

$$\chi(\gamma_2) \le \chi(\gamma_1),$$

$$\rho(\gamma_1, \gamma_2) \le \sqrt{\varepsilon},$$

and, for any $\gamma \in \Gamma \setminus \{\gamma_2\}$,

$$\chi(\gamma) > \chi(\gamma_2) - \sqrt{\varepsilon}.\rho(\gamma, \gamma_2). \tag{16.3}$$

Then there exists $s \in [0, 1]$ such that

$$c - \varepsilon \le f(\gamma_2(s))$$

and

$$\delta(f, \gamma_2(s)) \le \sqrt{\varepsilon}.$$

By contradiction, if not, considering the set

$$S = \{t \in [0, 1];\ c - \varepsilon \le f(\gamma_2(t))\},$$

we should have that for every $s \in S$, there exist $r(s) > 0$, $\alpha(s) > 0$, and a regularity mapping $\eta_s : B(\gamma_2(s), r(s)) \times [0, \alpha(s)] \to X$ satisfying

1. dist $(\eta_s(u, t)) \le t$, and
2. $f(u) - f(\eta_s(u, t)) \ge \sqrt{\varepsilon}.t$.

By continuity, for each $s \in S$ there is an interval $I(s)$ relatively open in $[0, 1]$ and containing s such that

$$\gamma_2(I(s)) \subset B(\gamma_2(s), r(s)/2).$$

Since S is compact, there is a finite subcovering $I(s_1), I(s_2), \ldots, I(s_k)$ of S. Define for $1 \le i \le k$,

$$\mu_i(t) = \begin{cases} \dfrac{\text{dist}(t, \complement I(s_i))}{\sum_{j=1}^{k} \text{dist}(t, \complement I(s_j))} & \text{if } t \in \bigcup_{1 \le i \le k} I(s_i), \\ 0 & \text{otherwise.} \end{cases}$$

And let $\phi : [0, 1] \to [0, 1]$ be a continuous function such that

$$\phi(t) = \begin{cases} 1 & \text{when } c \le f(\gamma_2(t)), \\ 0 & \text{when } f(\gamma_2(t)) \le c - \varepsilon. \end{cases}$$

Let

$$\tau = \min \left\{ \frac{1}{2} \min_{1 \le i \le k} r(s_i), \min_{1 \le i \le k} \alpha(s_i) \right\},$$

and define by induction a family of functions $y_i : [0, 1] \to X$ for $1 \le i \le k$ by

$$y_1 = \gamma_2,$$

and

$$y_{i+1} = \begin{cases} \eta_{s_i}(y_i(t))\tau\phi\mu_i & \text{if } \phi(t)\mu_i(t) \ne 0, \\ y_i(t) & \text{if } \phi(t)\mu_i(t) = 0, \end{cases}$$

for $1 \le i \le k$.

Then each y_i is well defined, and

$$\rho(\gamma_2, y_i) < \frac{r(s_i)}{2}. \tag{16.4}$$

Suppose that this is true for i, and take $t \in [0, 1]$ such that $\phi\mu_i \neq 0$. Then $t \in I(s_i)$, so that $\gamma_2(t) \in B(\gamma_2(s_i), r(s_i)/2)$. Then, by (16.4), we get that for each $t \in [0, 1]$

$$\text{dist}(y_i(t), \gamma_2(s_i)) \leq \text{dist}(y_i(t), \gamma_2(t)) + \text{dist}(\gamma_2, \gamma(s_i)),$$
$$< \frac{r(s_i)}{2} + \frac{r(s_i)}{2} = r(s_i).$$

So, $y_i(t) \in B(\gamma_2(s_i), r(s_i))$. But $\tau\phi\mu_i \leq \tau \leq \alpha(s_i)$. Thus y_{i+1} is well defined. The continuity of y_{i+1} can easily be verified using the first property of regularity mappings, which implies that

$$\text{dist}(y_{i+1}(t), y_i(t)) \leq \text{dist}\left(\eta_{s_i}(y_i(t), \tau\phi(t)\mu_i(t)), y_i(t)\right)$$
$$\leq \tau\phi(t)\mu_i(t)$$
$$\leq \tau\mu_i(t).$$

Hence,

$$\text{dist}(y_{i+1}(t), \gamma_2(t)) \leq \sum_{i=1}^{k} \text{dist}(y_{i+1}(t), y_i(t)),$$
$$\leq \tau \sum_{i=1}^{k} \mu_i(t).$$

But since $\sum_{i=1}^{k} \mu_i(t) \leq 1$, we get that

$$\text{dist}\left(y_{i+1}(t), \gamma_2(t)\right) \leq \tau \leq \frac{r(s_i)}{2}.$$

So, (16.4) is true for $i + 1$.

By the assumption $c - c_1 > \varepsilon$, we have that $f(\gamma_2(0)) = f(x_1) \leq c_1 < c - \varepsilon$, and $f(\gamma_2(1)) = f(x_2) \leq c_1 < c - \varepsilon$, but $\phi(0) = \phi(1) = 0$. So, for each $1 \leq i \leq k+1$, $y_i(0) = x_1$ and $y_i(1) = x_2$. Thus, $y_i \in \Gamma$ for $1 \leq i \leq k + 1$.

Now set $\overline{\gamma} = y_{k+1}$; by the second property of regularity mappings, we get for $1 \leq i \leq k$ that

$$f(y_{i+1}(t)) - f(y_i(t)) \leq f\left(\eta_{s_i}(y_i(t), \tau\phi\mu_i(t)) - f(y_i(t))\right).$$
$$\leq -\tau\sqrt{\varepsilon}\phi(t).$$

For t_0 such that $f(\overline{\gamma}(t_0)) = \chi(\overline{\gamma})$, $\phi(t_0) = 1$. Therefore,

$$f(\gamma_2(t_0)) - f(\overline{\gamma}(t_0)) \leq -\tau\sqrt{\varepsilon}.$$

Then

$$\chi(\overline{\gamma}) + \tau\sqrt{\varepsilon} \leq \chi(\gamma_2), \tag{16.5}$$

and $\gamma_2 \neq \overline{\gamma}$.

Equation (16.4) means that $\rho(\gamma_2, \overline{\gamma}) \leq \tau$. So, by (16.5), we conclude that

$$c(\overline{\gamma}) + \sqrt{\varepsilon}\rho(\gamma_2, \overline{\gamma}) \leq \chi(\gamma_1),$$

a contradiction with (16.3). □

Remark 16.2. We can easily see, by comparison to the proof of the nonsmooth MPT (see Chapter 15), that this theorem can be stated and proved with practically no change for a compact set K instead of $[0, 1]$ and a closed subset K_0 of K instead of $\{0, 1\}$.

The version attributed to Degiovanni and Marzocchi is stated in the general form suggested in the preceding remark and uses for its proof a *particular deformation lemma*. At this stage of abstraction (dealing with continuous functionals on metric spaces), no differential calculus (in the classical sense) existed yet.[1]

An immediate consequence of Ekeland's principle that will be useful in the sequel, and where the notion of weak slope appears, is the following.

Proposition 16.4. *Let X be a complete metric space and $f : X \to \mathbb{R} \cup \{+\infty\}$ a l.s.c. function. Let $r > 0$, $\sigma > 0$, and $E \subset X$ be such that $E \neq \varnothing$ and*

$$\inf_E f < \inf_X f + r\sigma.$$

Then, there exists $v \in X$ such that

$$f(v) < \inf_X f + r\sigma,$$
$$\text{dist}\,(v, E) < r, \qquad \text{and}$$
$$|df|(v) < \sigma.$$

Now comes the turn of a specific deformation lemma.

Lemma 16.5. *Let X be a metric space and $f : X \to \mathbb{R}$ be a continuous function. Consider a compact subset K of X and $\sigma > 0$ such that*

$$\inf\{|df|(u); \ u \in K\} > \sigma.$$

Then, there exists a neighborhood U of K in X, $\delta > 0$, and a continuous deformation $\eta : X \times [0, \delta] \to X$ such that

a. $\forall (u, t) \in X \times [0, \delta]$, $\text{dist}\,(\eta(u, t), u) \le t$,
b. $\forall (u, t) \in X \times [0, \delta]$, $f(\eta(u, t)) \le f(u)$,
c. $\forall (u, t) \in U \times [0, \delta]$, $f(\eta(u, t)) \le f(u) - \sigma t$.

Proof. For every $u \in K$ let us choose $\delta_u > 0$ and $\eta_u : B(u, \delta_u) \times [0, \delta_u] \to X$ according to the definition of the weak slope. Let $u_1, \ldots, u_n \in K$ be such that

$$K \subset \bigcup_{j=1}^{n} B\left(u_j, \frac{\delta_{u_j}}{2}\right).$$

Denote for simplicity $\delta_j = \delta_{u_j}$, $\eta_j = \eta_{u_j}$, and choose $s < \delta < \min\{\delta_1/2, \ldots, \delta_n/2\}$.

---

[1] See page 196 for a specific one.

Consider a neighborhood U of K in X and continuous functions $\theta_j : X \to [0, 1]$ $(1 \leq j \leq n)$ such that

$$\text{supp } \theta_j \subset B\left(u_j, \delta_j/2\right),$$

$$\forall v \in X, \ \sum_{j=1}^{n} \theta_j(v) \leq 1, \qquad \text{and}$$

$$\forall v \in U, \ \sum_{j=1}^{n} \theta_j(v) = 1.$$

Claim 16.1. For every $j = 1, \ldots, n$, there is a continuous map $\mathcal{K}_j : X \times [0, \delta] \to X$ such that for all $(u, t) \in X \times [0, \delta]$,

$$\text{dist}\left(\mathcal{K}_j(u, t), u\right) \leq \left(\sum_{h=1}^{n} \theta_h(u)\right) t,$$

and for all $(u, t) \in X \times [0, \delta]$,

$$f(\mathcal{K}_j(u, t)) \leq f(u) - \sigma\left(\sum_{h=1}^{n} \theta_h(u)\right) t.$$

First, set

$$\mathcal{K}_1(u, t) = \begin{cases} \eta_1(u, \theta_1(u)t) & \text{if } u \in \overline{B\left(u_1, 1/2\delta_1\right)}, \\ u & \text{if } u \notin \overline{B\left(u_1, 1/2\delta_1\right)}. \end{cases}$$

It is obvious that $\mathcal{K}_1$ satisfies the hypotheses.

Now take $2 \leq j \leq n$ and suppose by induction that $\mathcal{K}_{j-1}$ is defined. Since

$$\text{dist}\left(\mathcal{K}_{j-1}(u, t), u\right) \leq \left(\sum_{h=1}^{j-1} \theta_h(u)\right) t \leq \delta < \frac{1}{2}\delta_j,$$

we conclude that for all $u \in \overline{B\left(u_1, 1/2\delta_1\right)}$, $\mathcal{K}_{j-1}(u, t) \in B(u_j, \delta_j)$. Then, we define

$$\mathcal{K}_j(u, t) = \begin{cases} \eta_1(\mathcal{K}_{j-1}(u, t), \theta_j(u)t) & \text{if } u \in \overline{B\left(u_j, 1/2\delta_j\right)}, \\ \mathcal{K}_{j-1}(u, t) & \text{if } u \notin \overline{B\left(u_j, 1/2\delta_j\right)}. \end{cases}$$

By the induction hypothesis, it is easy to verify that $\mathcal{K}_j$ satisfies the conditions too, and the claim is proved.

To finish the proof of Lemma 16.5, it suffices to set $\eta = \mathcal{K}_n$. $\quad\square$

The following intermediate result is used by Degiovanni and Marzocchi to simplify the proof of their version of the MPT.

Lemma 16.6. *Let X be a metric space, $f : X \to \mathbb{R}$ be a continuous function, (D, S) a compact pair, and $\psi : S \to X$ a continuous map. Consider the set*

$$\Gamma = \left\{\gamma \in C(D; X); \ \gamma|_S = \psi\right\}$$

endowed with the uniform metric and define a continuous function $\Phi \colon \Gamma \to \mathbb{R}$ *by*

$$\Phi(\gamma) = \max_{\gamma(D)} f.$$

Let $\gamma \in \Gamma, \rho > 0, \sigma > 0$ *be such that*

$$\max_{\gamma(S)} \psi < \max_{\gamma(D)} f$$

and

$$\left(\text{for all } \xi \in D, \ f(\gamma(\xi)) \geq \max_{\gamma(D)} f - \rho \right) \quad \text{implies} \quad |df|(\gamma(\xi)) \geq \sigma.$$

Then $|d\Phi|(\gamma) \geq \sigma$.

Remark 16.3. Recall that (D, S) is a compact pair if D is compact and $S \subset D$ is a closed subset (and hence compact) of D.

Proof. Without loss of generality, we can assume that

$$\max_{\psi(S)} f < \max_{\gamma(D)} f - 3\rho.$$

Let $\sigma' \in \,]0, \sigma[$, U and $\eta \colon X \times [0, \delta] \to X$ be the deformation obtained by applying the previous deformation lemma to the compact set

$$\left\{ \gamma(\xi); \ f(\gamma(\xi)) \geq \max_{\gamma(D)} f - \rho \right\}$$

and to σ'. We can assume $f(u) > \max_{\gamma(D)} f - 2\rho$ for every $u \in U$.

We can also suppose that $\eta(u, t) = u$ whenever $f(u) < \max_{\gamma(D)} f - 3\rho$. Otherwise, we substitute $\eta(u, t)$ with $\eta(u, t\lambda(u))$, where $\lambda \colon X \to [0, 1]$ is a continuous function such that $\lambda(u) = 0$ for $f(u) \leq \max_{\gamma(D)} f - 3\rho$ and $\lambda(u) = 1$ for $f(u) \geq \max_{\gamma(D)} f - 2\rho$.

Let $\mathcal{N}$ be a neighborhood of γ in Γ such that

$$\text{for all } \tau \in \mathcal{N}, \ \max_{\tau(D)} f \geq \max_{\gamma(D)} f - \frac{2}{\rho},$$

and

$$\text{for all } \tau \in \mathcal{N}, \ \left\{ \tau(\xi); \ f(\tau(\xi)) \geq \max_{\gamma(D)} f - \rho \right\} \subset U.$$

Let $\delta' = \min\{\rho/(2\sigma'), \delta\}$ and let $\mathcal{K} \colon \mathcal{N} \times [0, \delta'] \to \Gamma$ be defined by

$$\mathcal{K}(\tau, t)(\xi) = \eta(\tau(\xi), t) \text{ for all } \xi \in D.$$

It is easy to verify that $\mathcal{K}$ is continuous and that $\rho(\mathcal{K}(\tau, t), \tau) \leq t$. Now take $\tau \in \mathcal{N}$

and $\xi \in D$. If $f(\tau(\xi)) \leq \max_{\gamma(D)} f - \rho$, then for every $t \in [0, \delta']$,

$$
\begin{aligned}
f(\mathcal{K}(\tau, t)(\xi)) &= f(\eta(\tau(\xi), t)), \\
&\leq f(\tau(\xi)), \\
&\leq \max_{\gamma(D)} f - \rho, \\
&\leq \max_{\tau(D)} f - \frac{\rho}{2}, \\
&< \Phi(\tau) - \sigma' t.
\end{aligned}
$$

Instead, if $f(\tau(\xi)) \geq \max_{\gamma(D)} f - \rho$, then for every $t \in [0, \delta']$,

$$
f(\mathcal{K}(\tau, t)(\xi)) = f(\eta(\tau(\xi), t)) \leq f(\tau(\xi)) - \sigma' t \leq \Phi(\tau) - \sigma' t.
$$

Therefore,

$$
\text{for all } t \in [0, \delta'], \ \Phi(\mathcal{K}(\tau, t)) \leq \Phi(\tau) - \sigma' t
$$

and $|d\Phi|(\gamma) \geq \sigma'$. The lemma is proved since $\sigma' \in {]}0, \sigma{[}$ was taken arbitrarily. $\square$

The statement of the MPT according to Degiovanni and Marzocchi [310] is the following.

Theorem 16.7 (Metric MPT, Degiovanni and Marzocchi Formulation). *Let X be a metric space, $f \colon X \to \mathbb{R}$ be a continuous function, (D, S) be a compact pair, $\psi \colon S \to X$ be a continuous map, and*

$$
\Gamma = \left\{ \gamma \in \mathcal{C}(D; X); \ \gamma|_S = \psi \right\}.
$$

Suppose that $\Gamma \neq \varnothing$ and that

$$
\text{for all } \gamma \in \Gamma, \ \max_{\psi(S)} f < \max_{\gamma(D)} f.
$$

If f satisfies the (PS) condition at the level

$$
c = \inf_{\gamma \in \Gamma} \max_{\gamma(D)} f,
$$

then c is a critical value for f.

Proof. By contradiction, let us suppose that c is not a critical value for f. Since $(\text{PS})_c$ holds, there exists $\sigma > 0$ such that

$$
u \in \mathcal{D}(f) \text{ and } c - \sigma \leq f(u) \leq c + \sigma \qquad \text{implies that} \qquad |df|(u) \geq \sigma. \quad (16.6)
$$

Define $\Phi \colon \Gamma \to \mathbb{R}$ as in the preceding lemma. Since Φ is bounded from below, by Proposition 16.4, there exists $\gamma \in \Gamma$ such that

$$
\Phi(\gamma) < c + \sigma \qquad \text{and} \qquad |d\Phi|(\gamma) < \sigma.
$$

But by (16.6), we have that

$$
\left(f(\gamma(\xi)) \geq \max_{\gamma(D)} f - \sigma \right) \qquad \text{implies that} \qquad |df|(\gamma(\xi)) \geq \sigma.
$$

By the preceding lemma, we get the contradiction $|d\Phi|(\gamma(\xi)) \geq \sigma$. $\square$

Comments and Additional Notes

The metric critical point theory is now a tangible reality. See, for example, [257,261,310, 494,495,516]. Among the tools there, two specific deformation lemmata were obtained in [310,494], but they both use the variational principle. Another one (homotopy based) in [494] is proved without Ekeland's principle, but it relies on the *paracompactness* of metric spaces.

◇ 16.I Strong Slope of Degiovanni and Marzocchi

The term weak in "weak slope" can be explained by the fact that Degiovanni and Marzocchi introduced before, in [305], a different notion they called *slope* and denoted by $|\nabla f|(x)$ to define a critical point theory for a certain class of functionals. The definition of this (strong) slope is the following.

Definition 16.4. Let $f: X \rightarrow \mathbb{R} \cup \{+\infty\}$ be a l.s.c. function and let $u \in \mathcal{D}(f)$. We define

$$|\nabla f|(x) = \begin{cases} \displaystyle\limsup_{v \to x} \frac{f(x) - f(v)}{\operatorname{dist}(x, v)} & \text{if } x \text{ is not a local minimum,} \\ 0 & \text{if } x \text{ is a local minimum.} \end{cases}$$

The number $|\nabla f|(x)$ is called the (strong) slope of f at x and may be infinite, too.

◇ 16.II A Specific Subdifferential Calculus

Many early applications of the metric theory to several problems in partial differential equations and variational inequalities where the functionals involved are not locally Lipschitz continuous were given [68,170–173,251,260,307,312,496] but this passes in general through a great deal of technicality. Recently, a *specific subdifferential calculus* has been developed by Campa and Degiovanni [166,307] when the underlying space is *normed* and has been used successfully in applications where Clarke's subdifferential seems inappropriate [64,67,69,166,314,856–860].

The beginning point of this work seems to be the remarks that $|df|(u)$, being a generalization of the norm of the derivative, may not have a rich calculus, and, also the fact mentioned earlier that applications to nonlinear problems that used it in a direct way involved many technicalities. Although, from its definition that uses "downhill" deformations, the weak slope is a good candidate for obtaining abstract "inf sup" critical point theorems.

This subdifferential calculus implies in particular that

$$|df|(u) < \infty \Rightarrow \left(\partial f(u) \neq \varnothing \text{ and } |df|(u) \geq \min\{\|\alpha\|;\ \alpha \in \partial f(u)\} \right). \quad (16.7)$$

So, the condition $|df|(u) = 0$ implies that $0 \in \partial f(u)$, and every critical point result in terms of $|df|$ implies a corresponding one in terms of the subdifferential $\partial f(u)$.

Equation (16.7) is satisfied for a locally Lipschitz functional for the Clarke subdifferential, but when f is only continuous, this is no longer true. It suffices to consider

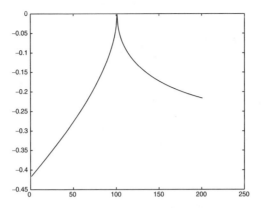

Figure 16.1. A mountain pass situation for a non-Lipschitz but continuous function, where Clarke's theory does not apply.

$f: \mathbb{R} \to \mathbb{R}$ defined by $f(u) = u - \sqrt{|u|}$ at $u = 0$. We have $|df|(0) = 0$. (It is so by the MPT for continuous functions; have a look at Figure 16.1.)

Nevertheless, Clarke's subdifferential of f at 0 is empty. This is to justify the introduction of the new subdifferential, when X is a normed space.

Definition 16.5. For any $u \in X$ such that $f(u) \in \mathbb{R}$, $v \in X$, and $\varepsilon > 0$, we define $f_\varepsilon^0(u; v)$ as the infimum of $r \in \mathbb{R}$ such that there exists $\delta > 0$ and a continuous mapping

$$v \colon \big(B_\delta(u, f(u)) \cap \mathrm{epi}\,(f)\big) \times]0, \delta] \to B_\varepsilon(v)$$

satisfying

$$f(w + tv((w, \mu), t)) \le \mu + rt$$

for all $(w, \mu) \in B_\delta(u, f(u)) \cap \mathrm{epi}\,(f)$ and $t \in]0, \delta]$, where $\mathrm{epi}\,(f) = \{(x, \lambda);\ f(x) \le \lambda\}$ is the *epigraph* of f.
Define also

$$f^0(u; v) = \sup_{\varepsilon > 0} f_\varepsilon^0(u; v).$$

The function $f^0(u; .)$ is convex, lower semicontinuous, and positively homogeneous of degree 1 (cf. [166]). The corresponding notion of subdifferential is defined as follows.

Definition 16.6. For all $u \in X$ such that $f(x) \in \mathbb{R}$, we set

$$\partial f(u) = \{\alpha \in X^*;\ \langle \alpha, v \rangle \le f^0(u; v),\ \forall v \in X\}.$$

If we drop the continuity condition on v, we get exactly the generalized directional derivative in the sense of Clarke and Rockafellar. Therefore, the real $f^0(u; v)$ is larger than the Clarke-Rockafellar generalized directional derivative. Hence, $\partial f(u)$ contains Clarke's subdifferential of f at u, which may be empty when $\partial f(u)$ is not, as we can see for the example given earlier for $f(u) = u - \sqrt{|u|}$ at $u = 0$ (there $|df|(0) = 0$). Nevertheless, the two notions agree when f is locally Lipschitz or directionally Lipschitz at u. Recall that f is directionally Lipschitz at u if the set of $v \in X$ such that

$$f^+(x; v) = \limsup_{\substack{(\xi,\mu)\to(x,f(x)) \\ (\xi,\mu)\in\text{epi}(f) \\ w\to v,\ t\to 0^+}} \frac{f(\xi + tw) - \mu}{t} < \infty$$

is nonempty. Corresponding notions of *normal and tangent cones* are also given in [166].

Proposition 16.8. *If $u \in X$ is such that $f(u) \in \mathbb{R}$, then*

   a. $|df|(u) < +\infty \iff \partial f(u) \neq \varnothing$,
   b. $|df|(u) < +\infty \Rightarrow |df|(u) \geq \min\{\|\alpha\|;\ \alpha \in \partial f(u)\}$.

Remark 16.4. Notice that condition b may be strict.

Proposition 16.9. *When $u \in X$ is such that $f(u) \in \mathbb{R}$ and $g\colon X \to \mathbb{R}$ is Lipschitz continuous, then*

$$\partial(f + g)(u) \subset \partial(f)(u) + \partial(g)(u),$$

if, moreover, g is of class $\mathcal{C}^1$, $\partial g(u) = \{g'(u)\}$, and we have equality.

◇ 16.III More Definitions of Critical Points on Metric Spaces

In connection with the metric MPT, we think that it would be useful to have a look at the work of Ioffe and Schwartzman [494, 495] who are also contributing to what is becoming a "metric critical point theory." You may also consult the paper [770], where a metric version of the intrinsic MPT appears (see Chapter 20) that uses the specific definition of Ioffe and Schwartzman.

In [494], it is reported that the roots of a metric critical point theory go back to Morse, who was the first to define some notions of critical points for continuous functions on metric spaces as early as 1937 [661].

Definition 16.7. A point $x \in X$ is said to be *Morse regular* if there are a neighborhood U of x and a continuous mapping $\eta\colon U \times [0, 1] \to X$ such that, for all $(u, t) \in U \times [0, 1]$,

   i. $\eta(u, 0) = u$, and
   ii. $\Phi(u) - \Phi(\eta(u, t)) > 0$ if $t > 0$.

The point x is said to be *Morse critical* if it is not Morse regular.

Always in [661], Morse defines an *ordinary point* as follows.

Definition 16.8. A point $x \in X$ is said to be an *ordinary point* if there are a neighborhood U of x, a continuous function $\eta\colon U \times [0, 1] \to X$, and a nondecreasing function $\sigma(t)$, positive for t positive, such that, for all $(u, t) \in U \times [0, 1]$,

   i. $\eta(u, 0) = u$, and
   ii.' $\Phi(u) - \Phi(\eta(u, t)) \geq \sigma(\text{dist}(u, \eta(u, t)))$.

The definition of Degiovanni and Marzocchi and Katriel of a δ-regular point is slightly different from that of Ioffe and Schwartzman, whose statement is the following.

Definition 16.9. A point $x \in X$ is said to be a δ-regular point if there are U and η as in definition 16.8, such that, for all $(u, t) \in U \times [0, 1]$,

  i. $\eta(u, 0) = u$, and
  ii.'' $\Phi(u) - \Phi(\eta(u, t)) \geq \delta \cdot \text{dist}(u, \eta(u, t))$, and $\eta(u, t) \neq u$ if $t > 0$.

The upper bound of such δ is denoted in [494] by $\delta(\Phi, x)$, like the notation adopted by Katriel. We keep this notation for the rest of this note since there is no ambiguity.

When Φ is of class $\mathcal{C}^1$, the deformation in the preceding definition of a δ-regular point can be taken either $u - t\Phi'(u)$ or $u - t\Phi'(x_0)$.

The relation $\|\Phi'(x)\| = \delta(\Phi, x)$ holds true there, too, and $\delta(\Phi, x)$ is l.s.c. in x.

The notions of δ-critical point and Morse critical point do not coincide even for smooth functions on $\mathbb{R}$. It suffices to consider the function $\Phi(x) = x^3$.

Property ii' (in the definition of an ordinary point) is somewhere between properties ii (in the definition of a Morse regular point) and ii'' (in the definition of a δ-regular point of Ioffe-Schwartzman).

It is clear that a δ-regular point in the sense of Degiovanni and Marzocchi and Katriel is also a δ-regular point in the sense of Ioffe and Schwartzman. (We neglect the assumption that $\eta(t, u) \neq u$ for $t > 0$.) But it is not clear that the inclusion should be strict.

When Φ is Lipschitz continuous in a neighborhood of u, then

$$|df|(u) \geq \text{dist}(0, \partial\Phi(u)). \tag{16.8}$$

So, a regular point using the terminology of nonsmooth (Lipschitz) analysis is also regular in the sense of Degiovanni and Marzocchi and, hence, it is regular in the sense of Ioffe and Schwartzman.

Indeed (using the subdifferential and the generalized gradient of Clarke), it suffices to notice that if

$$\delta < \text{dist}(0, \partial\Phi(u)),$$

then there is $h \in X$ such that

$$\Phi^\circ(u; h) = \limsup_{\substack{x \to u \\ t \to 0}} \frac{\Phi(x + th) - \Phi(x)}{t} < \delta.$$

So, we take $\eta(t, u) = u + th$. $\square$

Inequality (16.8) may be strict, as we can check in the examples seen earlier.

Remark 16.5. One more remark from [494] is that the concept of Morse regularity is topological and hence is invariant under local homeomorphisms, while the notion of a δ-regularity point is metric and then is invariant under locally Lipschitz homeomorphisms.

Another definition of the weak slope equivalent to that of Degiovanni and Marzocchi when the functional is continuous was given by Campa and Degiovanni. It is discussed in some detail in connection with the particular subdifferential calculus they introduced for continuous functionals on metric spaces.

◇ 16.IV A Critical Point Theory for l.s.c. Functionals

In [310], Degiovanni and Marzocchi developed a more general approach that requires the functional to be only l.s.c.

Definition 16.10. Let $f \colon X \to \mathbb{R} \cup \{+\infty\}$ be l.s.c. We define the function $\mathcal{G}_f \colon \operatorname{epi}(f) \to \mathbb{R}$ by $\mathcal{G}_f(u, \xi) = \xi$ where the epigraph of f is endowed with the metric

$$\operatorname{dist}((u, \xi), (v, \upsilon)) = \left(\operatorname{dist}(u, v)^2 + (\xi - \mu)^2\right)^{1/2}.$$

Then, $\operatorname{epi}(f)$ is closed in $X \times \mathbb{R}$ and $\mathcal{G}_f$ is Lipschitz continuous of constant 1. Therefore, $|d\mathcal{G}_f|(u, \xi) \leq 1$ for every $(u, \xi) \in \operatorname{epi}(f)$.

Proposition 16.10. Let $f \colon X \to \mathbb{R} \cup \{+\infty\}$ be a continuous function and let $u \in X$, $\xi \in \mathbb{R}$. Then

$$|d\mathcal{G}_f|(u, f(u)) = \begin{cases} \dfrac{|df|(u)}{\sqrt{1 + (|df|(u))^2}} & \text{if } |df|(u) < +\infty \\[2mm] 1 & \text{if } |df|(u) = +\infty \end{cases}$$

and

$$|d\mathcal{G}_f|(u, \xi) = 1 \quad \text{if } f(u) < \xi.$$

This proposition extends the definition of the weak slope to l.s.c. functionals too, but in an indirect way.

Definition 16.11. Let $f \colon X \to \mathbb{R} \cup \{+\infty\}$ be l.s.c. and pick $u \in \mathcal{D}(f)$ where

$$\mathcal{D}(f) = \{u \in X; \ f(u) < +\infty\} \text{ is the effective domain of } f.$$

Then

$$|df|(u) = \begin{cases} \dfrac{|d\mathcal{G}_f|(u, f(u))}{\left(1 + (|d\mathcal{G}_f|(u, f(u))^2\right)^{1/2}} & \text{if } |d\mathcal{G}_f|(u, f(u)) < 1, \\[2mm] +\infty & \text{if } |d\mathcal{G}_f|(u, f(u)) = 1. \end{cases}$$

Nevertheless, there is a criterion to get a lower estimate on $|df|(u)$ even in this l.s.c. case.

Proposition 16.11. Let $f \colon X \to \mathbb{R} \cup \{+\infty\}$ be a l.s.c. function and let $u \in \mathcal{D}(f)$. If there exist $\delta > 0$, $b > f(u)$, $\sigma > 0$, and a continuous deformation $\eta \colon (B(u, \delta) \cap f^b) \times [0, \delta] \to X$, such that for all $v \in B(u, \delta) \cap f^b$, and all $t \in [0, \delta]$,

$$\operatorname{dist}(\eta(v, t), v) \leq t \text{ and } f(\eta(v, t)) \leq f(v) - \sigma t,$$

then $|df|(u) \geq \sigma$.

There too, as it can checked immediately, the weak slope is l.s.c. with respect to the graph topology.

Proposition 16.12. *Let* $f: X \to \mathbb{R} \cup \{+\infty\}$ *be a l.s.c. function and let* $u \in \mathcal{D}(f)$. *If* $(u_n)_n \subset X$ *is such that* $u_n \to u$ *and* $f(u_n) \to f(u)$, *then*

$$|df|(u) \leq \liminf_n |df|(u_n).$$

Remark 16.6. We would like to emphasize the following facts concerning a possible extension of the theory as it is to l.s.c. functions reported before by Szulkin [891]:

> In the case of functions which are only l.s.c. such a construction [2] does not seem to be readily available, mainly because a non critical value c may be "semicritical" in the sense that there may exist a critical point $\bar{u}$ with $I(u) < c$ and a sequence $u_n \to \bar{u}$ with $I(u_n) \to c$.

This was confirmed later by Degiovanni and Marzocchi [310, p. 74].

> Let us point out that a general critical point theory for lower semicontinuous functionals seems not to be possible. Consider $f: \mathbb{R} \to \mathbb{R}$ defined by $f(x) = x + 1$ for $x < 0$ and $f(x) = x$ for $x \geq 0$. In this case $x = 0$ should not be considered as a mountain pass point, because the value $f(0)$ is not correct.

Nevertheless, we conclude that there is no reason that another approach would not be fruitful. The reader may consult the paper [409] by Frigon where the preceding example by Degiovanni and Marzocchi motivated a new critical point theory for multivalued mappings with a closed graph. This is also used to derive some results for l.s.c. functions that may be considered for a continuation of the aforementioned study. See also the note "the MPT for upper semicontinuous compact-valued mappings" in Chapter 25 for another multivalued approach that can be used for the study of l.s.c. functions.

◇ 16.V Pseudo-Gradient Vector Field for Continuous Functionals on Metric Spaces

In the very interesting paper [153] by Brézis and Nirenberg, the following result on the existence of a pseudo-gradient vector field for continuous functionals on metric spaces is given.

Lemma 16.13 ([153, Lemma 2]). *Let* E *be a metric space and* $f: E \to X^*$ *be a continuous function. Then, given* $\varepsilon > 0$, *there exists a locally Lipschitz function* $v: E \to X$ *such that for any* $\zeta \in E$:

    i. $\|v(\zeta)\| \leq 1$,
    ii. $\langle f(\zeta), v(\zeta) \rangle \geq \|f(\zeta)\| - \varepsilon$.

Applying this result with $f(\zeta) = \Phi'(\gamma(\zeta))$ where $\gamma \in \Gamma$, we obtain that, for all $\zeta \in E$,

---

[2] He means a compactness condition and a deformation result [891, p. 79].

i. $||v(\zeta)|| \leq 1$,
ii. $\langle \Phi'(\gamma(\zeta)), v(\zeta) \rangle \geq ||\Phi'(\gamma(\zeta))|| - \varepsilon$.

The lemma is proved using a partition of unity argument as in the usual construction of the pseudo-gradient vector field (see [748], for example).

This result is combined with Ekeland's principle in Chapter 17 to prove the MPT on convex domains.

◇ 16.VI Quantitative Deformation Lemmata and the Intrinsic MPT for Continuous Functionals

In an extension to the metric setting of the intrinsic MPT of Schechter [809], Corvellec shows in [257] how quantitative deformation properties can be used to obtain minimax results, even for the case when the usual geometric assumptions are not satisfied. This material is treated in detail in Chapter 20. See also [770, 771], where Ribarska et al. established another nonsmooth variant of the intrinsic MPT. Their proof is also based on a variant of the deformation lemma that uses the weak slope variant of Ioffe and Schwartzman.

◇ 16.VII Some Directions Where the Actual Metric Critical Point Theory May Be Extended

With the property of *uniform descent* in a neighborhood of a regular point, a local minimum is necessarily a critical point. But a local maximum is not necessarily critical, as we can see in the following example.

Consider the mapping $\Phi \colon \mathbb{R}^2 \to \mathbb{R}$ defined by $\Phi(x, y) = -A \operatorname{dist}((x_0, y_0), (x, y))$, where $A > 0$ is a constant, $u_0 = (x_0, y_0)$, and $\operatorname{dist}((x_0, y_0), (x, y)) = \sqrt{(x - x_0)^2 + (y - y_0)^2}$. Consider the *radial* deformation $\eta(t, .)$ that transforms a point $u = (x, y)$ on the cone described by Φ on the point $u' = (x', y')$ on the cone such that $\operatorname{dist}(u_0, u) = t \operatorname{dist}(u_0, u')$ for $t \in [0, 1]$ and u belongs to $[u_0, u']$. Then,

$$\Phi(\eta(t, u)) - \Phi(u) = -A \operatorname{dist}(u_0, \eta(t, u)) + A \operatorname{dist}(u_0, u),$$
$$= -A(\operatorname{dist}(u_0, u) + \operatorname{dist}(u, \eta(t, u))) + A \operatorname{dist}(u_0, u),$$
$$= -A \operatorname{dist}(u, \eta(t, u)), \text{ everywhere in } \mathbb{R}.$$

So, u_0 is A-regular. Hence, for the definition of regularity of Degiovanni and Marzocchi and Katriel, a local maximum is not necessarily critical. In particular, this implies that $\mathbb{K}(\Phi) \neq \mathbb{K}(-\Phi)$ in general.

Consider also the following example by Campa and Degiovanni [166] inspired by a result by Ribarska et al. [768]. Let $\Phi \colon \mathbb{R}^2 \to \mathbb{R}$ be defined by

$$\Phi(x, y) = a|y - m|x|| - \sigma x,$$

where $a, m, \sigma > 0$ and $am > \sigma$. Then, Φ is locally Lipschitz continuous and $0 \in \partial \Phi(0, 0)$ but $(0, 0)$ is regular in the sense of Degiovanni and Marzocchi.

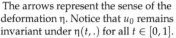

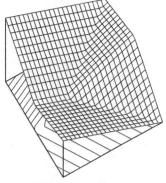

The arrows represent the sense of the deformation η. Notice that u_0 remains invariant under $η(t,.)$ for all $t \in [0,1]$.

This figure represents $\Phi(x,y) = 3|y - 3|x|| - 5x$, that is, $a = m = 3$ and $\sigma = 5$.

Figure 16.2. Some good candidates for critical points missed by the actual metric theory

◇ 16.VIII Morse Theory for Continuous Functionals on Metric Spaces

In [255, 259], Corvellec proves an analogue of the second deformation lemma for continuous functionals on metric spaces. As a consequence, he gets the Morse relation between the critical groups and Betti numbers.

◇ 16.IX Classification of the Critical Set of the Nonsmooth MPT

In [390] Fang extends the smooth theory of classification and localization seen in Chapter 12 to the case of continuous functionals. Moreover, as in his thesis [387], he also obtains similar results on the structure of the critical set generated by min-max principles. He considers homotopy, cohomotopy, and homology classes.

◇ 16.X Some Additional References

The lectures [173] by Canino and Degiovanni are devoted to the metric critical point theory for continuous functionals and to the existence of multiple solutions for quasilinear elliptic equations. The case in which f is invariant under the action of a compact Lie group is also considered. MPTs for continuous functionals and estimates of the number of critical points of f by means of the relative category are provided.

The paper [252] by Conti and Lucchetti is an overview of (smooth) critical point theory. In the final part, the case of continuous functionals defined on a metric space is presented.

In [261], Corvellec et al. prove a deformation lemma for continuous functionals on metric spaces. No use is made of Ekeland's variational principle.

◇ 16.XI On Uniform Descent Directions by Rockafellar

Compare the notion of metric stationary/regular point with the following notion from
[775, Chapter V]. (The following is a translation from french by the author with some
minor changes; see [776] for the English translation.)

The generalization which appears to be appropriate in this context may be described
in terms of "descent directions" with respect to a l.s.c. function f and a point x where
f takes a finite value. We will say that y is a *uniform descent direction* at x if, for some
$\rho > 0$, there is a neighborhood $X \in \mathcal{N}(x)$, $\delta > 0$, $\lambda > 0$ such that for all $t \in]0, \lambda[$ and
all $x' \in X$ with $f(x') \leq f(x) + \delta$ we have

$$f(x' + ty) \leq f(x') - t\rho. \tag{16.9}$$

This corresponds to the property $f^\circ(x; y) < 0$, where $f^\circ(x; y)$ is

$$f^\circ(x; y) = \limsup_{\substack{x' \to x \\ f(x') \to f(x) \\ t \downarrow 0}} \frac{f(x' + ty') - f(x')}{t}.$$

More generally, we will say that y is an *approximative uniform descent direction* at x
if, for some $\rho > 0$, for every neighborhood $Y \in \mathcal{N}(x)$ there is $X \in \mathcal{N}(x)$, $\delta > 0$, $\lambda > 0$
such that for all $t \in]0, \lambda[$ and all $x' \in X$ with $f(x') \leq f(x) + \delta$, we have

$$\inf_{y' \in Y} f(x' + ty') \leq f(x') - t\rho. \tag{16.10}$$

This corresponds to the property $f^\uparrow(x; y) < 0$, where $f^\uparrow(x; y)$, when f is l.s.c., is

$$f^\uparrow(x; y) = \limsup_{\substack{x' \to x \\ f(x') \to f(x) \\ t \downarrow 0}} \inf_{y' \to y} \frac{f(x' + ty') - f(x')}{t}$$

$$= \sup_{\substack{Y \in \mathcal{N}(y) \\ \delta > 0 \\ \lambda > 0}} \inf_{X \to \mathcal{N}(x)} \sup_{\substack{t \in]0, \lambda[\\ x' \to x \\ f(x') \leq f(x) - \delta}} \inf_{y' \to y} \frac{f(x' + ty') - f(x')}{t}.$$

A point x is said to be *substationary* for f if there is no approximative uniform descent
direction y to x.

The class of substationary points contains all minima, an important part of local
maxima, and saddle points of the form that appears in $x^2 - y^2$ on $\mathbb{R}^2$. For $x = (x_1, x_2) \in$
$\mathbb{R}^N = \mathbb{R}^{N_1} \times \mathbb{R}^{N_2}$, suppose that there are neighborhoods $X_1 \in \mathcal{N}(x_1)$, $X_2 \in \mathcal{N}(x_2)$
such that

$$f(x'_1, x_2) \geq f(x_1, x_2) \geq f(x_1, x'_2), \qquad \text{for all} \qquad x'_1 \in X_1, x'_2 \in X_2.$$

◇ 16.XII Extension of the Critical Point Theory for Continuous
Functionals on Metric Spaces to Multivalued Mappings with Closed Graph

The critical point theory for continuous functionals on metric spaces is extended to *multivalued mappings with closed graph* in [409] by Frigon (see Chapter 25 for the details).

V

Speculating about the Mountain Pass Geometry

17

The MPT on Convex Domains

The concept of convexity has far-reaching consequences in variational analysis. In the study of maximization and minimization, the division between problems of convex or nonconvex type is as significant as the division in other areas of mathematics between problems of linear or nonlinear type.

R.T. Rockafellar and R.J.-B. Wets, *Variational analysis*, Grundlehren der Mathematischen Wissenschaften, **317**, Springer, 1998.

A definition of the notions of critical points of differentiable functionals defined on convex sets is formulated by the means of variational inequalities. A corresponding version of the MPT is then given and proved twice. The first proof (which is only outlined) is based on an appropriate form of the deformation lemma, while the second uses Ekeland's variational principle.

When working on convex sets, extremal conditions may be expressed as variational inequalities. In this chapter, which may be considered at the midpoint between optimization and minimax methods, we will review some abstract variational results developed to deal with differentiable functionals on closed convex sets in Banach spaces. This type of functionals appears, for example, in problems with inequality constraints.

17.1 Variational Inequalities as Extremal Conditions

When working on convex sets, extremal conditions may be expressed as variational inequalities. Indeed, consider the minimization problem:

$$\text{Find} \qquad \min_{u \in M} \Phi(u), \qquad (17.1)$$

where $\Phi \colon M \subset X \to \mathbb{R}$ is a functional defined on the convex nonempty set M of a locally convex space X. If Φ is Gateaux differentiable, then any solution u of (17.1) satisfies the variational inequality

$$\langle \Phi'(u), v - u \rangle \geq 0, \qquad \text{for all } u \in M. \qquad (17.2)$$

If u is an interior point of M, then it is a critical point; that is, $\Phi'(u) = 0$.

Proposition 17.1. *When* Φ *is convex and Gateaux differentiable on* M, *then the minimization problem* (17.1) *and the variational inequality* (17.2) *are equivalent.*

Proof. It suffices to consider, for a fixed $v \in M$,

$$\varphi(t) = \Phi(u + t(v - u)).$$

If u is a solution of (17.1), then $\varphi(t) \geq \varphi(0)$ for all $t \in [0, 1]$. Therefore, $\varphi'(0) \geq 0$, that is, (17.2). Conversely, if u is a solution of (17.2), then $\varphi'(0) \geq 0$. Since φ is convex on $[0, 1]$, φ' is monotone, so

$$\varphi(1) - \varphi(0) = \varphi'(\theta) \geq \varphi'(0), \qquad 0 < \theta < 1.$$

Hence, $\Phi(v) \geq \Phi(u)$ for all $v \in M$, that is, (17.1).   □

We can show easily that when X is a reflexive Banach space, $M \subset X$ is a closed convex nonempty subset that is bounded, and $\Phi : M \to \mathbb{R}$ is weakly (sequentially) lower semicontinuous, the problem (17.1) admits a solution.

17.2 The MPT on Convex Domains

Consider a closed convex subset M of a Banach space X and suppose that $\Phi : M \to \mathbb{R}$. For $u \in M$, define the *steepest slope of* Φ *in* M,

$$g(u) = \sup_{\substack{v \in M \\ \|u-v\|<1}} \langle \Phi'(u), u - v \rangle,$$

as a measure for the slope of Φ in M. If $M = X$, $v - u$ covers all X, so we have that $g(u) = \|\Phi'(u)\|$. Moreover, if Φ is of class C^1, the function g is continuous in M.

Definition 17.1. A point $u \in M$ is said to be *critical* if $g(u) = 0$ and its value is said to be *critical*. Otherwise, u is *regular* and its value is *regular*.

This definition coincides with the usual one when $M = X$. So, the results obtained for such a notion constitute a generalization of similar ones on the whole space. Moreover, g is continuous in M when Φ is of class C^1.

We will say that the functional Φ satisfies the $(PS)_M$ condition if

> any sequence $(u_n)_n$ in M, such that
> $|\Phi(u_n)| \leq C$ uniformly and $g(u_m) \to 0$ as $m \to 0$,
> is relatively compact.

This condition implies in particular that the set $\mathbb{K}_c$ of critical points with some value c is compact.

17.2.1 A First Form Using a Deformation Lemma

The following form of the MPT on convex domains was proved by Struwe in his study of minimal surfaces.

Theorem 17.2 (MPT on Convex Domains, Struwe). *Suppose that M is a closed convex subset of a Banach space X, $\Phi \in C^1(X; \mathbb{R})$ satisfies (PS)$_M$ on M and admits two distinct relative minima u_1, u_2 in M. Then,*

   i. either $\Phi(u_1) = \Phi(u_2) = a$ and u_1, u_2 can be connected in any neighborhood of the set of relative minima $u \in M$ of Φ with $\Phi(u) = a$, or

  ii. there exists a critical point $\bar{u}$ of Φ in M that is not a relative minimizer of Φ.

This result was derived by Struwe in the study of minimal surface in the spirit of Morse and Tompkins and Shiffman (see the discussion at the end of the finite dimensional MPT).

The proof uses, in the standard way used with the classical MPT, a specific deformation lemma with a special version of pseudo-gradient vector fields adapted to this situation.

Lemma 17.3 (Deformation Lemma, Struwe). *Suppose $M \subset X$ is closed and convex, $\Phi \in C^1(X; \mathbb{R})$ satisfies (PS)$_M$ on M, and let $c \in \mathbb{R}$, $\bar{\varepsilon} > 0$ be two given real numbers. Then, for any neighborhood N of $\mathbb{K}_c$, there exist $\varepsilon \in]0, \bar{\varepsilon}[$ and a continuous deformation $\eta \colon M \times [0, 1] \to M$ such that*

    i. $\eta(u, t) = u$ if $g(u) = 0$, or if $t = 0$, or if $|\Phi(u) - c| \geq \bar{\varepsilon}$.

   ii. $\Phi(\eta(u, t))$ is nonincreasing in t for any $u \in M$.

  iii. $\eta(\Phi^{c+\varepsilon}, 1) \subset \Phi^{c-\varepsilon} \cup N$ and $\eta(\Phi^{c+\varepsilon} \setminus N, 1) \subset \Phi^{c-\varepsilon}$.

The proof of this deformation lemma, which is similar to that of the (standard) version, becomes straightforward once a new notion of pseudo-gradient vector flow is defined and its existence for C^1-functionals is proved.

Definition 17.2. A pseudo-gradient vector field $v \colon \tilde{M} \to X$ for Φ on M, where $\tilde{M} = \{u \in M; \, g(u) \neq 0\}$, is a locally Lipschitz vector such that for some $c > 0$ and for any $u \in \tilde{M}$,

    1. $u + v(u) \in M$,

    2. $\|v(u)\| < \min(1, g(u))$, and

    3. $\langle v(u), \Phi'(u) \rangle < -c \min(1, g(u))g(u)$.

As for the smooth case (Lemma 4.1), we have the existence of a pseudo-gradient vector field for Φ when it is of class C^1.

Lemma 17.4. *There exists a pseudo-gradient vector field $v \colon \tilde{M} \to M$ satisfying relation 3 in the preceding definition for $c = 1/2$. Moreover, v extends to a locally Lipschitz continuous vector field on the set of regular values $\tilde{X} = X \setminus \mathbb{K}$.*

17.2.2 A Second Form Using Ekeland's Principle

An extension of the MPT on closed convex sets M that does not require a priori the Palais-Smale condition, (PS)$_M$, is due to Ma [602]. While adopting the aforementioned definition of critical points and values by Struwe, Ma's results are based on the ideas and the spirit of the results appearing in the paper [153] by Brézis and Nirenberg. Suppose that $\Phi \colon M \to \mathbb{R}$ is a functional that admits a continuous *extension* to the whole space X which is Gâteaux differentiable and $\Phi' \colon X \to X^*$ is strong to weak∗-continuous. This implies in particular that the slope g is continuous on M.

The following form of a result on the existence of a pseudo-gradient vector field for continuous functionals on metric spaces, inspired from the paper [153] by Brézis and Nirenberg, is used in Ma's proof (cf. Note 16.V).

Lemma 17.5. *Let E be a metric space and $f \colon E \to X^*$ a strong-to-weak∗ continuous function. Then, for any $\varepsilon > 0$ and any continuous map $\gamma \colon E \to M$, there exists a locally Lipschitz function $v \colon E \to X$ such that, for any $x \in E$,*

a. $\gamma(x) - v(x) \in M$,
b. $\|v(x)\| \leq 1$,
c. $\langle f(x), v(x) \rangle \geq h(x) - \varepsilon$, *where*

$$h(x) = \sup_{\substack{y \in M \\ \|\gamma(x)-y\|<1}} \langle f(x), \gamma(x) - y \rangle.$$

The lemma may be proved using a partition of unity argument as in the usual construction of the pseudo-gradient vector field (as in [748], for example).

Let K be a compact metric space and let K^* be a closed nonempty subset of K such that $K^* \neq K$. Let

$$\Gamma = \left\{ \gamma \in \mathcal{C}(K; M), \ \gamma = \gamma^* \text{ on } K^* \right\},$$

where γ^* is a fixed continuous functional on K. Set

$$c = \inf_{\gamma \in \Gamma} \max_{x \in K} \Phi(\gamma(x)).$$

Then we have that

$$c \geq \max_{x \in K^*} \Phi(\gamma^*(x)).$$

The statement of the MPT on closed convex sets by Ma is the following.

Theorem 17.6 (MPT on Convex Domains, Ma). *Assume that for every $\gamma \in \Gamma$, $\max_{x \in K} \Phi(\gamma(x))$ is attained at some point in $K \setminus K^*$. Then, there exists a sequence $(u_n)_n$ on M such that*

$$\Phi(u_n) \to c \qquad \text{and} \qquad g(u_n) \to 0.$$

If in addition Φ satisfies (PS)$_M$, *then c is a critical value.*

Proof. For $x \in K$, denote

$$\rho(x) = \min\{\operatorname{dist}(x, K^*), 1\}$$

and consider for any fixed $\varepsilon > 0$ and any $\gamma \in \Gamma$ the perturbed functional

$$\Psi(\gamma, x) = \Phi(\gamma(x)) + \varepsilon\rho(x).$$

Define also

$$\Lambda(\gamma) = \max_{x \in K} \Psi(\gamma, x)$$

and

$$c_\varepsilon = \inf_{\gamma \in \Gamma} \Lambda(\gamma).$$

Then we can see easily that $c \le c_\varepsilon < c + \varepsilon$.

The functional Λ is lower semicontinuous with respect to the topology generated by the *usual* uniform convergence distance on Γ that will also be denoted "dist." By Ekeland's principle, there exists a $\gamma \in \Gamma$ such that

$$\Lambda(\gamma') - \Lambda(\gamma) + \varepsilon \operatorname{dist}(\gamma, \gamma') \ge 0, \qquad \text{for all } \gamma' \in \Gamma, \tag{17.3}$$

and

$$c \le c_\varepsilon \le \Lambda(\gamma) \le c_\varepsilon + \varepsilon \le c + 2\varepsilon. \tag{17.4}$$

By the assumption made in the theorem,

$$\Lambda(\gamma) > \max_{x \subset K^*} \Phi(\gamma(x)).$$

Let $B_\varepsilon = \{x \in K; \ \Psi(\gamma, x) = \Lambda(\gamma)\}$.

Claim 17.1. There exists $x_0 \in B_\varepsilon$ such that

$$g(\gamma(x_0)) \le 2\varepsilon. \tag{17.5}$$

Since $B_\varepsilon \subset K \subset K^*$, by Urysohn's lemma, there exists a function α in $C(K; [0, 1])$ satisfying

$$\begin{cases} \alpha(x) = 1 & \text{on } B_\varepsilon(\gamma), \\ \alpha(x) = 0 & \text{on } K^*. \end{cases}$$

Take for small $h > 0$, $\gamma' = \gamma_h$ for γ in (17.3) such that

$$\gamma_h(x) = \gamma(x) - hw(x)$$

with $w(x) = \alpha(x)v(x)$, where v is obtained by applying Lemma 17.5 with $E = K$, $f(x) = \Phi'(\gamma(x))$, and γ in (17.3).

It is clear that $\gamma_h \in \Gamma$. Moreover, the maximum

$$\Lambda(\gamma_h) = \max_{x \in K} \Psi(\gamma_h, x)$$

is attained at some point x_h in K. For a suitable sequence $h_n \to 0$, x_{h_n} converges to some $x_0 \in B_\varepsilon(\gamma)$. By (17.3) with $\gamma' = \gamma_h$ and by Lemma 17.5, we obtain

$$\Phi(\gamma(x_h) - hw(x_h)) + \varepsilon\rho(x_h) - \Lambda(\gamma) + \varepsilon h \ge 0.$$

By the mean value theorem, we have that

$$\Phi(\gamma(x_h)) + \varepsilon\rho(x_h) - \int_0^1 \langle \Phi'(\gamma(x_h) + thw(x_h)), hw(x_h)\rangle \, dt - \Lambda(\gamma) + \varepsilon h \ge 0.$$

And since $\Phi(\gamma(x_h)) + \varepsilon\rho(x_h) \leq \Lambda(\gamma)$, we have that, for $h = h_n$,

$$\int_0^1 \langle \Phi'(\gamma(x_{h_n}) + th_n w(x_{h_n})), h_n w(x_{h_n}) \rangle \, dt \leq 0.$$

As h_n tends to 0, we find

$$\langle \Phi'(\gamma(x_0)), v(x_0) \rangle \leq \varepsilon.$$

So that by Lemma 17.5 with $f = \Phi' \circ \gamma$, we get (17.5). And, hence, the MPT follows by choosing $\varepsilon = 1/n$ and $u_n = g(x_0)$ in (17.4) and (17.5).   □

In [213], Chang and Eells consider the following problem from mathematical biology:

$$\begin{cases} \Delta u + u^2 = f(x) & \text{in } \Omega, \\ \quad\quad u = 0 & \text{on } \partial\Omega, \end{cases} \tag{17.6}$$

where Ω is a piecewise smooth and bounded domain of $\mathbb{R}^N$ ($N < 6$), and f is a nontrivial nonnegative smooth and bounded function on Ω. They prove that (17.6) admits a negative solution and another one that is nontrivial. Ma used his version of the MPT to prove that (17.6) has a positive solution.

Struwe developed his version of the MPT to treat problems of the Plateau problem for minimal surfaces and for surfaces of constant mean curvature. The lecture notes [881] are devoted exclusively to this topic. See also [512, 880].

Comments and Additional Notes

◇ 17.I A Minimization Result on Closed Convex Subsets

We saw many times that a smooth functional bounded from below and satisfying (PS) has a minimum. Sometimes, in applications, we have to work only on some parts of the whole space, and many times, these are convex domains.

The following result was proved both by Hofer [478], using a version the deformation lemma adapted to closed convex subsets of Hilbert spaces, and by De Figueiredo and Solimini [302], using Ekeland's principle. The first approach is presented in the next chapter, which is consecrated to an MPT version in order intervals. We will now present the second approach.

Proposition 17.7 (De Figueiredo and Solimini). *Let X be a Hilbert space, $\Phi \in C^1(X; \mathbb{R})$ satisfying (PS). Let C be a closed convex subset of X. Suppose that the functional $K = \Phi' - I_d$ maps C into C and that Φ is bounded from below in C. Then, there is a point $u_0 \in C$ such that*

$$\Phi'(u_0) = 0 \quad\quad \text{and} \quad\quad \inf_C \Phi = \Phi(u_0).$$

Proof. By Ekeland's principle, given $\varepsilon > 0$ there is $u_\varepsilon \in C$ such that $\Phi(u_\varepsilon) \leq \inf_C \Phi + \varepsilon$ and

$$\Phi(u_\varepsilon) \leq \Phi(u) + \varepsilon \|u - u_\varepsilon\| \quad\quad \text{for all } u \in C. \tag{17.7}$$

Let's take $u = (1 - t)u_\varepsilon + tKu_\varepsilon$ with $0 \le t \le 1$ in (17.7) and use Taylor's formula to expand $\Phi(u_\varepsilon + t(Ku_\varepsilon - u_\varepsilon))$ about u_ε. We obtain then

$$t||\Phi'(u_\varepsilon)||^2 \le \varepsilon t||\Phi'(u_\varepsilon)|| + o(t).$$

It follows then, when t goes to 0, that $||\Phi'(u_\varepsilon)|| \le \varepsilon$. Finally, by (PS), we conclude that there exists $u_0 \in C$ such that $u_\varepsilon \to u_0$ for some sequence $\varepsilon \to 0$, and we conclude using the continuity of Φ and Φ'. $\quad\square$

Another criterion that can be of interest in handling some applications for which a particular form of topological degree has recently been developed in [669] is the following.

Definition 17.3. An operator $L \colon X \to X$, where X is a Hilbert space, is said to be of type S^+ if, for every sequence $(u_n)_n$ such that $u_n \rightharpoonup u$ and $\limsup(Lu_n, u_n - u) \le 0$, it follows that $u_n \to u$.

Then we have a similar result to the case when Φ' is a compact perturbation of the identity.

Proposition 17.8. Let $\Phi \in C^1(X; \mathbb{R})$ be such that Φ' is of type S^+. Suppose that Φ is bounded from below in the adherence of a ball $\overline{B}$. Then, there exists $v_0 \in \overline{B}$ such that $\Phi(v_0) = \inf_{\overline{B}} \Phi$ and $\Phi'(v_0) = \lambda v_0$ with $\lambda \le 0$.

Proof. By Ekeland's principle, given $\varepsilon > 0$, there is some $v_\varepsilon \in \overline{B}$ such that $\Phi(v_\varepsilon) \le \inf_{\overline{B}} \Phi + \varepsilon$ and

$$\Phi(v_\varepsilon) \le \Phi(v) + \varepsilon||v - v_\varepsilon||, \qquad \text{for all } v \in \overline{B}. \tag{17.8}$$

Take $u \in \overline{B}$ and set $v = v_\varepsilon + t(u - v_\varepsilon)$ in (17.8), with $0 \le t \le 1$. Use Taylor's formula again to expand $\Phi(v_\varepsilon + t(u - v_\varepsilon))$ about v_ε to obtain, tending t to 0,

$$0 \le \langle \Phi'(v_\varepsilon), u - v_\varepsilon \rangle + \varepsilon||u - v_\varepsilon||. \tag{17.9}$$

Taking now $u = v_0$ in (17.9), where v_0 is the weak limit of v_ε for some sequence $\varepsilon \to 0$, and making ε go to 0, we get that $\limsup\langle \Phi'(v_0), u - v_0 \rangle \le 0$. And since Φ' is of type S^+, it follows that $v_\varepsilon \to v_0$. We obtain readily that $\Phi(v_0) = \inf_{\overline{B}} \Phi$. On the other hand, from (17.9), it follows that $0 \le \langle \Phi'(v_0), u - v_0 \rangle$ for all $u \in \overline{B}$. This implies that there exists $\lambda \le 0$ such that $\Phi'(v_0) = \lambda v_0$. $\quad\square$

◇ 17.II On Amann's "Three Solution Problem" and the MPT on Convex Domains

Struwe applied his MPT on convex domains to treat variational inequalities [882]. He obtained again Amann's famous result on the "three solution problem" where the existence of *unstable* solutions of semilinear elliptic boundary value problems confined in an ordered interval between a sub- and a supersolution is obtained (for more, see Chapter 18).

Different variational approaches to the three solution problem can also be found in [204] where some regularizing properties of the flow are used to reduce the problem to a usual variational problem in a bounded open set of a Banach space to which the usual theory can be applied.

◇ 17.III An MPT on Closed Convex Subsets of Banach Spaces

In [208], Chang also proves an MPT on closed convex subsets of a Banach space.

Theorem 17.9 (MPT on Convex Domains, Chang). *Suppose that a C^1-functional Φ defined on a Banach space X satisfies (PS) with respect to a closed convex set C of X. Suppose that there exists $\alpha \in \mathbb{R}$ such that*

   i. $\sup_{x \in \partial Q} \Phi(x) \le \alpha < \inf_S f$,
   ii. $\sup_{x \in Q} \Phi(x) < +\infty$,

where ∂Q and S are two closed subsets of C that link with respect to C. Then, one of the following three alternatives occur:

   1. α is an accumulation point of critical values.
   2. α is an accumulation point with uncountable critical points.
   3. $c = \inf_{\gamma \in \Gamma} \sup_{x \in Q} \Phi(\gamma(x))$ is a critical value of Φ, where $\Gamma = \{\gamma \in C(Q; C); \; \gamma|_{\partial Q} = Id\}$.

The proof relies on an improved version of the deformation lemma contained in some of his earlier papers (see, in particular, [206, 947]).

◇ 17.IV MPT Versions on Closed Convex Sets for Some Particular Functions

A. A Critical Point Theory for Locally Lipschitz Functionals on Locally Convex Closed Subsets of Banach Spaces. In [942, 944], Wang establishes a minimax principle, including the MPT of course, for a locally Lipschitz functional Φ on *locally convex closed subsets S* of a Banach space (see Chapter 25 for some details).

B. An MPT on Closed Convex Sets for Functionals Satisfying the Schauder Condition. Suppose that H is a Hilbert space, $f : H \to \mathbb{R}$ is a C^1-functional that satisfies the (PS) condition, and M is a closed convex set in H. Let $f(x) = \|x\|^2 - h(x)$, $Ax = h'(x)$. If the condition $AM \subset M$ is satisfied, then we say that $f(x)$ satisfies the *Schauder condition* on M. In [887], Sun proves a version of MPT on closed convex sets for functionals satisfying the Schauder condition (see Chapter 25 for the details).

C. Yet Another MPT on Closed Convex Sets in Hilbert Spaces. In the article [460] (Some extensions of the mountain pass lemma), Guo et al. deal with some variants of the classical MPT. In the first part they prove an MPT with the limiting case, and in the second part an MPT for special functionals on *closed convex sets in Hilbert spaces* is improved.

◇ *17.V A Morse Theory on Convex Sets*

A Morse theory for functionals defined on convex sets appears in Struwe [868], in Chang and Eells [213], where the authors study the Plateau problem, and by Chang [208], who treats variational inequalities.

See also [205, Chapter I, Section 6.2], where Chang describes a concept of convexity to Banach manifolds and establishes a corresponding minimax theory and also a Morse theory for smooth functionals defined on these extended convex sets. The first and second deformation theorems, critical groups for isolated groups, and Morse relations for functions with isolated critical points do hold for locally convex subsets of C^2 paracompact Banach manifolds. According to Chang [205, Remark 6.2], the local convexity was first used in critical point theory in [943].

Compare the *alternative definition* of the functional g used to measure the regularity with the definition used in [215]. It is defined in Theorem 10.10 on page 113.

◇ *17.VI An MPT for Continuous Convex Functionals*

In [757] (Mountain-pass type theorems for nondifferentiable convex functions), Radulescu extends the MPT to *continuous convex functions*. The differentiability condition is replaced by subdifferentiability. A Palais-type pseudo-gradient lemma is established and the proof of the extended MPT uses Ekeland's variational principle.

18

MPT in Order Intervals

A natural instrument for the investigation of positive solutions are the methods of functional analysis in ordered spaces.

M.A. Krasnosel'skii

Some variational methods in ordered Banach spaces are investigated. In particular, we will see a variant of the MPT in order intervals in the spirit of some pioneering work by Hofer that exploited the natural *ordering*, intrinsic to semilinear elliptic problems.

Besides variational methods, many other methods are useful in the study of nonlinear problems. These methods exploit in general some additional information (of topological nature as in fixed point theory [983], or some monotonicity of the differential operator as in the theory of monotone operators [984], for example). It is natural then to expect to get better results if we combine these approaches to variational methods when the problem has a variational structure.

For example, consider the Dirichlet problem

$$(\mathcal{P}) \quad \begin{cases} -\Delta u = f(x, u) & \text{in } \Omega, \\ u = 0 & \text{on } \partial\Omega, \end{cases}$$

where $\Omega \subset \mathbb{R}^N$ is a bounded domain with smooth boundary $\partial\Omega$ and $f : \Omega \times \mathbb{R} \to \mathbb{R}$ is sufficiently smooth. The problem $(\mathcal{P})$ has some important features:

1. We have a *maximum principle* for $-\Delta$, in fact for a large class of second-order elliptic differential operators. This means that $-\Delta$ is *compatible* with the natural *ordering* of the underlying function space $H_0^1(\Omega)$ and makes $(\mathcal{P})$ equivalent to a fixed point equation for an order-preserving operator in a suitable (partially) ordered Banach space.
2. Moreover, when f satisfies a growth condition, we know that $(\mathcal{P})$ has a variational structure.

In the survey paper [33], without supposing the variational structure, Amann considered problems like $(\mathcal{P})$ by fixed point techniques in ordered spaces and got many interesting results. Few authors have treated $(\mathcal{P})$ and exploited both features 1 and 2 of the problem.

The first contribution in this direction seems to be the paper by Hofer [478]. He considered a potential operator $T: U \subset H \to H$ where H is an ordered real Hilbert space (see the definitions to follow) with an order given by a closed (proper) cone P, with T admitting the decomposition $T = I_d - K$ where K is compact and order preserving. The MPT we will see supposes similar conditions on the functional to those required in Hofer's presentation. This form of the MPT uses some advanced material with respect to the level we tried to address all the time in this text. So, our presentation will be somewhat sketchy in places. Nevertheless, we will clearly present the principles and hide the technicalities that should not bring "a plus" to the comprehension of the situation.

18.1 On Ordering and Variational Methods

We begin by recalling the necessary background to deal with the MPT in order intervals.

18.1.1 Ordering, Cones and Positive Operators

Let X be a nonempty set. An *ordering* in X, denoted in the sequel by "$\preceq$", is a relation in X that is reflexive, antisymmetric, and transitive. The pair $(X, \preceq)$ is called an *order set*.

For every $x, y \in X$, the set $[x, y] = \{z \in X; \ x \preceq z \preceq y\}$ is called the *order interval* between x and y. If X is a *real linear space*, an ordering *compatible* with the linear structure, that is, such that

i. $x \preceq y$ implies that $x + z \preceq y + z$ for all $z \in X$,
ii. $x \preceq y$ implies that $\alpha x \preceq \alpha y$ for all $\alpha \in \mathbb{R}_+ = [0, \infty[$,

is called a *linear ordering*. The pair $(X, \preceq)$ is called an *ordered vector space* (OVS).

Consider an OVS X and set $P = \{x \in X; \ 0 \preceq x\}$. The set P enjoys the following properties:

$C_1. \ P + P \subset P$,
$C_2. \ \mathbb{R}_+ P \subset P$, and
$C_3. \ P \cap (-P) = \{0\}$.

A nonempty subset P of a real vector space satisfying properties C_1–C_3 is called a *cone*. We can check immediately that every cone is convex. Moreover, every cone P in a real vector space X defines a linear ordering by

$$x \preceq y \qquad \text{if and only if} \qquad y - x \in P,$$

referred to as the *ordering induced by P*.

Writing $x \succ y$ if $y \preceq x$ and $y \neq x$, the set

$$P \setminus \{0\} = \{x \in X; \ x \succ 0\}$$

is called the *positive cone of the ordering*.

When $X = (X, \|.\|)$ is a Banach space ordered by a cone P, then X will be an *ordered Banach space* (OBS) if the order induced by P is compatible with the linear structure of X and with its topology; that is, "$\preceq$" is a linear ordering and the cone P is *closed*.

A cone is said to be *total* if $X = \overline{P - P}$ and it is *generating* if $X = P - P$.

Example 18.1.

1. Consider $X = \mathbb{R}^N$ and P any octant, that is,

$$x \preceq y \iff x_i \leq (-1)^{\varepsilon_i} y_i, \qquad i = 1, \ldots, N,$$

where $\varepsilon_i = \pm 1$ is fixed for each i.
2. Consider $X = C(\overline{\Omega}; \mathbb{R})$ with

$$P = \big\{ u \in C(\overline{\Omega}; \mathbb{R}); \ u(x) \geq 0, \ \forall x \in \overline{\Omega} \big\}.$$

This cone is generating.

An operator $T : U \subset X \to X$ is called *order preserving* (or *increasing*) if and only if

$$x \preceq y \qquad \text{implies that} \qquad Tx \preceq Ty,$$

and *strictly order-preserving* (or *strictly increasing*) if and only if

$$x \succ y \qquad \text{implies that} \qquad Tx \succ Ty,$$

and it is *strongly order-preserving* (or *strongly increasing*) if and only if

$$x \preceq y \qquad \text{implies that} \qquad Tx \ll Ty; \text{ that is, } Ty - Tx \in \text{int } P.$$

After the number of variants and extensions of the MPT we have seen up to now, you should certainly guess that someone, somewhere, has thought of a corresponding version of the MPT. That is indeed true, as we will see in the next section.

18.2 MPT in Order Intervals

In [568], Li and Wang proved an MPT in order intervals in which the position of the critical point (which is of mountain-pass type in this situation) is given precisely in terms of the ordering structure. With this variant of the MPT and using special flows, they proved the existence of multiple solutions and sign-changing solutions for semilinear elliptic Dirichlet problems.

Let H be a Hilbert space and $P_H \subset H$ be a closed convex cone. Let $X \subset H$ be a Banach space that is densely embedded in H. Let $P = X \cap P_H$ and assume that P has a nonempty interior, int $P \neq \emptyset$. Assume also that any order interval is finitely bounded. Suppose now that our functional $\Phi : H \to \mathbb{R}$ satisfies similar conditions to those used by Hofer [478].

Φ_1. $\Phi \in C^1(H; \mathbb{R})$, satisfies (PS) in H and the deformation property in X, and that Φ has only finitely many isolated critical points.

Φ_2. The gradient of Φ is of the form $\nabla\Phi = I_d - K_H$, where $K_H \colon H \to H$ is compact. The space X is stable by K_H, $K_H(X) \subset X$, and the restriction $K = K_H|_X \colon X \to X$ is continuous and strongly order preserving.

Φ_3. Φ is bounded from below on any order interval in X.

We recall first some technical results that will be needed in the sequel.

Lemma 18.1. *Suppose that Φ satisfies conditions Φ_1–Φ_3 and that $\underline{u} \preceq \overline{u}$ is a paired subsolution and supersolution of $\nabla\Phi = 0$ in X. Then there exists a negative pseudo-gradient vector field $\eta(t, .)$ such that $[\underline{u}, \overline{u}]$ is positively invariant under this vector field and $\eta(t, .)$ points inward in $[\underline{u}, \overline{u}]$. Moreover, if $\underline{u} \preceq \overline{u}$ is a paired strict subsolution and supersolution of $\nabla\Phi = 0$ in X, then*

$$\deg(I_d - K, [\underline{u}, \overline{u}], 0) = 1.$$

Proof of Lemma 18.1. Since $[\underline{u}, \overline{u}]$ is finitely bounded, the Leray-Schauder degree $\deg(I_d - K, [\underline{u}, \overline{u}], 0)$ is well defined. For any $x \in [\underline{u}, \overline{u}]$, we have

$$x - \nabla\Phi(x) = K(x) \gg K(\overline{u}) > \underline{u}, \tag{18.1}$$

$$x - \nabla\Phi(x) = K(x) \ll K(\underline{u}) < \overline{u}. \tag{18.2}$$

So, $x - \nabla\Phi(x) \in \operatorname{int}[\underline{u}, \overline{u}]$, the interior of $[\underline{u}, \overline{u}]$.

In the rest of the proof, we proceed as in the classical proof of the existence of a pseudo-gradient field for C^1-functionals. Set $\tilde{X} = \{x \in X; \ \nabla\Phi \neq 0\}$. For any $x_0 \in \tilde{X}$, there is $w \in X$, $||w|| = 1$ such that

$$\langle -\Phi'(x_0), w\rangle > \frac{2}{3}||\Phi'(x_0)||.$$

If $x_0 \in [\underline{u}, \overline{u}]$, by (18.1) and (18.2) we can require $x_0 + w$ to be in $\operatorname{int}[\underline{u}, \overline{u}]$. Let

$$v = \frac{3}{2}||\Phi'(x_0)||w.$$

Then,

$$\begin{cases} ||v|| < 2||\Phi'(x_0)|| \\ \langle -\Phi'(x_0), v\rangle > ||\Phi'(x_0)||^2. \end{cases}$$

From the continuity of Φ', for each x_0 there exists a neighborhood $\tilde{U}_{x_0}$ of x_0 such that, for all $x \in \tilde{U}_{x_0}$,

$$\begin{cases} ||v|| < 2||\Phi'(x)|| \\ \langle -\Phi'(x), v\rangle > ||\Phi'(x)||^2. \end{cases} \tag{18.3}$$

Take

$$U_x = \begin{cases} \tilde{U}_x & \text{if } x \in [\underline{u}, \overline{u}], \\ \tilde{U}_x \cap (\tilde{X} \setminus [\underline{u}, \overline{u}]) & \text{if } x \in \tilde{X} \setminus [\underline{u}, \overline{u}]. \end{cases}$$

Since $\tilde{X}$ is metrizable, it is paracompact. So, there is a locally finite $C^{1,0}$-partition of unity $(\beta_\alpha)_{\alpha \in \Lambda}$, where Λ is an index set. This gives a sense to the quantity

$$v(x) = \sum_{\alpha \in \Lambda} \beta_\alpha(x).v_\alpha, \tag{18.4}$$

where v_α is the point in U_{x_α} given by (18.3).

We check easily that

$$||v|| < 2||\Phi'(x)||,$$

and that

$$\langle -\Phi'(x), v \rangle > ||\Phi'(x)||^2.$$

Since $[\underline{u}, \overline{u}]$ is convex, by (18.4) and the definition of U_x, we know that $x + v(x) \in$ int $[\underline{u}, \overline{u}]$ for any $[\underline{u}, \overline{u}]$. Consider now the negative pseudo-gradient vector field $\eta(t, u)$ of Φ on X defined by

$$\begin{cases} \dfrac{d\eta(t, u)}{dt} = v(\eta(t, u)) \\ \eta(0, u) = u. \end{cases}$$

Then, $[\underline{u}, \overline{u}]$ is positively invariant under $\eta(t, u)$.

Remark 18.1. For the same pseudo-gradient vector field $\eta(t, u)$, for any subsolution $\underline{u}$ and any supersolution $\overline{u}$, we have the following stability (positive invariance of $[\underline{u}, \overline{u}]$ under η) property:

$$\begin{cases} \eta(t, \underline{u} + P) \subset \underline{u} + P, \\ \eta(t, \overline{u} + P) \subset \overline{u} + P. \end{cases}$$

Now, following standard arguments from [205, Theorems 1.4.2, 1.4.3, and 2.3.3] and using the fact that $[\underline{u}, \overline{u}]$ is positively invariant under $\eta(t, u)$, we have

$$\deg(I_d - K, [\underline{u}, \overline{u}], 0) = \sum_{q=0}^{\infty} (-1)^q \text{ rank } H_q([\underline{u}, \overline{u}]) = 1.$$

$\square$

Corollary 18.2. *If $\underline{u} \preceq \overline{u}$ is a paired strict subsolution and supersolution of $\nabla\Phi = 0$ in X, then*

$$\{u \in H; \nabla\Phi(u) = 0\} \cap \partial[\underline{u}, \overline{u}] = \varnothing.$$

Theorem 18.3 (MPT in Order Intervals, Li and Wang). *Suppose that Φ satisfies conditions Φ_1–Φ_3. Suppose also that there exist four points in X,*

$$\begin{cases} v_1 < v_2, \\ w_1 < w_2, \\ v_1 < w_2, \\ [v_1, v_2] \cap [w_1, w_2] = \varnothing \end{cases} \quad \text{with} \quad \begin{cases} v_1 \leq Kv_1, \\ v_2 > Kv_2, \\ w_1 < Kw_1, \\ w_2 \geq Kw_2. \end{cases}$$

Then Φ *has a mountain pass point* $u_0 \in [v_1, w_2] \setminus ([v_1, v_2] \cup [w_1, w_2])$. *More precisely, let* v_0 *be the maximal minimizer of* Φ *in* $[v_1, v_2]$ *and* w_0 *be the minimal minimizer of* Φ *in* $[w_1, w_2]$. *If* $v_0 < w_0$, *then* $v_0 \ll u_0 \ll w_0$. *Moreover, the critical group* $C_1(\Phi, u_0)$ *of* Φ *at* u_0 *is nontrivial.*

Proof. Since Φ is bounded from below on $[v_1, v_2]$ and satisfies the deformation property, Φ has at least one local minimizer in each interval. Let v_0 be the minimizer of Φ in $[v_1, v_2]$ and w_0 be the minimizer of Φ in $[w_1, w_2]$. Let

$$\Gamma = \begin{cases} \gamma(t) \in \mathcal{C}([0, 1]; [v_1, w_2]), & \text{such that} \\ \gamma(t) \in [v_1, w_2] \setminus ([v_1, v_2] \cup [w_1, w_2]) & \text{for } t \in]\frac{1}{3}, \frac{2}{3}[\\ \gamma(t) = \eta(\frac{1}{3} - t, \gamma(\frac{1}{3})) & \text{for } 0 \le t \le \frac{1}{3}, \ \gamma(\frac{1}{3}) \in \partial[v_1, v_2] \\ \gamma(t) = \eta(t - \frac{2}{3}, \gamma(\frac{2}{3})) & \text{for } \frac{2}{3} \le t \le 1, \ \gamma(\frac{2}{3}) \in \partial[w_1, w_2]. \end{cases} \tag{18.5}$$

It is easy to see that Γ is not empty and that it is a complete metric space for the distance of uniform convergence in X,

$$\text{dist}(x, y) = \max_{t \in [0,1]} \|x(t) - y(t)\|_X.$$

Set

$$c = \inf_{\gamma \in \Gamma} \sup_{t \in [0,1]} \Phi(\gamma(t)). \tag{18.6}$$

We will prove that c is a critical value of Φ. More precisely, there exists

$$u_0 \in \mathbb{K}_c \cap [v_1, w_2] \cap \{[v_1, w_2] \setminus ([v_1, v_2] \cup [w_1, w_2])\}.$$

The proof closely follows the arguments used by Shi to prove the nonsmooth MPT (see Chapter 15).

Step 1. Set for $\gamma(t) \in \Gamma$,

$$F(\gamma(t)) = \max_{t \in [0,1]} \Phi(\gamma(t)).$$

Then F is locally Lipschitz and bounded from below on Γ. Set

$$M(\gamma) = \left\{ s \in [0, 1]; \ \Phi(\gamma(t)) = F(\gamma(s)) = \max_{t \in [0,1]} \Phi(\gamma(t)) \right\}.$$

Step 2. From Ekeland's principle, for any positive sequence $(\varepsilon_n)_n$, $\varepsilon_n \searrow 0$, there exists a sequence $(\gamma_n)_n \subset \Gamma$ such that, for all $n \in \mathbb{N}$,

$$\begin{cases} c \le F(\gamma_n) \le c + \varepsilon_n, \\ F(\gamma) > F(\gamma_n) - \varepsilon_n \, \text{dist}(\gamma, \gamma_n), & \text{for any } \gamma \ne \gamma_n. \end{cases} \tag{18.7}$$

Thus for any $\gamma \in \Gamma$, we have

$$F^\circ(\gamma_n, h) = \limsup_{\lambda \searrow 0} \frac{F(\gamma_n + \lambda \gamma) - F(\gamma_n)}{\lambda} \ge -\varepsilon_n \max_{t \in [0,1]} \|\gamma(t)\|_X. \tag{18.8}$$

Step 3. From (18.8), by Proposition 15.11 (cf. Chapter 15), there exists $t_n \in M(\gamma_n)$ such that

$$\langle \nabla \Phi(\gamma_n(t)), u \rangle \geq -\varepsilon_n \|u\|, \quad \forall u \in X. \tag{18.9}$$

This implies that $\nabla \Phi(\gamma_n(t_n)) \to 0$ as $n \to \infty$.

Step 4. From the definition of Γ and Lemma 18.1, we get $v_0 \ll u_0 \ll w_0$. Then, by Lemma 18.1 and Corollary 18.2, we have

$$\inf_{u \in \partial[v_1, v_2]} \Phi(u) > \Phi(v_0), \qquad \inf_{u \in \partial[w_1, w_2]} \Phi(u) > \Phi(w_0). \tag{18.10}$$

Therefore,

$$c > \max \big\{ \Phi(v_0), \Phi(w_0) \big\}.$$

Using arguments given by Hofer [480], u_0 is a mountain pass point.

By a similar argument, we know that $(\Phi^c \setminus \{u_0\}) \cap W$ is neither nonempty nor path-connected.

From the definition of the critical group and since $(\Phi^c \setminus \{u_0\}) \cap W$ is nonempty, we get that

$$C_0(\Phi, u_0) = H_0((\Phi^c \setminus \{u_0\}) \cap W, (\Phi^c \setminus \{u_0\}) \cap W) = 0.$$

The fact that $(\Phi^c \setminus \{u_0\}) \cap W$ is not path-connected implies then that $H_0((\Phi^c \setminus \{u_0\}) \cap W) \neq 0$.

Claim 18.1. $C_1(\Phi, u_0) \neq 0$.

Indeed, consider the exact sequence

$$H_1(\Phi^c \cap W, (\Phi^c \setminus \{u_0\}) \cap W) \xrightarrow{\partial} H_0(\Phi^c \cap W, (\Phi^c \setminus \{u_0\}) \cap W)$$
$$\to H_0(\Phi^c \cap W) \to H_0(\Phi^c \cap w, (\Phi^c \setminus \{u_0\}) \cap W).$$

If $C_1(\phi, u_0) = H_1(\Phi^c \cap W, (\Phi^c \setminus \{u_0\}) \cap W)$, then $H_0((\Phi^c \setminus \{u_0\}) \cap W) \simeq H_0(\Phi^c \cap W)$, but $H_0(\Phi^c \cap W) \simeq 0$, so we get a contradiction. $\square$

Multiplicity results involving order structures have been investigated by Amann using fixed point techniques. The method of Li and Wang is variational with the order structure built in. They get more information on the third critical point u_0 than in Amann's monograph [35, Theorem 14.2], u_0 is of mountain-pass type, and the critical group $C_1(\Phi, u_0) \neq 0$. Moreover, $v_0 \ll u_0 \ll w_0$ was not known in Amann's result. Some applications are given to several classes of elliptic boundary value problems with Dirichlet conditions.

Comments and Additional Notes

◇ 18.I Variational Results in OBS

In [478], Hofer proved many interesting existence and multiplicity results obtained by a classification of isolated critical points of a particular type of $\mathcal{C}^2$-functionals in an ordered Hilbert space. In particular, he obtains a special variant of the *Krein-Rutman theorem* (see the notes below and the references cited there) for the linear eigenvalue problem

$$\Phi''(v).u = \lambda u.$$

This enables Hofer to prove that the local Leray-Schauder degree of a critical point given by the MPT is equal to -1 for the class of functionals he considered. This will not be discussed here and the interested reader is referred to [478] and to Chapter 12 for the details.

We begin by fixing the terminology and some notations. A Φ-*family* on C is the set

$$D = D(\Phi, C) = \big\{ \gamma \in C([0,1] \times C; C); \ \gamma(0,.) = I_d$$
$$\text{and } t \mapsto \Phi(\gamma(t, u)) \text{ is nonincreasing for all } u \in C \big\}.$$

We can check easily that the operation "$*$" defined on $D \times D$ by

$$\gamma_1 * \gamma_2(t, u) = \begin{cases} \gamma_2(2t, u) & t \in [0, 1/2] \\ \gamma_1(2t - 1, \gamma_2(1, u)) & t \in [1/2, 1] \end{cases}$$

maps $D \times D$ into D. The first of the two results by Hofer we wanted to present is a specific variant of the deformation lemma.

Theorem 18.4 (Deformation Lemma, Hofer). *Let H be a real Hilbert space and C be a nonempty closed convex set such that $C \subset U$ where U is open. Suppose that $\Phi \in C^1(U; \mathbb{R})$ satisfies $(PS)_c$ and $\nabla\Phi = I_d - K$ where $KC \subset C$.*
Then the family $D(\Phi, C)$ on C has the following property.

For given real numbers $d \in \mathbb{R}$, $\varepsilon_0 > 0$ and a relative neighborhood $W \subset C$ of $\mathbb{K}_d \cap C$, there exist $\varepsilon \in]0, \varepsilon_0]$ and a deformation $\eta \in D$ such that

$$\eta(\{1\} \times ((\{\Phi(x) \le d + c\} \cap C) \subset W)) \{\Phi(x) \le d + c\} \subset \cap C.$$

Sketch of the Proof. The classical proof by Rabinowitz works. However, we have to approximate K sufficiently well by a locally Lipschitz continuous function $\tilde{K} : C \to C$.

We choose then a "cutoff" function $\beta : C \to [0, 1]$ being locally Lipschitz continuous such that $u \mapsto \beta(u)(\tilde{K}(u) - u)$ has a bounded range. Then, consider the positive semiflow associated with the differential equation:

$$\begin{cases} \dot{u} = \beta(u)(\tilde{K}u - u) \\ u(0) = u_0. \end{cases}$$

The set C is shown to be positively invariant. Moreover, one has the global existence on $\mathbb{R}_+$.

Define $\eta = \sigma|_{[0,1]\times C}$. It has the desired properties. $\square$

Remark 18.2. If $\bigcup_{e \in [c,d]} \mathbb{K}_e \cap C = \varnothing$ and the hypotheses of the former deformation lemma hold, there exists $\eta \in D = D(\Phi, C)$ such that $\eta(\{1\} \times (\{\Phi \le d\} \cap C) \subset \{\Phi \le d\} \cap C$.

Indeed, by the compactness of $[c, d]$, there exist

$$c \le d_1 < d_2 < \cdots < d_k \le d \quad \text{and} \quad \varepsilon_i > 0, \text{ for } i = 1, \ldots, k$$

such that $([d_i - \varepsilon_i, d_i + \varepsilon_i])_{i=1}^{k}$ is a covering of $[c, d]$ and corresponding $\eta_i \in D$ with $\eta_i(\{1\} \times (\{\Phi \le d\} \cap C)) \subset \{\Phi \le d\} \cap C$. We define $\eta \in D$ by $\eta = (\cdots((\eta_1 * \eta_2) * \eta_3) * \cdots * \eta_k)$. □

Without using any order structure, this deformation lemma implies easily the following minimization result proved directly by De Figueiredo and Solimini [302] using Ekeland's principle presented in the preceding chapter (on page 212).

Proposition 18.5 (Hofer). *Let H be a real Hilbert space, $C \subset H$ be nonempty, closed and convex, and U be an open set containing C. Assume that $\Phi \in C^1(U; \mathbb{R})$ satisfies* $(PS)_c$ *and that* $\nabla\Phi = I_d - K$ *such that* $KC \subset C$ *and* $\Phi|_C$ *is bounded from below.*
Then Φ attains its infimum d in C; that is, $\mathbb{K}_d \cap C \ne \varnothing$ where $d = \inf_C \Phi$.

Proof. Suppose by contradiction that $\mathbb{K}_d \cap C = \varnothing$. Then there exist $\varepsilon > 0$ and $\sigma \in D = D(\Phi, C)$ such that $\sigma(\{1\} \times \{\Phi \le d\}) \cap C = \varnothing$, which is a contradiction since $\{\Phi \le d\} \cap C \ne \varnothing$. So, $\{x \in C; \ \Phi(x) = d\} \supset \mathbb{K}_d \cap C \ne \Phi$. Let's show the opposite inclusion. Suppose that there is a $w \in \{x \in C; \ \Phi(x) = d\} \setminus (\mathbb{K}_d \cap C)$; then there would exist a relative neighborhood $W \subset C$ of $\mathbb{K}_d \cap C$ such that $w \notin W$. By the deformation lemma there are $\eta \in D$ and $\epsilon > 0$ with

$$\eta(1, W) \in \{\Phi \le d\} \cap C = \varnothing,$$

which is again a contradiction.   □

◇ *18.II Krein-Rutman Theorem*

A well-known result of Perron and Forbenius for matrices is the following. If $M = (a_{ij})_{ij}$ is a square matrix of order n such that

$$a_{ij} > 0, \qquad i, j = 1, 2, \ldots, n,$$

then the linear mapping associated to M is compact and the spectral radius $\rho(M)$ is a simple eigenvalue of M with an associated eigenfunction in $(\mathbb{R}^*)^n$. The famous Krein-Rutman theorem extends this result to a class of positive operators.

Let X be a Banach space. We recall that for every continuous linear operator $T \in \mathcal{L}(E; E)$, the limit

$$\rho(T) = \lim_{k \to \infty} \|T^k\|^{1/k}$$

exists and is called the *spectral radius* of T (cf. [977]).

Theorem 18.6 (Krein-Rutman [539]). *Let (X, P) be an OBS with a total positive cone. Suppose that $T \in \mathcal{L}(E; E)$ is compact and has a positive spectral radius $\rho(T)$. Then $\rho(T)$ is an eigenvalue of T with eigenvectors in P.*

As an interesting application, consider a bounded domain in $\mathbb{R}^N$ with smooth boundary. Consider also the Dirichlet linear problem:

$$(\mathcal{P}) \qquad \begin{cases} -\Delta u = f \in L^2(\Omega), \\ u \in H_0^1(\Omega). \end{cases}$$

It is known that $(\mathcal{P})$ has a unique solution in $H^2(\Omega) \cap H_0^1(\Omega)$. Moreover, if $f \geq 0$ almost everywhere, then $u \geq 0$ almost everywhere. Denote $\Gamma^+ = \{u \in L^2(\Omega); u(x) \geq 0 \text{ a.e. }\}$. Then

$$B = (-\Delta)^{-1} \in \mathcal{L}(L^2(\Omega); L^2(\Omega))$$

and

$$B(\Gamma^+) \subset \Gamma^+.$$

Since Ω is bounded, it follows that B is compact. So, we can apply the Krein-Rutman theorem to B, $L^2(\Omega)$, and $P = \Gamma^+$. We conclude then that the Laplacian with Dirichlet boundary conditions has a "smallest" eigenvalue with an associated eigenfunction that does not change sign. The *simplicity* of this smallest eigenvalue is also discussed in [33, 630–632].

◇ *18.III Some Complementary Information Using Morse Theory*

Under quite similar conditions on the partial order used and on the functional appearing in the MPT in order intervals, Bartsch et al. [99] investigated the Morse index and critical groups, in some abstract critical point theorems, for functionals on partially ordered Hilbert spaces.

These results are used to look for multiple sign-changing solutions of a nonlinear elliptic Dirichlet problem.

19

The Linking Principle

Geometry may sometimes appear to take the lead over analysis, but in fact precedes it only as a servant goes before his master to clear the path and light him on the way.

J.J. Sylvester

The notion of *linking* is very important in critical point theory. It expresses in an elegant way the geometric conditions that appear in all the abstract results seen until now. Various definitions in many contexts (homotopical, homological, local, isotopic, etc.) were given. Some linking notions are presented and their respective forms of linking theorems, containing the MPT, are stated.

After the MPT in 1973, in the late 1970s two new parents came to consolidate the family of minimax theorems in critical point theory. The *multidimensional MPT* by Rabinowitz [737] in 1978 (Chapter 8 is entirely devoted to this result) and the *saddle point theorem* (see the additional notes to follow), also by Rabinowitz [738] in 1978. The three results have something in common (we will see soon) and are proved similarly using a stereotyped method. Indeed, as we could see in the chapters devoted to the MPT and the multidimensional MPT, the proof is always done in two steps. This is also true for the case of the saddle point theorem.

1. An *intersection* property on the sets that define the geometric condition is proved.
2. Supposing that the "inf max" value c is regular, we get a contradiction by the deformation lemma in exactly the same way.

Linking in the Sense of Benci and Rabinowitz

In 1979, Benci and Rabinowitz [113] formulated the geometric conditions that appear in the three theorems in a way that expresses some linking between a set S and the boundary ∂Q of a manifold Q. Consider a real Banach space $E = E_1 \oplus E_2$, where both E_1 and E_2 may be infinite dimensional. Let P_1 and P_2 be respective projectors of

E onto E_1 and E_2 associated with splitting of E. Set

$$S = \big\{ \gamma \in \mathcal{C}([0, 1] \times E; E); \ \gamma(0, u) = u$$
$$\text{and } P_2\gamma(t, u) = P_2 u - \kappa(t, u) \text{ where } \kappa : [0, 1] \times E \to E \text{ is compact} \big\}.$$

Let $S, Q \subset E$ with $Q \subset \tilde{E}$ a given subspace of E. The notation ∂Q will refer to the boundary of Q in $\tilde{E}$.

Definition 19.1 (Linking, Benci and Rabinowitz). The sets S and ∂Q *link*, in the sense of Benci and Rabinowitz, if any $\gamma \in S$ such that $\gamma(t, \partial Q) = \varnothing$ for all $t \in [0, 1]$ satisfies $\gamma(t, Q) \cap S \neq \varnothing$ for all $t \in [0, 1]$.

This particular form of the set S is due to the fact that the tool used to prove the topological property of intersection expressed earlier as a linking of S and ∂Q is the Leray-Schauder degree. This degree requires the functional to have the particular form of a *compact perturbation of the identity*.

For heuristic purposes, to quote Rabinowitz [748], one may think of the sets S and ∂Q as linking if every manifold modeled on Q sharing the same boundary intersects S.

Benci and Rabinowitz used this notion of linking to prove the following result.

Theorem 19.1 (Linking Theorem, Benci and Rabinowitz). *Let E be a real Hilbert space with $E = E_1 \oplus E_2$ and $E_2 = E_1^\perp$. Suppose that $\Phi \in \mathcal{C}^1(E; \mathbb{R})$ satisfies* (PS) *and is such that*

Φ_1. *$\Phi(u) = \frac{1}{2}(Lu, u) + b(u)$, where $Lu = L_1 P_1 u + L_2 P_2 u$ and $L_i : E_i \to E_i$ is bounded and self-adjoint, $i = 1, 2$, and*

Φ_2. *b' is compact.*

Φ_3. *There exists a subspace $\tilde{E} \subset E$, two sets $S \subset E$, $Q \subset E$, and two constants $\alpha > \omega$ such that S and ∂Q link,*

   i. $S \subset E_1$ and $\Phi|_S \geq \alpha$, and

   ii. Q is bounded and $\Phi|_{\partial Q} \leq \omega$.

Then Φ possesses a critical value $c \geq \alpha$.

The result is stated for a Hilbert space H as we can see and requires a particular form of the functional Φ. Nevertheless, it may be considered a generalization of the MPT, the generalized MPT, and the saddle point theorem because it succeeds in solving the variational problems treated before by these critical point theorems, where the energy indeed has this form.

19.1 The Linking Principle

There is another more abstract formulation of the notion of linking that is widely known and used nowadays, so that it may be called *standard* or *usual linking* or only *linking* to avoid confusion with the other definitions. It is just the one by Benci and Rabinowitz, where the particular forms of Φ and E have been hidden (see [114, 882, 956], for example).

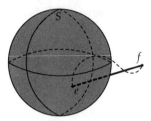

Figure 19.1. The MPT linking situation.

Definition 19.2 (Standard Linking). Let S, $Q \subset E$ where E is a real Banach space and Q is a subset of a subspace $\tilde{E}$ of E with boundary ∂Q in $\tilde{E}$. We say S and ∂Q *link* if

  i. $S \cap \partial Q = \varnothing$, and
  ii. for any $\gamma \in C(E; E)$ such that $\gamma|_{\partial Q} = I_d$, we have $\gamma(Q) \cap S \neq \varnothing$.

Moreover, if Γ is a subset of $C(E; E)$, then S and ∂Q are said to link with respect to Γ if conditions i and ii are satisfied for any $\gamma \in \Gamma$.

The following two examples yield, respectively, the geometries of the MPT and that of the saddle point theorem and the generalized MPT.

Example 19.1. Consider two points e and f in X, and B a neighborhood of e in X, such that $f \notin \overline{B} = S$ (see Figure 19.1). Let Γ be the set of all continuous paths joining e and f and set $Q = [e, f]$. Then, S and ∂Q link.

Example 19.2.
  1. Let $X = X_1 \oplus X_2$ be such that $\dim X_2 < \infty$.
     Let $S = X_1$ and $Q = B_R(0) \cap X_2$ with relative boundary
     $$\partial Q = \{u \in X_2; \ \|u\| = R\}.$$
     Then S and ∂Q link.

  2. Let $X = X_1 \oplus X_2$ with $\dim X_2 < \infty$ and let $e \in X_1$ with $\|e\| = 1$ be given. Suppose $0 < \rho < R_1, 0 < R_2$, and let
     $$S = \{u \in X_1; \ \|u\| = \rho\},$$
     $$Q = \{se + u_2; \ 0 \leq s \leq R_1, u_2 \in X_2 \text{ and } \|u_2\| \leq R_2\}$$
     with relative boundary
     $$\partial Q = \{se + u_2; \ s \in \{0, R_1\} \text{ or } \|u_2\| = R_2\}.$$
     Then S and ∂Q link (see Figure 19.2.)

The notion of linking was used to formulate the following abstract critical point theorem that contains the earlier ones by Rabinowitz. We adopt here a nice form attributed to Willem [956]. Many other forms exist in the literature.

Then S and ∂Q link.

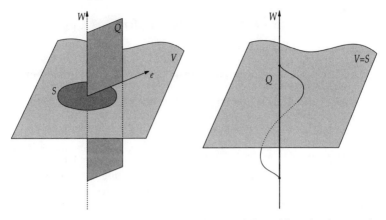

Figure 19.2. The linking in the multidimensional MPT and the saddle point theorem situations, respectively.

Theorem 19.2 (Linking Principle, Willem). *Suppose that* $\Phi \in C^1(E; \mathbb{R})$, $S \subset E$ *is closed, and* $Q \subset E$ *satisfy the following:*

a. *S and ∂Q link.*
b. *There exists $a, b \in \mathbb{R}$ such that*

$$a = \sup_{\partial Q} < b = \inf_S \Phi.$$

c. *$d = \sup_{u \in Q} \Phi(u) < \infty$.*

Let

$$c = \inf_{\gamma \in \Gamma} \sup_{u \in Q} \Phi(\gamma(u)).$$

Moreover, if Φ satisfies $(PS)_c$, *then c defines a critical value of Φ.*

Assumption c is satisfied, for example, when Q is compact.

Proof. Using assumptions a and b, we get that $b \le c$, while condition c ensures that $c \le d$.

By contradiction, suppose that our conclusions were false; then there would exist ε, δ, and γ such that for $\mathcal{O} = \gamma(Q)$,

$$u \in \Phi^{-1}([c - 2\varepsilon, c + 2\varepsilon]) \cap \mathcal{O}_{2\delta} \qquad \text{implies} \qquad \|\phi'(u)\| \ge 4\varepsilon/\delta,$$

since by assumption,

$$c - 2\varepsilon > a, \tag{19.1}$$

and

$$\gamma(Q) \subset \Phi^{c+\varepsilon}. \tag{19.2}$$

Set $\beta(u) = \eta(1, \gamma(u))$, where η is the deformation given by the quantitative deformation lemma of Willem (Theorem 4.2, p. 38).

Relations ii from Theorem 4.2 and (19.1) imply that, for any $u \in \partial Q$,

$$\beta(u) = \eta(1, \gamma(u)) = \eta(1, u) = u.$$

Hence, $\beta \in \Gamma$. By relation iii from Theorem 4.2 and (19.2), we get the contradiction

$$c \leq \sup_{u \in Q} \Phi(\beta(u)) \leq c - \varepsilon.$$

$\square$

The linking principle is still true in the limiting case (see Chapter 9). This has been proved by many authors [153, 423, 498, 956]. We continue with Willem's approach and present his localization theorem, which *adds some precision to Ghoussoub's result* (Theorem 9.6).

Theorem 19.3 (Localization Theorem, Willem). *Suppose that* $\Phi \in C^1(X; \mathbb{R})$, *where* X *is a Banach space and* $S, Q \subset X$ *are closed subsets of* X.
Let

$$c = \inf_{\gamma \in \Gamma} \sup_{u \in Q} \Phi(\gamma(u)),$$

where $\Gamma = \{\gamma \in C(Q; X); \ \gamma(u) = u \text{ on } \partial Q\}$.
Suppose that

    i. *S and ∂Q link,*
    ii. *dist $(S, \partial Q) > 0$, and*
    iii. *$\inf_S \Phi = c > -\infty$.*

Then, for any $\varepsilon > 0$, $\delta \in]0$, dist $(S, \partial Q)/2[$ and $\gamma \in \Gamma$ such that

$$\sup_{u \in Q} \Phi(\gamma(u)) < c + \varepsilon,$$

there exists $u \in X$ such that

    a. *$c - \varepsilon \leq \Phi(u) \leq c + 2\varepsilon$,*
    b. *dist $(u, S \cap [\gamma(Q)]_\delta) \leq 2\delta$, and*
    c. *$\|\Phi'(u)\| < 4\varepsilon/\delta$.*

If, moreover, Φ satisfies $(PS)_c$, then $\mathbb{K}_c \cap S \neq \emptyset$.

Proof. By contradiction, suppose that these conclusions were false; then there would exist ε, δ, and γ such that, for $\mathcal{O} = S \cap [\gamma(Q)]_\delta$,

$$u \in \Phi^{-1}([c - 2\varepsilon, c + 2\varepsilon]) \cap \mathcal{O}_{2\varepsilon}) \qquad \text{implies} \qquad \|\Phi'(u)\| \geq 4\varepsilon/\delta,$$

for Φ replaced by $\tilde{\Phi} = -\Phi$ and $\tilde{c} = c$ replaced by $-c$. Let η be the corresponding deformation given by the quantitative deformation lemma, and let β be the mapping defined on Q by

$$\gamma(u) = \eta(1, \beta(u)).$$

Notice that $\text{dist}\,(\mathcal{O}, \partial Q) \geq \text{dist}\,(S, \partial Q) > 2\delta$. By assumption ii of Theorem 4.2, for any $u \in \partial Q$,

$$\eta(1, u) = u = \gamma(u) = \eta(1, \beta(u)).$$

So, $\beta(u) = u$ on ∂Q and hence $\beta \in \Gamma$. Then, by property a, there exists $u \in Q$ such that $\beta(u) \in S$, and property b implies that

$$\beta(u) \in S \subset \tilde{\Phi}^{\tilde{c}}.$$

So, from property iv, we get

$$\text{dist}\,(\beta(u), \gamma(Q)) \leq \|\beta(u) - \gamma(u)\| \leq \delta.$$

But then, $\beta(u) \in S \cap \tilde{\Phi}^{\tilde{c}}$, and by property vii we get that

$$\gamma(u) = \eta(1, \beta(u)) \in \tilde{\Phi}^{\tilde{c}-\varepsilon}.$$

So that we get the contradiction

$$c + \varepsilon \leq \Phi(\gamma(u)) < c + \varepsilon.$$

$\square$

19.2 Linking of Deformation Type

Silva [843] introduced a notion of *linking of deformation type with respect to a given functional*, with the aim of classifying "linkings" as presented by Benci and Rabinowitz. This allowed him it to establish new forms of the multidimensional MPT (and hence the MPT) and the saddle point theorem. The novelty in his results was that the geometric conditions were global and porting on the whole spaces that appear in the splitting of the underlying space on which the problem is treated.

Definition 19.3 (Linking of Deformation Type, Silva). Consider two linking sets S and ∂Q. The linking between them is said to be of deformation type with respect to Φ if there exists $\alpha \in \mathbb{R}$ and $\gamma \in C([0, 1] \times E; E)$ such that $\gamma(0, .) = I_d$ and

   i. $\gamma(t, \partial Q) \cap S = \varnothing$, for all $t \in [0, 1]$,
   ii. $\gamma(1, \partial Q) \subset \Phi^\gamma$,
   iii. $\Phi(u) > \alpha$.

This notion is used to provide a new linking theorem.

Theorem 19.4 (Linking Theorem, Silva). *Let E be a real Banach space and S, Q be two subsets of E. Suppose that $\Phi \in C^1(E; \mathbb{R})$ and S and ∂Q have a linking of deformation type with respect to Φ. If Φ satisfies* (PS)$_d$ *for every $d \geq \alpha$, and α is given in condition iii, then Φ possesses a critical value $c \geq \alpha$ characterized by*

$$c = \inf_{\gamma \in \Gamma} \max_{u \in Q} \Phi(\gamma(1, u)),$$

where $\Gamma = \{\gamma \in C([0, 1] \times E; E); \ \gamma(0, .) = I_d$ and γ satisfies conditions i and ii $\}$. *Furthermore, if $c = \alpha$, then $\mathbb{K}_c$ possesses at least one nonisolated critical point of Φ on E.*

The main tool used by Silva is the following special form of the deformation lemma, especially modeled to handle linkings of deformation type.

Lemma 19.5 (Deformation Lemma, Silva). *Let E be a real Banach space and $\Phi \in C^1(E; \mathbb{R})$. If $\alpha < \beta$ and Φ satisfies $(PS)_c$ for every $c \in [\alpha, \beta]$, then, given $r, \bar{\varepsilon} > 0$, there exist $c_1 > 0$, $R_0 > r$, $\varepsilon \in]0, \bar{\varepsilon}[$, and $\eta \in C([0, 1] \times E; E)$ such that*

η_1. $\eta(0, u) = u$ for every $u \in E$,
η_2. $\eta(0, u) = u$ for every $u \in \Phi^{\alpha - \bar{\varepsilon}} \cup B_r(0)$, $t \in [0, 1]$,
η_3. $\eta(1, \Phi^\beta \setminus B_{R_0}(0)) \subset \Phi^{\alpha - \varepsilon}$,
η_4. $\|\eta(t, u) - u\| \leq c_1 t$ for every $u \in X$, $t \in [0, 1]$,
η_5. $\Phi(\eta(t, u)) \leq \Phi(u)$ for every $u \in X$, $t \in [0, 1]$.

Corollary 19.6. *Let $X = X_1 \oplus X_2$ be a real Banach space such that $\dim X_1 < \infty$. Suppose that $\Phi \in C^1(X; \mathbb{R})$ satisfies the following:*

Φ_1. *There exists $\beta \in \mathbb{R}$ such that $\Phi(u) \leq \beta$, for every $u \in X_1$, and*
Φ_2. *there exists $\alpha \in \mathbb{R}$ such that $\Phi(u) \geq \alpha$, for every $u \in X_2$.*

If Φ satisfies $(PS)_c$ for every $c \in [\alpha, \beta]$, then Φ possesses a critical value in $[\alpha, \beta]$.

Remark 19.1.
   i. If $X_1 = \{0\}$, condition Φ_1 is trivially true and $\inf_X \Phi$ is a critical value of Φ on X.
   ii. Note that, since $\alpha \leq \Phi(0) \leq \beta$, we have that $\alpha \leq \beta$.
   iii. In the classical form of the saddle point theorem [738], condition Φ_1 is replaced by the following:
   R. There exists $\bar{\beta} < \alpha$ and a bounded neighborhood U of 0 in X_1 such that

$$\Phi(u) \leq \bar{\beta}, \qquad \text{for every } u \in \partial U.$$

Corollary 19.7 (Multidimensional MPT, Silva). *Let $X = X_1 \oplus X_2$ be a real Banach space, $\dim X_1 < \infty$. Suppose that $\Phi \in C^1(E; \mathbb{R})$ and satisfies*

Φ_1. $\Phi(u) \leq 0$, for every $u \in X_1$.
Φ_2. *There exists $\rho > 0$ such that*

$$\Phi(u) \geq 0 \qquad \text{for every } u \in \partial B_\rho(0) \cap X_2,$$

   and
Φ_3. *there exist $e \in \partial B_1(0) \cap X_2$ and $\beta \in \mathbb{R}$ such that*

$$\Phi(u) \leq \beta \qquad \text{for every } u \in X_1 \oplus \mathbb{R}^+ e,$$

where $X_1 \oplus \mathbb{R}^+ e = \{v + te \in X_1 \oplus \mathbb{R}e; \ v \in X_1, \ t \geq 0\}$.

If Φ satisfies $(PS)_c$ *for every* $c > 0$ *and* (PS) *for bounded sequences, then* Φ *possesses a critical point other than* 0.

Example 19.3 (A new linking situation discovered by Silva). Consider $X = X_1 \oplus \mathbb{R}e \oplus X_2$ to be a real Banach space, such that $\dim X_1 < \infty$ and $e \in X$, $||e|| = 1$. Given $0 < \rho < R$, we get that

$$S = S_\rho = \overline{(X_2 \setminus B_\rho(0))} \cup ((\mathbb{R}^+ e \oplus X_2) \cap \partial B_\rho(0))$$

and

$$D_R = \partial B_R(0) \cap (X_1 \oplus \mathbb{R}e).$$

Defining

$$\Gamma_1 = \left\{ \gamma \in C^1([0, 1] \times X; X); \ \gamma(0, .) = I_d, \ \gamma(t, .)|_{\partial D_R} = I_d|, \ t \in [0, 1] \right\},$$

where ∂D_R is the boundary of D_R as a subset of $X_1 \oplus \mathbb{R}e$, we obtain the following.

Lemma 19.8. *Let* $X = X_1 \oplus \mathbb{R}e \oplus X_2$ *be a real Banach space,* $\dim X_1 < \infty$ *and* $e \in X$, $||e|| = 1$. *If* $0 < \rho < R$ *and* S_ρ *and* D_R *are as defined earlier, then*

$$\gamma(t, D_R) \cap S_\rho \neq \varnothing \qquad \text{for every } \gamma \in \Gamma \text{ and all } t \in [0, 1].$$

19.3 Newer Extensions of the Notion of Linking

The notion of linking seemed to be an optimal formulation of the topological relation of intersection that appears in minimax theorems for a while, until the paper [818] appeared where a new notion of *linking of two sets* was given. It permitted Schechter to get many interesting extensions and variants of the MPT (see the *intrinsic MPT* in Chapter 20, for example).

Consider a family $\mathcal{H}$ of homeomorphisms $\Gamma(t) \in C([0, 1] \times E; E)$, which contract the whole of E into single points. More precisely, $\mathcal{H}$ is the set of all continuous maps $\Gamma \colon E \times [0, 1] \to E$ such that

a. $\Gamma(0) = I_d$.
b. There is $u_0 \in E$ such that $\Gamma(1)u = u_0$, for all $u \in E$.
c. $\Gamma(s)u \to u_0$ as $s \to 1$ uniformly on bounded subsets of E.
d. For each $s \in [0, 1[$, $\Gamma(s)$ is a homeomorphism of E onto itself.
e. $\Gamma(s)^{-1}$ is continuous from $E \times [0, 1[$ to E.

Definition 19.4 (Linking of Two Sets, Schechter and Tintarev). *For* $A, B \subset E$, A *links* B *if*

$$\begin{cases} A \cap B = \varnothing, \\ \forall \Gamma(t) \in \mathcal{H}, \ \exists t_0 \in [0, 1] \qquad \text{such that } \Gamma(t_0)A \cap B \neq \varnothing. \end{cases}$$

Roughly speaking, A links B if it cannot be continuously contracted to a point without intersecting B. Moreover, this notion is "almost symmetric" if A and B are both closed and bounded and $E \setminus A$ is path-connected; then

$$A \text{ links } B \quad \text{implies that} \quad B \text{ links } A.$$

A limitation in the definition of the standard linking of a closed set with the boundary ∂Q of a submanifold is that it requires the second set to be the boundary of a submanifold. Indeed, this has the drawback that if, for example, $Q = M \cap B(0, R)$, where M is a closed infinite dimensional subspace of E, we can construct a map γ_0 in $\Gamma = \{\gamma \in \mathcal{C}(E; E); \ \gamma|_{\partial Q} = I_d\}$ that maps Q into ∂Q. Such a map cannot satisfy $\gamma_0(Q) \cap B \neq \varnothing$ for any set B satisfying $A \cap B = \varnothing$. **So, it is hopeless to prove a linking theorem where ∂Q is not finite dimensional.** In particular, the second form of the saddle point theorem of Rabinowitz, by Silva, which is still true if any of the two spaces that appear in the splitting of E is finite dimensional, as shown by Schechter (see, for example, [808]), cannot be handled this way.

Many interesting examples of linking theorems built on the notion of linking of two sets have been given in many papers by Schechter (see, for example, [807, 811, 813]) and also in his book [816]. To avoid making this chapter excessively long, these will not be given there. Nevertheless, we prefer to give some illustrative examples and only another (somewhat old) from the MPT that allow extension of some results on superlinear problems. Schechter, who is very "fecund," has developed many extensions of the MPT. Some of them are detailed in this monograph.

Example 19.4. Let $E = E_1 \oplus E_2$, where E_1 and E_2 are closed subspaces of E, one of them being finite dimensional. Then $E_1 \cap B(0, R)$ links E_2 for each $R > 0$. This is useful to get the form of the saddle point theorem indicated earlier.

Example 19.5. Let $E = E_1 \oplus E_2$, where E_1 and E_2 are closed subspaces of E, one of them being finite dimensional. Let $w_0 \in E_1 \setminus \{0\}$ and $0 < r < R$;

$$A = \{w \in E_1; \ \|w\| = R\},$$
$$B = \{v \in E_2; \ \|v\| \geq r\} \cup \{u = v + sw_0; \ v \in E_2, s \geq 0, \|u\| = r\}.$$

Then A links B.

Example 19.6. Let A and $(B_n)_n$ be subsets of E such that A is bounded and links B_n for each n. Suppose that

$$B_n = B_n' \cup B_n'' \quad \text{where } \operatorname{dist}(B_n'', 0) \to 0 \text{ as } n \to \infty,$$

and that there is a set $B \subset E$ such that

$$A \cap B = \varnothing, \qquad B_n' \subset B, \qquad n = 1, 2, \ldots$$

Then A links B.

In [257], while extending the intrinsic MPT to the context of critical point theory of continuous functionals on metric spaces, Corvellec found himself constrained to refor-mulate the definition of the linking of two sets, the problem being that the deformations appearing in the metric theory are not homeomorphisms. He stayed faithful to the spirit of the original definition, as we can see.

Definition 19.5 (Linking of Two Sets, Corvellec). Let E be a metric space and $A, B \subset X$. We say that A and B *link* if

    i. B is contractible in X,
   ii. $B \cap A = \varnothing$, and
  iii. for any contraction Ψ of b in X, $\Psi(B \times [0, 1]) \cap A \neq \varnothing$.

Recall that a nonempty subset B of X is *contractible* in X_1 if there exists a continuous $\Psi : B \times [0, 1] \to X$ and some $u_0 \in X$ such that

$$\begin{cases} \Psi(u, 0) = u, & \forall u \in B, \\ \Psi(u, 1) = u_0, & \forall u \in B. \end{cases}$$

Ψ is called a *contraction* of B in X.

The reformulation of the geometry of the generalized MPT (see Chapter 8) with the newer notion of linking is the following. Consider $E = Y \oplus Z$, where $\dim Y < \infty$ and $v_0 \in Z \setminus \{0\}$. For $\rho > 0$, $R > 0$, set

$$A_\rho = \partial B(0, \rho) \cap Z,$$

$$B'_R = \big(B(0, R) \cap \mathbb{R}^+ v_0 \oplus Y\big) = \{sv_0 + u; \ s \geq 0, \ u \in Y, \ \|sv_0 + u\| = R\}, \text{ and}$$

$$B_R = B'_R \cup (B(0, R) \cap Y) \subset Y \oplus \mathbb{R} v_0.$$

Then B_R links A_ρ if $R > \rho$.

A very important point there is that *the inf sup values given by the standard linking and by the notion of linking of two sets, in the case of the multidimensional MPT coincide.* Indeed, Corvellec [257] proved the following result.

Theorem 19.9. *Let X be a Banach space that splits into the direct sum $X = Y \oplus Z$, with Y finite dimensional. Set*

$$A = Z, \qquad B_R = \partial B(0; R) \cap Y, \qquad R > 0.$$

Then B_R links A. Moreover, if $\Phi : X \to \mathbb{R}$ is continuous and

$$c = \inf_{\psi \in \Phi_{B_R}} \sup_{B_R \times [0,1]} \Phi \circ \psi, \qquad \tilde{c} = c = \inf_{\gamma \in \Gamma} \sup_{\overline{B(0;R)} \cap Y} \Phi \circ \gamma,$$

where Φ_{B_R} is the set of contractions of B_R in X and

$$\Gamma = \{\gamma : \overline{B}(0; R) \cap Y \to X; \ \gamma \text{ is continuous and } \gamma|_{B_R} = Id\},$$

then $c = \tilde{c}$.

Proof. The fact that B_R links A follows from a standard application of Brouwer's topological degree, similar to the proof that $\gamma(\overline{B}(0; R) \cap Y) \cap A \neq \varnothing$ for any $\gamma \in \Gamma$

(see, e.g., [748, Chapter 5]). If $\psi \in \Phi_{B_R}$ with $\psi(u, 1) = u_0$, we can define $\gamma \in \Gamma$ by

$$\gamma(u) = \begin{cases} \psi\left(\dfrac{Ru}{\|u\|}, 1 - \dfrac{\|u\|}{R}\right) & \text{if } u \neq 0 \\ u_0 & \text{if } u = 0, \end{cases}$$

so that $\gamma(\overline{B}(0; R) \cap Y) = \psi(B_R \times [0, 1])$ and $\tilde{c} \leq c$.

Conversely, for $\gamma \in \Gamma$ define $\psi \in \Phi_{B_R}$ by

$$\psi(u, t) = \gamma((1 - t)u),$$

so that $\psi(B_R \times [0, 1]) = \gamma(\overline{B}(0; R) \cap Y)$ and $c \leq \tilde{c}$. $\square$

Comments and Additional Notes

◇ 19.I Duality to Homotopy-Stable Families of Sets

Let B be a closed subset of X. Ghoussoub [425] calls a class $\mathcal{F}$ of compact subsets of X a *homotopy stable family with boundary B* if

a. every set in $\mathcal{F}$ contains B, and

b. for any set in $\mathcal{F}$ and any continuous deformation of the identity $\eta \in C([0, 1] \times X; X)$, $\eta(t, x) = x$ for $(t, x) \in (\{0\} \times X) \cup ([0, 1] \times B)$ that leaves B invariant, we have $\eta(\{1\} \times A) \in \mathcal{F}$.

The situation of the MPT corresponds to the *mountain pass family* $\mathcal{F}_{u_0}^{u_1}$, the class of all continuous paths joining u_0 to u_1, which is homotopy stable with boundary $B = \{u_0, u_1\}$.

Ghoussoub expresses the notion of linking as a duality to a homotopy-stable class $\mathcal{F}$ with boundary B.

A set M is said to be *dual* to a homotopy-stable family $\mathcal{F}$ with boundary B if

$$M \cap B = \varnothing \qquad \text{and} \qquad M \cap A \neq \varnothing \text{ for every } A \in \mathcal{F}.$$

Dual sets are very important in locating critical points, especially in the "limiting case" situation (see Chapter 9). In the situation of the MPT, any sphere S centered at u_0 with radius $\rho < \|u_0 - u_1\|$ is dual to the mountain-pass class $\mathcal{F}_{u_0}^{u_1}$.

◇ 19.II Saddle Point Theorem of Silva

Theorem 19.10. *Let $E = X_1 \oplus X_2$ be a real Banach space, where X_1 is finite dimensional. Suppose $\Phi \in C^1(E; \mathbb{R})$ and satisfies*

i. $\Phi(u) \leq 0$, *for every $u \in X_1$.*

ii. *There exists $\rho > 0$ and $\gamma > 0$ such that*

$$\Phi(u) \geq \gamma \text{ for every } u \in \partial B(0, \rho) \cap X_2.$$

iii. There exists $e \in \partial B(0, 1) \cap X_1$ and $\beta \in \mathbb{R}$ such that

$$\Phi(u) \leq \beta \text{ for every } u \in X_1 \oplus \mathbb{R}^+ e,$$

where $X_1 \oplus \mathbb{R}^+ e = \{v + te \in X_1 \oplus \mathbb{R}e; \ v \in X_1, \ t \geq 0\}$.

If Φ satisfies (PS)$_c$ for every $c \in [\gamma, \beta]$, then Φ possesses a critical point $u \in E$ characterized by a minimax argument.

Concerning these results, notice that if $X_1 = \{0\}$, then $E = X_2$ and conditions ii and iii are the usual geometry of the MPT. Silva permits $\beta \geq \gamma$, unlike Rabinowitz [737]; however, he requires a global estimate in condition iii.

If $\gamma = 0$, he concludes that Φ possesses critical points other than 0. However, he cannot conclude that the corresponding critical value is obtained by a minimax argument.

Silva also studies functionals with an infinite dimensional splitting. Of course, he is unable to prove the intersection result expressing the linking directly. He uses approximation arguments through an appropriate Palais-Smale condition, stronger than the usual one. This way he gets a relation between the *local linking condition* of [587] and the methods developed by Benci and Rabinowitz [113]. For the details, the interested reader is referred to [842, 843].

◇ *19.III Isotopic Linking*

We pass now to another notion of linking, the *isotopic linking* of Tintarev [921]. Consider the class $\mathcal{I}$ of *isotopies* on a Banach space E defined as the class of functions $\gamma \in C([0, 1] \times E; E)$ such that

    i. γ is bounded on every bounded set,
    ii. $\gamma(t, .)$ is a homeomorphism for each $t \in [0, 1]$, and
    iii. $\hat{\gamma}(t, .) = \gamma^{-1}(t, .)$ is bounded on bounded sets.

Notice that condition iii implies that $\hat{\gamma} \in \mathcal{I}$. Tintarev defines the isotopic linking in the following way.

Definition 19.6 (Isotopic Linking, Tintarev). Let B be a nonempty subset of E. We shall say that a collection $\mathcal{L}_B$ of nonempty bounded subsets of E is a *linking class* of B if

1. $A \in \mathcal{L}_B \Rightarrow A \cap B = \varnothing$,
2. $\gamma \in \mathcal{I}, A \in \mathcal{L}_B, \gamma([0, 1] \times A) \cap B = \varnothing$ implies that for all $t \in [0, 1], h(t, A) \in \mathcal{L}_B$, and
3. there exists $C_B > 0$, such that for all $A \in \mathcal{L}_B$, $\text{dist}(A, B) \leq C_B$.

This notion is used to prove the following linking theorem.

Theorem 19.11 (Isotopic Linking Theorem, Tintarev). Let $\Phi \in C^1(E; \mathbb{R})$ and assume that Φ' is uniformly continuous on all bounded sets. Let $B \subset E$ and $A \in \mathcal{L}_B$ be *nonempty bounded sets, such that*

$$\sup_{A} \Phi < \inf_{B} \Phi.$$

Then

$$c = \inf_{A' \in \mathcal{L}_B} \sup_{x \in A'} \Phi(x)$$

is finite and there is a sequence $(x_n)_n$ satisfying

$$\Phi'(x_n) \to 0, \qquad \Phi(x_n) \to c.$$

◇ *19.IV Homological and Cohomological Linkings*

See the final chapter, Chapter 25.

◇ *19.V Local Linking*

Let $X = X_1 \oplus X_2$ be a Banach space. In 1984, Li and Liu [567] introduced a notion of *local linking* that generalizes the notions of local minimum and local maximum. This notion was and is still used to obtain theorems with multiple critical points. We cite, for example, [567] in 1984, Li and Willem [569] in 1995, and Perera [713] in 1999, where Morse index estimates for critical points produced by local linking are given.

Definition 19.7 (Local Linking at Zero, Li and Willem). A function $\Phi \in C^1(X; \mathbb{R})$ has a *local linking at* 0 with respect to X_1 and X_2 if, for some $r > 0$,

$$\Phi(u) \geq 0 \text{ for } u \in X_1, \ ||u|| \leq r,$$

$$\Phi(u) \leq 0 \text{ for } u \in X_2, \ ||u|| \leq r.$$

It is clear then that 0 is a critical point of Φ. When 0 is a nondegenerate critical point of a C^2-functional defined on a Hilbert space and $\Phi(0) = 0$, then Φ has a local linking at 0.

Li and Liu supposed the stronger assumptions that

$$\Phi(u) \geq c > 0 \text{ for } u \in X_1, ||u|| = r,$$

and $\dim X_2 < +\infty$.

Concerning the definition of local linking at 0, similar global geometric conditions were supposed by Silva. (See Section 19.2 to get a little idea on the work of Silva.)

◇ *19.VI Generalized Linking Theorem*

Before closing this chapter, we want to state another linking theorem where the splitting does not require that one of the two spaces is finite dimensional. Let Y be a separable subspace of a Hilbert space H and let $Z = Y^\perp$. Let $P: H \to Y$, $Q: H \to Z$ be the orthogonal projections onto Y and Z, respectively. The τ-topology on H is generated by the norm

$$|||u||| = \max\left(||Qu||, \sum_{k=1}^{\infty} \frac{1}{2^{k+1}}|(Pu, e_k)|\right),$$

where $(e_k)_k$ is a total orthonormal sequence in Y.

Let $\rho > r > 0$ and $z \in Z$ be such that that $||z|| = 1$. Define

$$\begin{aligned} M &= \{y + \lambda z; \; ||u|| \le \rho, \; \lambda \ge 0, \; y \in Y\}, \\ M_0 &= \{y + \lambda z; \; y \in Y, \; ||u|| = \rho, \; \text{and } \lambda \ge 0 \text{ or } ||u|| \le \rho \text{ and } \lambda = 0\}, \\ N &= \{u \in Z; \; ||u|| = r\}. \end{aligned}$$

Theorem 19.12 (Generalized Linking Theorem, Kryszewsky-Szulkin [542]). *Let* $\Phi \in C^1(H; \mathbb{R})$ *be such that* Φ *is* τ-*u.s.c. and* $\nabla\Phi$ *is weakly sequentially continuous and let*

$$b = \inf_N \Phi > 0 = \sup_{M_0} \Phi, d = \sup_M \Phi < \infty. \tag{19.3}$$

Then there exists $c \in [b, d]$ *and a sequence* $(u_n)_n \subset H$ *such that*

$$\Phi(u_n) \to c, \qquad \Phi'(u_n) \to 0.$$

Corollary 19.13. *Let* $\Psi \in C^1(H; \mathbb{R})$ *be weakly sequentially l.s.c., bounded from below and such that* $\nabla\Psi$ *is weakly sequentially continuous. If*

$$\Phi(u) = \frac{||Qu||^2}{2} - \frac{||Pu||^2}{2} - \Psi(u)$$

satisfies (19.3), *then there is* $c \in [b, d]$ *and a sequence* $(u_n)_n \subset H$ *such that*

$$\Phi(u_n) \to c, \qquad \Phi'(u_n) \to 0.$$

20

The Intrinsic MPT

Intrinsic *adj* inherent, born, built-in, congenital, connate, constitutional, deep-seated, elemental, essential, inborn, inbred, indwelling, ingenerate, ingrained, innate, intimate.

From Webster's Electronic Thesaurus

Using the concept of *linking of two subsets A and B*, seen in Chapter 19, Schechter proved an *intrinsic* version of the MPT where an estimate for $\|\Phi'(u)\|$ appears, as a function of the difference between the supremum of Φ on A and its infimum on B, and of the distance between B and the proper subset of A where Φ assumes greater values than on B.

We will present Schechter's result and some of its immediate consequences, but we will focus on its metric extension due to Corvellec, which presents nicely and clearly its principles and basic ideas.

The main references for the subject of this chapter are the papers [257, 770, 771, 808]. You may also consult the chapter of notes and remarks at the end of Schechter's book [816].

The aim of Schechter, in [808], was a new statement of the MPT without the aid of "auxiliary sets" (the local minimum and the lower point e in the statement of the original MPT or the compact set K and its closed subset K^* that appear in the statements of [153, 623, 628, 835], for example). He wanted also to investigate what happens if some of the geometric assumptions fail to hold. This ran into a new form, which was reconsidered again by Corvellec [257] and Ribarska et al. [770, 771], who obtained metric variants requiring less smoothness on the functional, as we will see.

Schechter systematically uses the *linking of two sets* of E (Definition 19.4 on page 233), a notion he introduced with Tintarev in [818], in the numerous results he obtains in [808].

Let E be a Banach space and let Ψ be the set of all continuous maps $\Gamma(t) \colon E \times [0, 1] \to E$, $t \in [0, 1]$ such that

a. $\Gamma(0) = I_d$,
b. there is an $x_0 \in E$ such that $\Gamma(1)x = x_0$ for each $x \in E$,
c. $\Gamma(t)x \to x_0$ as $t \to 1$ uniformly on bounded subsets of E, and

d.   for each $t \in [0, 1[$, $\Gamma(t)$ is a homeomorphism of E onto itself and Γ^{-1} is continuous on $E \times [0, 1[$.

Recall that a subset A of E links a subset B of E if $A \cap B = \varnothing$ and for each $\Gamma \in \Psi$ there is a $t \in]0, 1]$ such that $\Gamma(t)A \cap B \neq \varnothing$.

The notations Γ and Ψ will refer in the sequel to the objects appearing in this particular definition.

20.1 The Intrinsic MPT

So, consider two subsets A and B of a Banach space E that link and let $\Phi : E \to \mathbb{R}$ be a C^1-functional. Set

$$a = \inf_{\substack{\Gamma \in \Psi \\ u \in A}} \sup_{0 \le s \le 1} \Phi(\Gamma(s)u) \qquad \text{and} \qquad b_0 = \inf_{B} \Phi. \tag{20.1}$$

Assume also that

$$\operatorname{dist}(A, B) > 0. \tag{20.2}$$

Let $B' = B \cap \Phi^a = \{v \in B; \ \Phi(v) < a\}$ and $d' = \operatorname{dist}(A, B')$. Note that since A links B, we have $b_0 \le a$ and $B' = \varnothing$ if and only if $b_0 = a$.

Let α, T be two positive numbers such that

$$a - b_0 < \alpha T \qquad \text{and} \qquad T < d'. \tag{20.3}$$

Theorem 20.1 (Intrinsic MPT, Schechter). *Assume in addition that*

$$-\infty < b, \qquad a < \infty. \tag{20.4}$$

Then for every $\delta > 0$ sufficiently small there exists $u \in E$ such that

$$b_0 - \delta \le \Phi(u) \le a + \delta, \qquad ||\Phi'(u)|| \le \alpha, \tag{20.5}$$

and the following alternative holds. Either

$$\operatorname{dist}(u, B') < T \tag{20.6}$$

or

$$\operatorname{dist}(u, B \setminus B') < \delta/\alpha. \tag{20.7}$$

Supposing by contradiction that the conclusion of the theorem does not hold, Schechter constructed a flow generated by a vector field in the usual way followed in the proof of the deformation lemma and got a contradiction. We do not give the proof here because we will state and prove a metric extension of this result by Corvellec.

As a consequence of Theorem 20.1, when setting $B' = \varnothing$, $d' = \infty$ and taking for each n, $T_n = 1$, $\delta_n = 1/n^2$ and $\alpha_n = 1/n$, we have the following result.

Corollary 20.2. *If $b_0 = a$, then there is a sequence $(u_k)_k \subset E$ such that*

$$\Phi(u_k) \to a, \qquad \Phi'(u_k) \to 0, \qquad and \qquad \text{dist}\,(u_k, B) \to 0. \tag{20.8}$$

In a similar way we get also the following interesting consequence.

Corollary 20.3. *Let $(A_n)_n$ and $(B_n)_n$ be sequences of subsets of E and define*

$$a_n = \inf_{\substack{\Gamma \in \Psi \\ u \in A_n}} \sup_{0 \le s \le 1} \Phi(\Gamma(A)u), \qquad b_{0n} = \sup_{B_n} \Phi, \tag{20.9}$$

$$a = \liminf_n a_n, \qquad b_0 = \limsup_n b_{0n}, \tag{20.10}$$

and

$$B'_n = B_n \cap \Phi^{a_n} = \left\{ v \in B_n;\ \Phi(v) < a_n \right\} \tag{20.11}$$

$$d'_n = \text{dist}\,(A_n, B'_n). \tag{20.12}$$

Assume that A_n links B_n, $\text{dist}\,(A_n, B_n) > 0$, and that

$$-\infty < b_0 \le a < \infty \tag{20.13}$$

$$d'_n \to \infty. \tag{20.14}$$

Then there exists a sequence $(u_k)_k$ such that

$$\Phi(u_k) \to c, \qquad b_0 \le c \le a, \qquad and \qquad \Phi'(u_k) \to 0. \tag{20.15}$$

20.2 A Metric Extension

Corvellec unified and extended in [257] the intrinsic MPT and the various results of Schechter discussed earlier to the context of the critical point theory for continuous functions on metric spaces, in the spirit of the work of Degiovanni and Marzocchi, and Katriel. He used a specific *quantitative deformation lemma*, similar to Willem's one in the smooth case (Lemma 4.2 on page 38).

> While working on metric spaces with functions that are only continuous, one ought not expect to get deformations that are homeomorphisms at each time, and the way to recover this property is not yet known. This seems to us to be because of the *very* definition of the metric regularity of continuous functions at some point u, which supposes only a uniform *descent property* in a neighborhood of u and not a similar uniform *ascension property*.

The first thing Corvellec does is to reformulate the definition of the linking of two sets to remove the homeomorphism property from the deformation $\Psi(., t)$, in a way that makes it simpler (see Definition iii).

Linking of Two Sets (Corvellec). *Let E be a metric space and $A, B \subset X$. We say that A and B link if*

  i. *B is contractible in X,*
  ii. *$B \cap A = \emptyset$, and*
  iii. *For any contraction Ψ of b in X, $\Psi(B \times [0, 1]) \cap A \ne \emptyset$.*

The main tools used by Corvellec are a quantitative deformation lemma for continuous functions on metric spaces, which is interesting for itself (compare it, for example, to those that appear in Chapter 16), and a *quantitative noncritical interval theorem*.

Lemma 20.4 (Quantitative Nonsmooth Deformation Lemma, Corvellec). *Let* (X, dist) *be a metric space and let* $\Phi \colon X \to \mathbb{R}$ *be a continuous function,* $c \in \mathbb{R}$, A *a closed subset of* X, *and* $\delta, \sigma > 0$ *such that*

$$c - 2\delta \le \Phi(u) \le c + 2\delta \qquad \text{implies that} \qquad |d\Phi|(u) > 2\sigma.$$

Then there exists $\eta \colon X \times [0, 1] \to X$ *continuous with*

   a. $\text{dist}(\eta(u, t), u) \le (\delta/\sigma)t$,
   b. $\eta(u, t) \ne u \implies \Phi(\eta(u, t)) < \Phi(u)$, *and*
   c. $u \in A \cap \Phi_{c-\delta}^{c+\delta}$ *implies that* $\Phi(\eta(u, t)) \le \Phi(u) - (\Phi(u) - c - \delta)t$.

Lemma 20.5 (Quantitative Noncritical Interval Theorem, Corvellec). *Let* (X, dist) *be a metric space and let* $\Phi \colon X \to \mathbb{R}$ *be a continuous function,* $a, b \in \mathbb{R}$, *with* $a < b$, $\delta > 0$, *and* $\sigma > 0$ *such that*

$$a - \delta \le \Phi(u) \le b + \delta \qquad \text{implies that} \qquad |d\Phi|(u) > \sigma.$$

Then there exists $\eta \colon X \times [0, 1] \to X$ *continuous with*

   a. $\text{dist}(\eta(u, t), u) \le \left(\dfrac{b - a}{\sigma} \right) t$,
   b. $\eta(u, t) \ne u$ *implies that* $\Phi(\eta(u, t)) < \Phi(u)$, *and*
   c. $u \in \Phi_a^b$ *implies that* $\Phi(\eta(u, t)) \le \Phi(u) - (\Phi(u) - a)t$.

Coming back now to the intrinsic MPT, consider two subsets A and B of X such that B links A. Set

$$a = \inf_A \Phi, \qquad b_0 = \sup_B \Phi,$$

and denote by Ψ_B the set of contractions of B in X. Let

$$b = \inf_{\psi \in \Psi_B} \sup_{B \times [0,1]} (\Phi \circ \psi).$$

Since B links A, then $b_0 \le b$ and $a \le b$.

Theorem 20.6. *Assume that* $b \in \mathbb{R}$, $b_0 \le a$, *and* $b - a \le \delta/2$ *for some* $\delta > 0$. *Then, for* $\sigma > 0$, *there exists* $u \in X$ *with*

$$b - 2\delta \le \Phi(u) \le b + 2\delta, \qquad \text{dist}(u, A) \le 2\delta/\sigma, \qquad \text{and} \qquad |d\Phi|(u) \le 2\sigma.$$

Proof. By contradiction, suppose that for some $\sigma > 0$, we have $u \in X$,

$$b - 2\delta \le \Phi(u) \le b + 2\delta, \ \text{dist}(u, A) \le 2\delta/\sigma \qquad \text{implies that} \qquad |d\Phi|(u) > 2\sigma.$$

Let $\eta \colon X \times [0, 1] \to X$ be the continuous deformation given by the quantitative deformation lemma with A replaced by $\{u \in X; \ \text{dist}(u, A) \le \delta/\sigma\}$ such that

$$\text{dist}(\eta(u, t), u) \le (\delta/\sigma)t,$$

and

$$\eta(u, t) \neq u \quad \text{implies that} \quad \Phi(\eta(u, t)) < \Phi(u),$$

$$\left(\text{dist}(u, A) \leq \delta/\sigma, u \in \Phi_{b-\delta}^{b+\delta}\right) \quad \text{implies that} \quad \Phi(\eta(u, t)) \leq \Phi(u) - (\Phi(u) - b + \delta)t.$$

Let $\psi_0 \in \Phi_B$ with $\sup_{B \times [0,1]}(\Phi \circ \psi_0) \leq b + \delta$ and define $\psi \in \Phi_B$ by

$$\psi(u, t) = \begin{cases} u & \text{if } 0 \leq t \leq 3/4, \\ \psi_0(u, 4t - 3) & \text{if } 3/4 \leq t \leq 1. \end{cases}$$

So that $\sup_{B \times [0,1]}(\Phi \circ \psi) \leq b + \delta$. Finally, define $\tilde{\psi} \in \Phi_B$ by

$$\tilde{\psi}(u, t) = \eta(\psi(u, t), t).$$

Let $(u, t) \in B \times [0, 1]$ such that $\tilde{\psi}(u, t) \in A$. Then, $t > 3/4$. Indeed, if t was in $[0, 3/4]$, we would have $\tilde{\psi}(u, t) = \eta(u, t)$, so that if $\tilde{\psi}(u, t) \neq u$, $\Phi(\tilde{\psi}(u, t)) < \Phi(u) \leq b_0 \leq a$ and $\tilde{\psi}(u, t) \notin A$. On the other hand, we have dist$(\psi(u, t), A) \leq \delta/\sigma$. Hence,

$$\begin{aligned} \Phi(\tilde{\psi}(u, t)) &= \Phi(\eta(\psi(u, t), t), \\ &\leq \Phi(\psi(u, t)) - (\Phi(\psi(u, t)) - b + \delta/2)t, \\ &\leq b - \delta/2, \\ &\leq a, \end{aligned}$$

and this contradicts the fact that $\tilde{\psi}(u, t) \in A$.   □

Theorem 20.7 (Intrinsic MPT, Corvellec). *Let (X, dist) be a complete metric space, $\Phi: X \to \mathbb{R}$ a continuous function, and A, B two subsets of X such that B links A. Set*

$$a = \inf_A \Phi, \qquad b_0 = \sup_B \Phi, \qquad b = \inf_{\psi \in \Psi_B} \sup_{B \times [0,1]} (\Phi \circ \psi),$$

and assume that $a, b \in \mathbb{R}$.

a. *In the case $b_0 \leq a$, there exists a sequence $(u_n)_n \subset X$ such that*

$$\Phi(u_n) \to b \qquad \text{and} \qquad |d\Phi|(u) \to 0.$$

If, moreover, $b = a$, we can choose $(u_n)_n$ such that dist$(u_n, A) \to 0$. *So, if Φ satisfies* (PS)$_b$, *there exists $u \in X$ such that*

$$\Phi(u) = b, \qquad |d\Phi|(u) = 0, \qquad \text{and} \qquad u \in \overline{A}.$$

b. *In the case $b_0 > a$, then if either*
    i. *$0 < \alpha \leq$ dist$(B \cap \Phi_a, A)$, or*
    ii. *$0 < \alpha \leq$ dist$(B \cap \Phi^{b_0}, A)$ and if $\sigma > (b_0 - a)/\alpha$,*
    then for any $\delta > 0$ there exists $u \in X$ with

$$a - \delta \leq \Phi(u) \leq a + \delta \qquad \text{and} \qquad |d\Phi|(u) \leq \sigma.$$

Proof. For case a, we apply Theorem 20.6 with $\delta = \delta_h = 1/h^2$, $\sigma = \sigma_h = 1/h$, and either the given A if $b = a$, or A replaced by $\tilde{A} = \Phi_{b-1/(2h^2)}$ if $b > a$.

In the latter case, it is clear that for large h, $B \cap \tilde{A} = \varnothing$ because $b_0 \leq a < b$, while by the definition of b, $\psi(B \times [0, 1]) \cap \tilde{A} \neq \varnothing$ for every $\psi \in \Phi_B$.

For case b, if $b > b_0$, the conclusion follows from case a.

So, assume that $b = b_0$. Notice that the strict inequality $b_0 > a$ means that $B \cap \Phi_a$ and $B \cap \Phi^{b_0}$ are nonempty, so that $\alpha < +\infty$.

By contradiction, suppose that for some $\delta > 0$

$$u \in X, \ a - 2\delta \leq \Phi(u) \leq b + 2\delta \qquad \text{implies that} \qquad |d\Phi|(u) > \sigma,$$

and that δ is so small that $b - a + 2\delta \leq \sigma\alpha$. Let $\eta\colon X \times [0, 1] \to X$ be given by the quantitative noncritical interval theorem (Theorem 20.5) (with a replaced by $a - \delta$ and b by $b + \delta$) such that

$$\text{dist}\,(\eta(u, t), u) \leq \frac{b - a + 2\delta}{\sigma} t,$$

$$\eta(u, t) \neq u \qquad \text{implies that} \qquad \Phi(\eta(u, t)) < \Phi(u), \ \text{and}$$

$$u \in \Phi_{a-\delta}^{b+\delta} \qquad \text{implies that} \qquad \Phi(\eta(u, t)) \leq \Phi(u) - (\Phi(u) - a + \delta)t.$$

Let $(b - a + \delta)/(b - a + 2\delta) < \varepsilon < 1$ be fixed. Then,

$$\Phi(u) \leq b + \delta, \quad \text{and} \quad \varepsilon \leq t \leq 1 \quad \text{implies that} \quad \Phi(\eta(u, t)) < a. \qquad (20.16)$$

Indeed, if $a \leq \Phi(u) \leq b + \delta$ and $t \in [\varepsilon, 1]$, we have that

$$\Phi(\eta(u, t)) \leq (1 - t)\Phi(u) + (a - \delta)t \leq b + \delta - (b - a + 2\delta)\varepsilon < a,$$

while whenever $\Phi(u) < a$, (20.16) holds.

Now, let $\psi_0 \in \Phi_B$ with $\sup_{B \times [0,1]}(\Phi \circ \psi_0) \leq b + \delta$ and define $\psi \in \Phi_B$ by

$$\psi(u, t) = \begin{cases} u & \text{if } 0 \leq t \leq \varepsilon, \\ \psi_0\left(u, \dfrac{t - \varepsilon}{1 - \varepsilon}\right) & \text{if } \varepsilon \leq t \leq 1, \end{cases}$$

and define $\tilde{\psi}(u, t) = \eta(\psi(u, t), t)$.

Claim 20.1. $\tilde{\psi}(B \times [0, 1]) \cap A = \varnothing$.

Since the claim contradicts the fact that B links A, the proof would be complete once we would have proved the claim. The claim is true. Indeed, let $u \in B$, then for $t \in [0, \varepsilon]$ we have $\tilde{\psi}(u, t) = \eta(u, t)$, so that

$$\text{dist}\,(\tilde{\psi}(u, t), u) = \text{dist}\,(\eta(u, t), u) \leq \frac{b - a + 2\delta}{\sigma}\varepsilon < \alpha, \qquad (20.17)$$

while

$$\tilde{\psi}(u, t) = \eta(u, t) \neq u \qquad \text{implies that} \qquad \Phi(\tilde{\psi}(u, t)) < \Phi(u). \qquad (20.18)$$

In case i, whenever $\Phi(u) > a$ we have $\tilde{\psi}(u, t) \notin A$ by (20.17), while (20.18) shows that $\tilde{\psi}(u, t) \notin A$ whenever $\Phi(u) \leq a$. (Recall that $A \cap B = \varnothing$.)

In case ii, (20.17) and (20.18) show that $\tilde{\psi}(u, t) \notin A$ since $\Phi(u) \leq b_0$. For $t \in [\varepsilon, 1]$ we have, by (20.16),

$$\Phi(\tilde{\psi}(u, t)) = \Phi\left(\eta\left(\psi_0\left(u, \frac{t - \varepsilon}{1 - \varepsilon}\right), t\right)\right) < a,$$

so that again $\tilde{\psi}(u, t) \notin A$. □

Case a is similar to the "usual" minimax principle in the presence of linking.

The preceding theorem can be used to obtain the existence of critical points of Φ, even in the case where the *usual geometry*,

$$-\infty < b_0 = \sup_B \Phi \leq \inf_A \Phi = a < +\infty, \tag{20.19}$$

is *not satisfied*.

Corollary 20.8. *Let X be a complete metric space, $\Phi\colon X \to \mathbb{R}$ be continuous, and $(A_n)_n$, $(B_n)_n$ be two sequences of subsets of X such that, for each n,*

$$\operatorname{dist}(B_n, A_n) > 0 \qquad \text{and} \qquad B_n \text{ links } A_n.$$

Set

$$a_n = \inf_{A_n} \Phi, \qquad b_{n0} = \sup_{B_n} \Phi, \qquad b_n = \inf_{\psi \in \Psi_B} \sup_{B_n \times [0,1]} (\Phi \circ \psi),$$

where Ψ_B is the set of the contractions of B_n in X,

$$d_n = \operatorname{dist}(B_n \cap \Phi_{a_n}, A_n), \qquad d_{n0} = \operatorname{dist}(B_n, \Phi^{b_{n0}} \cap A_n).$$

Assume that b_n, $a_n \in \mathbb{R}$ and that either

$$\limsup_{n \to \infty} \frac{b_{n0} - a_n}{d_n} \leq 0 \qquad \text{or} \qquad \limsup_{n \to \infty} \frac{b_{n0} - a_n}{d_{n0}} \leq 0$$

(with the convention $1/+\infty = 0$). And, let

$$b = \liminf_{n \to \infty} b_n, \qquad a = \liminf_{n \to \infty} a_n,$$

and assume that $a, b \in \mathbb{R}$ and that Φ satisfies $(\text{PS})_c$ for all $c \in [a, b]$.
Then there exists $u \in X$ such that $|d\Phi|(u) = 0$ and $\Phi(u) \in [a, b]$.

Comments and Additional Notes

◇ 20.I A Version of the Intrinsic MPT with No Regularity Assumptions at All

In [770], the intrinsic MPT is extended once again. This time, the authors use the definition of weak slope by Ioffe and Schwartzman [494] (which is close to the weak slope of Degiovanni and Marzocchi and Katriel, used by Corvellec; see Chapter 16 for the exact definitions and some comments) and they do not require *any regularity conditions* on Φ, thus allowing a possible treatment of the lower semicontinuous case if needed. This may open the door to some new applications in nature. These results were already announced in [771].

The following geometric language, related to the MPT, is used in the papers consecrated to the intrinsic MPT. The local minima 0 and e are called the *boundary* while the sphere surrounding 0 is the *barrier*.

So, in the situation of the MPT, the values of the functional are higher on the barrier than in the boundary, while in [808], the barrier is split into two parts: a "high" part where the values of Φ are greater or equal than its values on the boundary and a "low" part where this is not true. Schechter requires only that the low part is sufficiently far from the boundary.

In the generalizations by Corvellec [257] and Ribarska et al. [770], a result with a split boundary is also given. This is interesting because, as noticed by Ribarska et al., the boundary and the barrier do not play symmetric roles.

21

Geometrically Constrained MPT

This resembles the situation of a traveler trying to cross a mountain range without climbing higher than necessary. If we can find a continuous path connecting the two points which does not take the traveler higher than any other such path, it is expected that this path will produce a critical point.

However, there is a difficulty which must be addressed. One must allow the competing path to roam freely, and conceivably they can take the traveler to infinity while he is trying to cross some local mountains.

M. Schechter, A bounded mountain pass lemma without the (PS) condition and applications. *Trans. Am. Math. Soc.*, **331** (1992)

We will see in this chapter the so-called bounded MPT of Schechter and the mountain impasse theorem of Tintarev. They correspond to the situation where the functional does not necessarily satisfy the Palais-Smale condition but still has the geometry of the MPT. The peculiarity of these two results is that they both require the continuous paths appearing in the minimaxing procedure of the MPT to be within a level set of some auxiliary function. This adds enough compactness to give some interesting results.

Consider the general situation of a functional Φ having the geometry of the MPT but no critical point of level $c = \inf_{\gamma \in \Gamma} \sup_{t \in [0,1]} \Phi(\gamma(t))$. This happens, for example, when (PS) fails to hold as we saw in Examples 10.1 and 10.2 by Brézis and Nirenberg. We know that, by either Ekeland's principle or the quantitative deformation lemma, there exists a (PS) sequence tending to c. But as the approximation paths go farther and farther from the origin, the sole mountain pass geometry does not suffice in general to provide critical points.

Since then, (PS) is not satisfied in many concrete problems. Many attempts were made to give new results where they are not needed. In this chapter, we will see situations where the (PS) condition is not required, but some constraints on the paths in the minimaxing argument are used instead. The idea is the following. Since the role of the (PS) condition is to allow us to deal with unbounded regions in a uniform way (for a detailed discussion, see Chapter 13), a possible different approach, as done in [114, 189], for example, is to require (PS) on bounded regions and to control the growth of $||\Phi'(u_n)||^{-1}$ near infinity. This permits us, as with (PS), to deal with unbounded regions in a uniform way. Schechter and Tintarev [800, 817, 917, 918, 920] propose another approach. They restrict competing paths to a bounded region.

We will describe in this chapter the bounded MPT without the (PS) condition [800] by Schechter and the mountain impasse theorems of Tintarev.

21.1 A Bounded MPT without the (PS) Condition

To assure that the competing paths will not leave the bounded region they are restricted to, as they approach the optimal one, Schechter [800] imposes a boundary condition. He first begins by generalizing the notion of pseudo-gradient.

Lemma 21.1. *Let H be a Hilbert space, and let $\Phi(u)$ be a continuous mapping of H into itself such that $\Phi(u) \neq 0$ for all u. Let $v_i(u)$ be continuous mappings such that $v_i(u)$ does not vanish on a closed set Q_i. Assume that*

$$(v_i(u), v_j(u)) = 0, \qquad i \neq j,$$

and that there are numbers $\Theta_i \geq 0$ such that

$$\Theta^2 = \sum_i \Theta_i^2 < 1,$$

and

$$(\Phi(u), v_i(u)) \leq \Theta_i \|\Phi(u)\|\, \|v_i(u)\|, \qquad u \in Q_i.$$

If $\alpha < 1 - \Theta$, then there is a locally Lipschitz map $\Psi(u)$ such that $\|\Psi(u)\| \leq 1$ and

$$(\Phi(u), \Psi(u)) \geq \alpha \|\Phi(u)\|, \qquad u \in H,$$

$$(\Psi(u), v_i(u)) < 0, \qquad u \in Q_i.$$

The "bounded" version of the MPT, derived by Schechter to study some semilinear elliptic problems, is the following.

Theorem 21.2 (Bounded Mountain Pass Theorem, Schechter). *Let $\Phi(u) \in C^1(H; \mathbb{R})$ satisfying the following form of the Palais-Smale condition:*

$$\begin{array}{ll} u_k \rightharpoonup u \text{ weakly}, & |\Phi(u_k)| \leq C, \\ \Phi'(u_k) \to 0 & \text{imply that } \Phi'(u) = 0. \end{array}$$

Assume that $H = N \oplus M$, where N, M are orthogonal subspaces with $\dim N < \infty$. Assume that there are constants $R \geq R_0 > 0$ such that

$$c_0 = \max_{\partial B_0} \Phi < c_1 = \inf_B \Phi \leq c_2 = \max_{B_0} \Phi,$$

where

$$B_0 = \{v \in N; \|v\| \leq R_0\}, \qquad B = \{w \in M; \|w\| \leq R\},$$

and that for each c satisfying $c_1 \leq c \leq c_2$ there are constants $\sigma > 0$, $\Theta < 1$ such that

$$(\Phi'(u), u) \geq -\Theta R \|\Phi'(u)\|$$

holds for all u satisfying

$$c - \sigma \le \Phi(u) \le c + \sigma, \qquad ||u|| = R.$$

Then there exists $u \in H$ satisfying $\Phi'(u) = 0$.

Notice that the critical value obtained is not characterized by a minimax argument. Indeed, in the proof, setting $Q = B(0, R)$, Schechter supposes by contradiction that Φ admits no critical point in Q. Then, for $S = \{\gamma : B_0 \to Q \text{ continuous}; \gamma|_{\partial B_0} = I_d\}$, he proves that $\gamma(B_0) \cap B \ne \varnothing$, for all $\gamma \in S$ and that $c_1 \le c \le c_2$, where

$$c = \inf_{\gamma \in S} \sup_{v \in B_0} \Phi(\gamma(v)).$$

He uses Theorem 21.1 and gets a contradiction with the definition of c.

We pass now to the *mountain impasse theorem* of Tintarev. He continued alone the adventure begun with Schechter, in the same spirit, and obtained nice results, which we will have the opportunity to see now.

21.2 Mountain Impasse Theorem

The situation described earlier (the mountain pass geometry without supposing (PS)) is what Tintarev calls the *mountain impasse theorem* (MIT). He studied this situation in a series of papers [917,918,920] and with Schechter [817]. We will focus our attention on the result contained in [920], where he generalizes the results of [817,917,918] to a general Banach space setting and nonsmooth functionals while the paths are allowed to be *noncompact* but have to lie in some closed set.

The nonsmooth form of this result allows one to deal with problems with discontinuous nonlinearities. The proofs are based on the nonsmooth techniques (cf. [243]), like those used in Chapter 15, and where a critical point ($\Phi'(u) = 0$) is such that $0 \in \partial\Phi(u)$, the subgradient corresponding to u. As a consequence, the proofs are shorter than in the previous papers.

Consider a reflexive Banach space E and two real-valued locally Lipschitz functionals f and Φ defined on E such that $\sup_{u \in E} f(u)$. Let K be a Haussdorff space, K^* a closed subset of K such that $K \ne \varnothing$ and $K^* \ne K$, and let $\gamma^* \in C(K; E)$.

The *constrained mountain pass geometry* for Φ will refer to the following conditions:

$$c_0 = \sup_{\zeta \in K^*} \Phi(\gamma^*(\zeta)) < \inf_{\gamma \in \Gamma_t} \sup_{\zeta \in K} \Phi(\gamma(\zeta)) = c(t), \tag{21.1}$$

where

$$\Gamma_t = \{\gamma \in C(K; f^t); \gamma = \gamma^* \text{ on } K^*\},$$

which is assumed to be nonempty for t greater than some $t_0 \in \mathbb{R}$.

Consider also the set of *unconstrained paths* used in the classical form of the MPT,

$$\Gamma_\infty = \{\gamma \in C(K; E); \gamma = \gamma^* \text{ on } K^*\},$$

and the corresponding *relaxed* inf sup *value* $c(\infty)$.

When (21.1) is satisfied for $t = \infty$ (and K is compact), we would be in the context of the nonsmooth MPT variant of Shi, seen in Chapter 15, where we obtained a sequence $(x_n)_n \subset E$ such that (see Chapter 15 for the details)

$$\Phi(x_n) \to c(\infty) \qquad \text{and} \qquad \inf_{x^* \in \partial \Phi(x_n)} \|x^*\| \to 0. \tag{21.2}$$

Consider a Banach space E. We will need the following technical results in the sequel.

Lemma 21.3. *Let $C \subset E^*$ be a closed convex set such that*

$$\text{dist}(C, 0) > 0.$$

Then there exists $x_0 \in E$ such that $\|x_0\| = 1$ and

$$\inf_{\xi \in C} \langle x_0, \xi \rangle = \text{dist}(C, 0). \tag{21.3}$$

It suffices to consider a *supporting* cotangent vector to C at the closest point of C to the origin. Recall that a point $x_0 \in C$ is a *supporting point*, where C is a nonempty closed convex subset of X, if there exists a bounded linear functional $f \in X^*$ such that

$$f(x_0) = \sup_{x \in C} f(x).$$

Lemma 21.4. *Let $u \mapsto C_u$ be a mapping from a closed subset D of E into closed convex subsets of E^*. Suppose, moreover, that it is uniformly bounded on bounded subsets of D and satisfies the following continuity property:*

$$\begin{cases} (u_n)_n \subset D, \ (\xi_n)_n \subset C_{u_n} \\ u_n \to u, \qquad \xi_n \rightharpoonup \xi \end{cases} \qquad \text{implies that} \quad \xi \in C_u. \tag{21.4}$$

Assume that there exists a $\delta_0 > 0$ such that

$$\text{dist}(C_u, 0) \geq \delta_0 \qquad \text{for } u \in D. \tag{21.5}$$

Then, for every $\delta < \delta_0$, there exists a locally Lipschitz continuous mapping $Z(u)$ of D into E such that

$$\|Z(u)\| \leq 1, \qquad u \in D, \tag{21.6}$$

$$\langle Z(u), \xi \rangle \geq \delta, \qquad \xi \in C_u, \ u \in D. \tag{21.7}$$

Compare to Lemma 15.12 by Choulli et al. in the Notes of the nonsmooth MPT.

Proof. By Lemma 21.3, for every $u \in D$ there exist $x(u) \in E$ such that $\langle x(u), \xi \rangle \geq \delta_0$ for all $\xi \in C_u$. Then, there exists a neighborhood $N(u)$ such that

$$\langle x(u), \xi \rangle \geq \delta, \qquad \text{for all } \xi \in C_u, \ v \in N(u).$$

Otherwise, there would exist a sequence $(v_n)_n$ tending to u and a sequence $(\xi_n)_n \in C_{v_n}$ such that $\langle x(u), \xi_n \rangle < \delta$. Consider a weakly convergent subsequence, still denoted

$(\xi_n)_n$; we have by (21.4) that

$$\xi_0 = w - \lim_n \xi_n \in C_u$$

and also

$$\langle x(u), \xi_n \rangle < \delta_0,$$

a contradiction.   □

Denote by

$$\text{conv}\,(X, Y) = \{\alpha x + (1 - \alpha)y;\ x \in X, y \in Y \text{ and } \alpha \in [0, 1]\}$$

the union of all the intervals $[a, b]$ whose endpoints are such that $a \in X$ and $b \in Y$. This is not the convex hull of $X \cup Y$, which we can check easily in simple examples.

The main result in this section is the following.

Theorem 21.3 (Mountain Impasse Theorem, Tintarev). *Suppose* (21.1)*. Then the following alternative holds.*

  a. *Either there exists a sequence $(u_n)_n \subset f^t$ such that*

$$\Phi(u_n) \to c(t) \qquad and \qquad \text{dist}\,(\partial\Phi(u_n), 0) \to 0, \qquad (21.8)$$

  b. *or there is a sequence $(u_n)_n \subset f^t$ such that*

$$\Phi(u_n) \to c(t) \qquad and \qquad \text{dist}\,(\text{conv}\,(\partial\Phi(u_n), \partial f(u_n)), 0) \to 0. \quad (21.9)$$

Proof. Assume that neither assumption holds. Then there would exist $\varepsilon > 0$ and $\delta_1 > 0$ such that

$$\begin{aligned}
|\Phi(u) - c(t)| \le 3\varepsilon, \text{ and } u \in \partial f^t & \quad \text{implies that} \quad & \text{dist}\,(\text{conv}\,(\partial\Phi(u), \partial f(u)), 0) \ge \delta_1, \\
|\Phi(u) - c(t)| \le 3\varepsilon, \text{ and } u \in f^t & \quad \text{implies that} \quad & \text{dist}\,(\partial\Phi(u), 0) \ge \delta_1.
\end{aligned}$$
$$(21.10)$$

Set $Q = f^t \cap \Phi_{c(t)-2\varepsilon}^{c(t)+2\varepsilon}$, $Q_1 = f^t \cap \Phi_{c(t)-\varepsilon}^{c(t)+\varepsilon}$, $Q_2 = f^t \setminus Q$ and let

$$\eta(u) = \frac{\text{dist}\,(u, Q_2)}{\text{dist}\,(u, Q_1) + \text{dist}\,(u, Q_2)}.$$

Then, $\eta(u)$ is a Lipschitz continuous functional on E that vanishes on $\overline{Q_2}$ and is equal to 1 on Q_1. Take $\delta_0 < \delta_1$ and let $\alpha(u) \in C(Q; \mathbb{R})$ be such that $0 \le \alpha(u) \le 1$ and

$$\begin{cases} \alpha(u) = 1 & \text{for } u \text{ in } f^t, \\ \text{dist}\,(\text{conv}\,(\partial\Phi(u), \partial f(u)), 0) \ge \delta_0 & \text{in the support of } \alpha(u). \end{cases}$$

Let $D = Q$ and $C_u = \text{conv}\,(\partial\Phi(u), \alpha(u)\partial f(u))$. We know that subgradients of Lipschitz functions satisfy the continuity property (21.4) (see [243], for example). And since $\alpha(u)$ is continuous, C_u also satisfies (21.4). The relation (21.5) is also satisfied because of (21.10), the definitions of Q and Q_1, and since dist (conv $(\partial\Phi(u), \theta$ ·

$\partial f(u)), 0) \geq \text{dist}(\text{conv}(\partial \Phi(u), f(u)), 0)$ for $\theta \in [0, 1]$. Thus for $\delta < \delta_0$, by Lemma 21.4, there exists a locally Lipschitz mapping $Z(u)\colon Q \to E$ such that

$$\|Z(u)\| \leq 1, \quad u \in Q, \tag{21.11}$$

$$\langle Z(u), \xi \rangle \geq \delta, \quad \xi \in C_u, \ u \in Q. \tag{21.12}$$

Now set $W(u) = -\eta(u)Z(u)$. It is locally Lipschitz continuous and bounded on the whole of f^t because $\Phi(u)$ vanishes outside Q, which is a closed subset of f^t. We can solve

$$\begin{cases} \dfrac{d\sigma(t)}{dt} = W(\sigma(t)) \\ \sigma(0) = u \end{cases} \tag{21.13}$$

uniquely in $[0, \infty[$ for each $u \in f^t$ provided $\sigma(t)$ does not exit $f(t)$. Indeed, the solution $\sigma(t, u)$ of (21.13) *does not exit* f^t for $t \geq 0$.

Since $\|W(u)\| \leq 1$, we have that

$$\|\sigma(t, u) - u\| \leq t, \quad t \geq 0. \tag{21.14}$$

Moreover,

$$\begin{aligned} \sup\{\partial_t \Phi(\sigma(t, u))\} &\leq \sup_{\xi \in \partial \Phi(\sigma(t,u))} \langle W(\sigma(t, u)), \xi \rangle \\ &\leq \sup_{\xi \in C_{\sigma(t,u)}} -\eta(\sigma(t, u))\langle Z(\sigma(t, u)), \xi \rangle \\ &\leq -\delta\eta(\sigma(t, u)) \leq 0. \end{aligned} \tag{21.15}$$

By the definition of $c(t)$, there is a $\gamma \in \Gamma_t$ such that

$$\Phi(\gamma(\zeta)) < c(t) + \varepsilon, \quad \zeta \in K. \tag{21.16}$$

If for some $\zeta \in K$ and $t_1 \geq 0$ and since (21.15) and (21.16) exclude $\Phi(\sigma(t_1, \gamma(\zeta))) > c(t) + \varepsilon$, $\sigma(t_1, \gamma(\zeta)) \notin Q_1$, then we must have

$$\Phi(\sigma(t_1, \gamma(\zeta))) < c(t) - \varepsilon. \tag{21.17}$$

Hence,

$$\Phi(\sigma(t, \gamma(\xi))) < c(t) - \varepsilon \quad \text{for all } t > t_1. \tag{21.18}$$

On the other hand, by (21.15),

$$\Phi(\sigma(t_1, \gamma(\zeta))) - g(\gamma(\zeta)) \leq -\delta t_1 < -2\varepsilon, \tag{21.19}$$

which implies that $t_1 > \delta/2\varepsilon$. Consequently,

$$\Phi(\sigma(\delta/\varepsilon, \gamma(\zeta))) < c(t) - \varepsilon \quad \text{for all } \zeta \in K. \tag{21.20}$$

Moreover, η vanishes in the neighborhood of $\gamma^*(K^*)$. Hence, $\sigma(\delta/\varepsilon, \gamma(\xi)) \in \Gamma_t$, which together with (21.3) contradicts the inequality (21.1). $\square$

If we require an additional convergence condition that is weaker than the Palais-Smale condition and turns out to be a *verifiable* absolute continuity condition in applications, we get the following result.

Theorem 21.4. *Assume* (21.1) *and let* $(u_n)_n \in E$ *a sequence such that either*

$$\Phi(u_n) \to c(t), \quad f(u_n) \le t, \quad and \quad \operatorname{dist}(\partial\Phi(u_n), 0) \to 0, \tag{21.21}$$

or

$$\Phi(u_n) \to c(t), \quad f(u_n) \in \partial f^t, \quad and \quad \operatorname{dist}(\operatorname{conv}(\partial f(u_n), \partial\Phi(u_n)) \to 0 \tag{21.22}$$

have a convergent subsequence. Then the following alternative holds.
Either there exists $u \in f^t$ *such that*

$$\Phi(u) = c(t), \quad 0 \in \partial\Phi(u), \tag{21.23}$$

or there exists $u \in \partial f^t$, $\xi \in \partial\Phi(u)$, $\eta \in \partial f(u)$, *and* $\theta \in [0, 1]$ *such that*

$$\Phi(u) = c(t), \quad \theta\xi + (1 - \theta)\eta = 0. \tag{21.24}$$

Proof. By the MIT, we can find $\theta_j \in [0, 1]$, $\xi_i \in \partial f(u_j)$, and $\eta_j \in \partial\Phi(u_j)$ such that $\theta_j\xi_j + (1 - \theta_j)\eta_j \to 0$ in E^*. Since $\cup_j \partial f(u_j)$ is a bounded set (and the same for $\partial\Phi$), we may consider renamed convergent subsequences $\xi_j \rightharpoonup \xi$, $\eta_j \rightharpoonup \eta$, $\theta_j \to \theta$. Moreover, by a closedness property of the pseudo-gradient, $\xi \in \partial f(u)$, $\eta \in \partial\Phi(u)$ where u is the limit of u_j. So, (21.24) holds. The proof of (21.23) is similar. □

Comments and Additional Notes

◇ 21.1 Mountain Impasse Theorem for Strictly Differentiable Functionals

A functional $f: X \to Y$ is strictly differentiable at x, with strict derivative $D_s f(x)$, if f is Lipschitz near x and for any $v \in X$

$$\lim_{\substack{x' \to x \\ t \downarrow 0}} \frac{f(x' + tv) - f(x')}{t} = \langle D_s f(x), v \rangle.$$

This notion is weaker that $f \in C^1$. When f is Lipschitz continuous and ∂f is a singleton, then f is *strictly differentiable*.

In the situation when both ∂f and $\partial\Phi$ are singletons, we get the following more usual form of the MPT where critical points with Lagrange multipliers appear.

Corollary 21.7 (Mountain Impasse Theorem for Strictly Differentiable Functionals, Tintarev). *Assume* (21.1) *and let* f, Φ *be Lipschitz continuous and strictly differentiable. Assume that any sequence* $(u_n)_n \subset E$ *such that either*

$$\Phi(u_n) \to c(t), \quad f(u_n) \le t, \quad and \quad \Phi(u_n) \to 0, \tag{21.25}$$

or

$$\Phi(u_n) \to c(t), \quad (u_n) \in \partial f^t, \quad and \quad (1 - \theta_j)f'(u_n + \theta_j\Phi'(u_n) \to 0, \tag{21.26}$$

with some $\theta_j \in [0, 1]$, *has a convergent subsequence. Then the following alternative holds.*

Either there exists $u \in E$ *such that*

$$\Phi(u) = c(t), \qquad \Phi'(u) = 0, \qquad f(u) \le t, \tag{21.27}$$

or there exists $u \in E$ *and* $\theta \in [0, 1]$ *such that*

$$\Phi(u) = c(t), \qquad f(u) = t, \qquad \theta\Phi'(u) + (1 - \theta)f'(u) = 0. \tag{21.28}$$

The conditions (21.25) (21.26) are weaker than the (PS) condition. In applications, these are absolute continuity conditions and not (PS).

In a version where better versions of (21.24) and (21.28) are used, only a variation of constants in (21.21) and (21.22) is allowed. As a result, *critical points at infinity* are found.

Theorem 21.8. *Assume* (21.1) *for all t large enough. Assume that for every* $c_1, c_2 \in \mathbb{R}$ *and every sequence* $(u_n)_n \in E$ *that either*

$$\Phi(u_n) \to c_2, \quad f(u_n) \le c_1, \quad and \quad \mathrm{dist}\,(\partial\Phi(u_n), 0) \to 0, \tag{21.29}$$

or

$$\Phi(u_n) \to c_2, \quad f(u_n) \to c_1, \quad and \quad \mathrm{dist}\,(\mathrm{conv}\,(\partial f(u_n), \partial\Phi(u_n)) \to 0 \tag{21.30}$$

have a convergent subsequence. Then, for every $\varepsilon > 0$, *the following alternative holds. Either there exists* u *such that*

$$c(\infty) \le \Phi(u) \le c(\infty) + \varepsilon, \quad 0 \in \partial\Phi(u), \tag{21.31}$$

or for every $\lambda > 0$, *there exists* u_λ, $\xi_\lambda \in \partial\Phi(u_\lambda)$, *and* $\mu(\lambda) \in [0, \lambda]$ *such that*

$$\xi_\lambda + \mu_\lambda(\lambda)\eta_\lambda = 0 \tag{21.32}$$

and

$$\Phi(u_\lambda) \to c(\infty), \qquad f(u_\lambda) \to \infty \qquad as\ \lambda \to 0. \tag{21.33}$$

In particular, if f *and* Φ *are strictly differentiable, then* (21.32) *means that*

$$\Phi'(u_\lambda) + \mu(\lambda)f'(u) = 0, \qquad \mu(\lambda) \to 0 \qquad as\ \lambda \to 0. \tag{21.34}$$

The interested reader is encouraged to get a look at the series of papers by Schechter and Tintarev [800, 817, 917, 918, 920].

VI

Technical Climbs

22

Numerical MPT Implementations

> The particular form obtained by applying an analytical integration method may prove to be insuitable for practical purposes. For instance, evaluating the formula may be numerically instable (due to cancellation, for instance) or even impossible (due to division by zero).
>
> A.R. Krommer and C.W. Ueberhuber, *Computational integration*, 1998.

This chapter is devoted to some numerical implementations of the MPT. We first present the "mountain pass algorithm" by Choi and McKenna. Then we describe a partially interactive algorithm, also based on the MPT, by Korman for computing unstable solutions. A third algorithm used in quantum chemistry by Liotard and Penot [584] is described in the final notes.

The fact that the solutions obtained by the MPT are *unstable* makes them very interesting candidates for solving particular problems whose solutions are unstable in nature. But this instability, in addition to the *global* character of the inf max characterization (we "inf max" on functional spaces that are infinite dimensional), makes them hard to capture numerically. Moreover, the available discretization algorithms are not very useful with domains whose boundaries present some curvature. This may explain, in our opinion, the delayed appearance of a numerical algorithm for computing the solutions obtained by the MPT.

We will begin with the mountain pass algorithm of Choi and McKenna, an algorithm that is beginning to become known by nonlinear analysts and seems to be destined for a *bright* future. It combines the finite element method with the steepest descent method to reproduce the MPT in a constructive way.

22.1 The Numerical Mountain Pass Algorithm

Consider the semilinear elliptic problem

$$(\mathcal{P}) \quad \begin{cases} -\Delta u = f(x, u) & \text{in } \Omega, \\ u = 0 & \text{on } \partial\Omega, \end{cases}$$

259

where Ω is a smooth bounded domain of $\mathbb{R}^N$. Suppose that the nonlinearity satisfies the following conditions:

 i. $f(x, s)$ is locally α-Holder continuous in $\overline{\Omega} \times \mathbb{R}$ with $0 < \alpha < 1$.
 ii. There are constants $a_1, a_2 \geq 0$ such that if $N > 2$,

$$|f(x, s)| \leq a_1 + a_2 |s|^p,$$

where $0 \leq p < (N + 2)/(N - 2)$, and if $N = 2$,

$$|f(x, s)| \leq a_1 \exp(\varphi(s)),$$

where $\varphi(s)/s^2 \to 0$ as $|s| \to \infty$. If $N = 1$, condition ii is not needed. Choi and McKenna proved [228] that a careful implementation of the ideas of the MPT lead to the construction of a *robust* and *globally convergent* algorithm, which always converges to a solution with the *mountain pass property*. The general scheme of the algorithm is the following.

1. A piecewise linear path joining the local minimum w to some point e whose altitude $\Phi(e)$ is lower than that of the minimum is considered in a finite dimensional approximating subspace.
2. The maximum of Φ along that path is calculated.
3. The path is deformed by pushing the point at which the local maximum is located in the direction of *steepest descent*.
4. Repeat the second step, stopping only when a critical point is reached. It is identified by the fact that lowering the local maximum is no longer possible.

This algorithm has been used since then with many boundary value problems. The solutions of different types of symmetry can successfully be computed by restricting the initial guess in particular symmetry subspaces.

22.1.1 Steepest Descent Direction

The *steepest descent method* seems to be the most appropriate method for finding the local minima of a given functional without requiring a good *initial guess*. The way it is implemented in the mountain pass algorithm is the following. Recall that the *steepest descent direction* at a point $u \in E$ is the direction at which Φ decreases most rapidly "per unit distance" in E.

> When different norm structures are assigned to E, the distance and hence the steepest descent direction are altered. And even for two different equivalent norms on E, the steepest descent directions may be different.

Choi and McKenna chose the H^1-norm in [228], which seems natural when treating semilinear elliptic equations with Dirichlet data on the boundary of a bounded domain Ω. This led, for that specific norm, to the fact that the steepest descent direction points toward the unique minimum point if the equation is linear.

Computing the steepest descent direction of Φ at $u \in H_0^1(\Omega)$ corresponds to finding $v \in H_0^1(\Omega)$ such that $\|v\|_{H^1} = 1$, and

$$(\Phi(u + \varepsilon v) - \Phi(u))/\varepsilon$$

is as negative as possible when $\varepsilon \to 0$. This is equivalent to finding the minimum of the Fréchet derivative of Φ at u on v, that is, minimizing $\langle \Phi'(u), v \rangle$ subject to the the constraint $\|v\|_{H^1} = 1$.

This constrained problem is equivalent to the unconstrained one of minimizing the functional $\Psi \colon H_0^1(\Omega) \to \mathbb{R}$ defined by

$$\Psi(v) = \int_\Omega \left[\nabla u . \nabla v - f(x, u)v + \lambda |\nabla v| \right] dx,$$

where a Lagrange multiplier $\lambda \neq 0$ appears.

Indeed, for any $w \in H_0^1$,

$$\langle \Psi'(v), w \rangle = \int_\Omega \left[\nabla u . \nabla w - f(x, u)w + 2\lambda \nabla v \nabla w \right] dx;$$

a critical point of Ψ corresponds to a weak solution of the linear elliptic equation with the right-hand side in $H^{-1}(\Omega)$,

$$2\lambda \Delta v = -\nabla u - f(x, u). \tag{22.1}$$

If u is not a solution of $(\mathcal{P})$, then a nontrivial solution $2\lambda v$ exists and, hence, so does the minimum of Ψ. This will define the steepest descent once the value of λ is known. So solving the linear problem (22.1) gives us $2\lambda v$, then $|\lambda|$ is determined by the fact that $\|v\|_{H^1} = 1$. It's easy to see that the sign of λ is positive by considering a weak solution v of (22.1) and computing

$$\frac{\Phi(u + \varepsilon v) - \Phi(u)}{\varepsilon} = \int_\Omega [\nabla u . \nabla v - f(x, u)v] \, dx + O(\varepsilon)$$
$$= \int_\Omega -2\lambda \nabla v . \nabla v \, dx + O(\varepsilon)$$
$$= -2\lambda + O(\varepsilon).$$

So, as $\varepsilon \to 0$, -2λ dominates, and λ is chosen to be positive so that v represents the steepest descent direction.

But Practically, How Is It Implemented?

We seek a numerical approximation of the weak solution v to (22.1) with u known at the mesh points using the finite element method. Setting $\bar{v} = 2\lambda v + u$, the steepest descent method can be computed once we have found a weak solution to the equation

$$-\Delta \bar{v} = f(x, u). \tag{22.2}$$

This is equivalent to minimizing on $H_0^1(\Omega)$

$$\Phi(\bar{v}) = \int_\Omega \left[\frac{1}{2} |\nabla \bar{v}|^2 - f(x, u)\bar{v} \right] dx.$$

Choi and McKenna restricted themselves, for the purpose of numerical computations, to semilinear elliptic equations on the unit square and used triangular elements on a uniform mesh with size $h = 1/(M+1)$. Given such a triangulation, piecewise linear basis functions φ_j are used [864], and φ_j is 0 at all mesh points except at x_j, where it is 1. With such a finite dimensional subspace, for any approximation $\bar{v}^h = \sum_j q_j \varphi_j$ of v,

$$\Phi(\bar{v}^h) = \frac{1}{2} q^T K q - \int_\Omega f(x, u) \sum_j q_j \varphi_j \, dx,$$

where K is the $M^2 \times M^2$ *stiffness matrix*, given by

$$K = \begin{pmatrix} B & -I \\ -I & B & -I \\ & & \ddots \\ & & -I & B & -I \\ & & & -I & B \end{pmatrix}. \tag{22.3}$$

Here I is the $M \times M$ identity matrix and B is the $M \times M$ matrix,

$$B = \begin{pmatrix} 4 & -1 \\ -1 & 4 & -1 \\ & & \ddots \\ & & -1 & 4 & -1 \\ & & & -1 & 4 \end{pmatrix}. \tag{22.4}$$

Approximating the integral in the expression of $\Phi(\bar{v}^h)$ gives

$$\Phi(\bar{v}^h) = \frac{1}{2} q^T K q - q^T \Theta(x, u) h^2, \tag{22.5}$$

where $\Theta = (f(x_1, u(x_1)), \ldots, f(x_{M^2}, u(x_{M^2})))^T$. The coefficients q and, hence, $\bar{v}^h$ are calculated by solving the system of linear algebraic equations

$$Kq = h^2 \Theta(x, u). \tag{22.6}$$

From the theory of finite element approximation, it is known that $\bar{v}^h = \sum_j q_j \varphi_j \to \bar{v}$ with an error of $O(h^2)$ in H^1-norm and that the error due to the numerical quadrature does not affect the order of convergence. The H^1-norm of $\bar{v}$, which is approximated by $\bar{v}^h$, is then computed by

$$\|\bar{v}^h\|_{H^1} = \left(\int_\Omega \nabla \bar{v}^h . \nabla \bar{v}^h \right)^{1/2} (q^T K q)^{1/2}. \tag{22.7}$$

The distance between two points of H^1 are computed similarly.

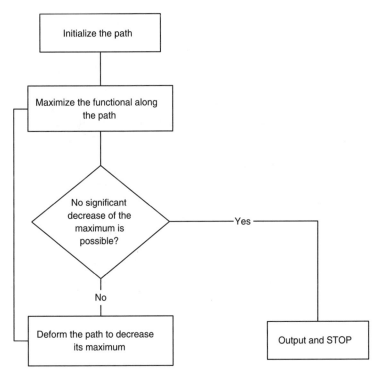

Figure 22.1. The mountain pass algorithm flow chart.

22.1.2 The Numerical Algorithm

We will now describe the algorithm in detail. As said in the beginning, there are mainly four main points in the algorithm (the flow chart is in Figure 22.1):

1. A path from the local minimum w to the point of lower altitude e is initialized.
2. The maximum of Φ along this path is found.
3. The path is deformed in such a way to make the maximum along the path decrease as fast as possible (in the steepest descent direction).
4. A decision whether to stop or not is made depending on whether or not the maximum was actually a critical point and no significant decrease was possible.

Step 1. Initializing the Path. Fix a number N to be used as the number of segments in the initial (piecewise linear) path Γ_0. Each segment is uniquely identified by its endpoints. The straight line from w to e could be a possible starting path. However, the endpoint e is not necessarily optimal. So, the algorithm begins with a little improvement of the path by choosing

$$\gamma(i) = w + \frac{i}{N}(e - w), \qquad 1 \le i \le N$$

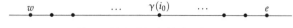

Figure 22.2. The initial path Γ_0.

and then continues evaluating $\Phi(\gamma(i))$ for $i \geq 1$ until we find $i = i_0$ such that $\Phi(\gamma(i_0)) \leq \Phi(w)$. We then replace the original e by $\gamma(i_0)$, thereby shortening the distance between w and e but retaining the geometry of the MPT. The initial path Γ_0 is then chosen to be the straight line between w and the new e, and the corresponding $\gamma(i)$ to be the equally spaced points along this path (see Figure 22.2).

Step 2. Maximizing Φ along the Path. Given the discretized path $\gamma(i)$, the value of $i = i_m$ with the maximum of $(\Phi(\gamma(i)))_i$ is computed. Note that since w is a local minimum and $\Phi(w) \leq \Phi(e)$, if N is large enough then $1 \leq i_m \leq N - 1$. The next thing to try is then to locate a higher maximum between $i = i_m$ and $i = i_{m+1}$. Consider first the segment joining $\gamma(i_m)$ and $\gamma(i_{m+1})$ and seek an $\alpha \geq 0$ such that

$$\Phi(\gamma(i_m) + \alpha.(\gamma(i_{m+1}) - \gamma(i_m)))$$

stops increasing. Proceed by dichotomy (dividing the interval by 2 each time) and test

$$\gamma_1 := \tfrac{1}{2}(\gamma(i_m) + \gamma(i_{m+1})),$$
$$\gamma_2 := \tfrac{1}{2}(\gamma(i_m) + \gamma(i_1)),$$
$$\gamma_3 := \tfrac{1}{2}(\gamma(i_m) + \gamma(i_2)), \dots$$

until you find

a. γ_i such that $\Phi(\gamma_i) > \Phi(\gamma(i_m))$, or the norm
b. $\|\gamma_i - \gamma(i_m)\|_{H^1} = \frac{1}{2^i}\|\gamma(i_{m+1}) - \gamma(i_m)\|_{H^1}$ is smaller than a prescribed value.

In case a, the value $\gamma_{\text{mid}} = \tfrac{1}{2}(\gamma(i_m) + \gamma_i)$ is computed and a quadratic polynomial interpolation is performed to find a better value for the maximum of Φ, which will be assigned to $\gamma(i_m)$.

In case b, $\gamma(i_m)$ is not changed. Then, consider the segment joining $\gamma(i_{m-1})$ and $\gamma(i_m)$ and perform a similar task to what has just been done for the segment joining $\gamma(i_m)$ and $\gamma(i_{m+1})$.

Then refine the path in the neighborhood of the local maximum by moving some of the nearby closer nodes. If $\|\gamma(i_m) - \gamma(i_{m-1})\|$ is larger than a prescribed value, a number of mesh points (5 in [228]) are moved around this local maximum to avoid the "missing" of the saddle point and still keep the total number of points N in the discretized path. So, in [228] the old $\gamma(i_{m+1})$ becomes $\gamma(i_{m+5})$ and new $\gamma(i_{m+1}), \dots, \gamma(i_{m+4})$ are added, equally spaced, on the line between $\gamma(i_m)$ and the old $\gamma(i_{m+1})$ (the new $\gamma(i_{m+5})$). The same task is also performed for $\gamma(i_m)$ and $\gamma(i_{m-5})$ (see Figure 22.3). If m is too close to 0 or N, this adjustment is not performed.

Step 3. Deforming the Path. The steepest descent direction at the local maximum is computed by solving (22.6). Thus, a numerical approximation of $2\lambda v = \bar{v} - u$ is computed, and the local maximum is moved "downhill" in the direction of the steepest

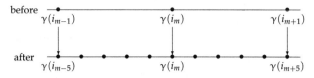

Figure 22.3. Refining the path to avoid missing the saddle point.

descent for a certain distance determined by a procedure similar to Step 2. If this distance or the difference in Φ before and after the descent is smaller than a prescribed value, then we have a numerical approximation of a critical point and stop the algorithm. If not, we repeat Steps 2 and 3.

22.2 A Partially Interactive Algorithm

It is well known that a nonlinear problem

$$\begin{cases} -u''(x) = f(x, u(x)) & \text{in }]a, b[\\ u(a) = u(b) = 0 \end{cases} \tag{22.8}$$

has a *subsolution* $\underline{u}$ and a *supersolution* $\overline{u}$, which are ordered. Then (22.8) possesses at least one solution $u(x)$. Moreover, two approximating sequences converging, respectively, to minimal and maximal solutions of (22.8) can be constructed (see [33]). However, this approach works only for stable solutions that are local extrema. The question of solvability of (22.8) and computation of the corresponding solutions need to be investigated when the subsolution $\underline{u}$ lies above the supersolution $\overline{u}$ when

$$\underline{u}(x) > \overline{u}(x) \text{ for all } x \in]a, b[.$$

In the general case, the answer is known to be negative, as shown by Kazdan and Warner (see the discussion in [33, p. 653]). Consider the following *Emden-Fowler equation* that occurs in nuclear physics, gas dynamics, and so on (cf. [599]):

$$\begin{cases} -u'' = a(x)u^p & \text{in }]a, b[, \ p > 0 \\ u(a) = u(b) = 0 \end{cases} \tag{22.9}$$

Its positive solutions are necessarily *unstable* in the sense that if $u(x) > 0$ is a solution of (22.9), then it follows immediately that $\alpha u(x)$ is a subsolution of (22.9) for any constant $\alpha > 1$, and it is a supersolution if $0 < \alpha < 1$. Unstable solutions are characterized as unstable steady states for the corresponding parabolic equation

$$\begin{cases} u_t - u_{xx} = f(x, u) & \text{for } a < x < b, \ t > 0 \\ u(a, t) = u(b, t) = 0. \end{cases} \tag{22.10}$$

Stability for the Dirichlet problem (22.8) means that the solutions of the corresponding evolution equations (22.10) tend to $u(x)$. It is well known that the numerical solutions of (22.10) with initial data being subsolution (supersolution) of (22.8) will tend increasing (decreasing) to the steady state, which is a solution of (22.8). This fact is used in [488] to develop some schemes with monotone convergence.

Computation of unstable solutions is not generally easy. For example, the *monotone iteration method* (see the notes following) *cannot possibly work* (cf. [791], for example). The typical variational result to look for such unstable solutions, as you must guess, again and again, is the MPT.

In [526], Korman proposes a "partially interactive algorithm" where the unstable solutions of (22.8) are the saddle points of the associated energy Φ on $H_0^1(]a, b[)$ to be found using the MPT.

So, assume that

i. $f(x, 0) = 0$ for all $a < x < b$, so that $\Phi(0) > 0$ and 0 is also a trivial solution of (22.8).

ii. $\displaystyle\lim_{u \to 0} \frac{F(x, u)}{u^2} = 0$ uniformly in $]a, b[$.

Poincaré inequality implies that $\Phi > 0$ in a small neighborhood of 0. If we suppose, moreover, that there is a point $e \in H_0^1(]a, b[)$ such that $\Phi(e) < 0$ and that Φ satisfies the (PS) condition, the MPT yields the existence of a critical point.

So, problem (22.8) will be considered under conditions i and ii and $e(x)$ such that $\Phi(e) < 0$ will be supposed to be known. In the examples given for illustration in [526], $e(x)$ has the form $Ax(1 - x)$ for a positive constant A.

Dividing the interval $[a, b]$ into N equal parts of length $h = (b - a)/N$ and denoting $x_0 = a, x_k = x_0 + kh$ for $k = 1, \ldots, N, u_k = u(x_k)$, and $U = (u_0, u_1, \ldots, u_N)$. First, the equation is replaced by its standard finite difference

$$\begin{cases} -\dfrac{u_{k+1} - 2u_k + u_{k-1}}{h^2} = f(x_k, u_k), \quad k = 1, \ldots, N-1 \\ u_0 = u_N = 0. \end{cases} \tag{22.11}$$

Define

$$d_k = -\frac{u_{k+1} - 2u_k + u_{k-1}}{h^2} - f(x_k, u_k), \qquad k = 1, \ldots, N-1,$$

$$\mathfrak{p} = \mathfrak{p}(U) = \sum_{k=1}^{N-1} d_k^2, \qquad \text{defmax}(U) = \max_{1 \le k \le N-1} |d_k|. \tag{22.12}$$

Solving (22.11) is equivalent to finding a zero of $\mathfrak{p}$, that is, to minimizing the non-negative function $\mathfrak{p}(U)$.

22.2.1 Description of the Algorithm

The idea used to find the saddle point is to go close to the nontrivial solution (0 is a trivial solution of (22.8) that minimizes $\mathfrak{p}$, which must be avoided) and then to perform a minimization algorithm for $\mathfrak{p}(U)$.

The algorithm is divided into two parts.

1. First, compute an approximation of the minimax of Φ to be near the saddle point.

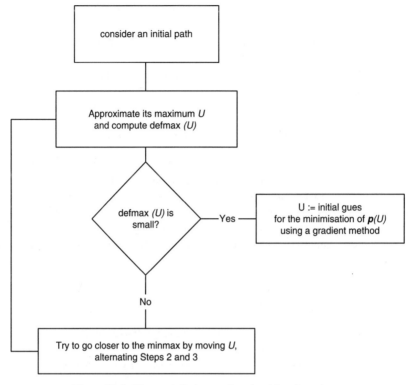

Figure 22.4. The partially interactive algorithm flow chart.

2. Then use this approximation to minimize the residual of the difference equation.

We will describe them in some detail now (see Figure 22.4).

Part I. Interactive Search for the Minimax

Step 1. For an integer parameter N_1, set $\tau = 1/(2N_1)$, $E = (e_0, \ldots, e_N)$, where $e_i = e_i(x_i)$. Join E with the origin of $\mathbb{R}^{N+1}$ by a straight line and set

$$V_i = (1 - i.\tau)E, \qquad i = 0, 1, \ldots, 2N_1.$$

Assume that

$$\Phi(V_{i_0}) = \max_{0 \le i \le 2N_1} \Phi(V_i).$$

Then, $U^1 = V_{i_0}$ is taken to be a first guess and, using (22.11), defmax (U^1) is computed. If defmax (U^1) is considered small enough (*smaller than one* in [526]), then U^1 is passed as the initial guess for Part II. If a smaller value of defmax is desired, we try to get closer to the saddle point on a path minimizing Φ.

Step 2. Computing the Fréchet derivative of Φ and using integration by parts, we get that

$$\Phi'(u)v = \int_a^b (u'v' - f(x,u)v)\,dx = -\int_a^b (u'' + f(x,u))v\,dx.$$

Hence, $\Phi'(u)v < 0$ for $v = -\tau_1(-u'' - f(x,u))$, where $\tau_1 > 0$.

So, we define

$$\begin{cases} U^{1,0} = U^1 \\ U_k^{1,i+1} = U_k^{1,i} - \tau_1.d(U_k^{1,i}) \quad i = 1, \dots, m_1, \quad k = 1, \dots, N-1, \end{cases} \tag{22.13}$$

with $U_0^{1,i+1} = U_N^{1,i+1} = 0$. This allows us to compute $U_k^{1,i+1}$ given $U_k^{1,i}$. The functional Φ is expected to decrease for τ_1 small. So, if the $(U_k^n)_n$ are bounded from below, they must converge to a solution of (22.11), where τ_1 is a constant parameter and m_1 is the number of steps. We keep iterating (22.13) and monitor defmax$(U_k^{1,i})$. If it does not decrease, m_1 is kept small and we pass to the next step.

Step 3. We take τ and N_1 as in Step 1 and join $U^{1,m}$ by straight lines to E and to the origin,

$$\begin{cases} V_{1,i} = (1 - 2i\tau)E + 2i\tau U^{1,m_1} & i = 0, 1, \dots, N_1 \\ \tilde{V}_{1,i} = (1 - 2i\tau)E & i = 0, 1, \dots, N_1. \end{cases}$$

Among all $V_{1,i}$ and $\tilde{V}_{1,i}$, we select the one maximizing Φ and call it U^2. We then return to Step 2 to produce $U^{2,1}, \dots, U^{2,m_2}$.

We continue alternating Steps 2 and 3, terminating the process when defmax(U^{s,m_s}) is suitably small.

Part II. Minimization of p(U) *Using a Gradient Descent*

We continue our description of the way Korman proceeded in [526]. So, we compute

$$P_{u_1} = -2\left(\frac{-2d_1 + d_2}{h^2} + f_u(x,u_1)d_1\right),$$

$$P_{u_k} = -2\left(\frac{-2d_{k-1} - 2d_k + d_{k+1}}{h^2} + f_u(x,u_k)d_k\right), \quad k = 2, \dots, N-2,$$

$$P_{u_{N-1}} = -2\left(\frac{-2d_{N-2} + 2d_{N-1}}{h^2} + f_u(x,u_{N-1})d_{N-1}\right).$$

If we define $d_0 = d_N = 0$, $d = (d_0, d_1, \dots, d_N)$, we can write

$$-\nabla_p = 2\Delta_h d + 2f_u(x,u)d,$$

$\Delta_h d$ being the discrete version of d''.

Starting with $U^0 = (u_{01}, \dots, u_{0(N-1)}) = (U_1^{s,m_s}, \dots, U_{N-1}^{s,m_s})$, we iterate

$$U_{(i+1)k} = U_{ik} - \tau_2 \nabla p(x_k, U_k), \quad i = 1, \dots, M, \ k = 1, \dots, N-1,$$

with suitably chosen τ_2 and M. This permits the amelioration of the solution at the interior mesh points, keeping $u_0 = u_N = 0$.

The particular descent algorithm used in Part II may be slow in some particular cases and can be replaced by a faster one. The reader is referred to [526] for some numerical examples.

Comments and Additional Notes

◇ 22.I   A Second Algorithm from Quantum Chemistry

The algorithm proposed by Liotard and Penot [584] finds a *critical point* of Φ and a *path* joining the local minimum 0 and the point e with lower level. This is possible because we are in a finite dimensional setting (cf. Chapter 5, dedicated to the finite dimensional MPT).

In quantum chemistry, the functional Φ represents the energy of a molecule; 0 and e that are supposed to be local minima correspond to two *stable molecular states* (0 is the *chemical reactant* and e is the *chemical product*). The problem is to find the *reaction path* joining them. Such a path is required to make the least possible change in the variation of the energy Φ. The highest point on the path being a saddle point (mountain pass point) whose knowledge is (said to be) of utmost importance in reaction rates theory [439].

Description of the Algorithm

We consider a C^2-function $\Phi : \mathbb{R}^N \to \mathbb{R}$, with two minima M_1 and M_2, that satisfies the (PS) condition (for example, Φ may be coercive, as in Chapter 5). The following algorithm finds a saddle point S "between" M_1 and M_2 and an approximating path joining M_1 to M_2 and passing through S.

A practical constraint in chemical applications seems that the direct evaluation of the Hessian is too costly. So, Liotard and Penot required that their algorithm must only require the explicit knowledge of Φ and its gradient $\nabla \Phi$ at a given point P. Of course, a path is approximated at the ith iteration by a *chain* of points $\{P_1, \ldots, P_{n_i}\}$ with $M_1 = P_1$ and $M_2 = P_{n_i}$, as in the former algorithm. An iteration consists of replacing the highest point of the chain P_H with a new one, P^*, according to

$$P^* = P_H + \lambda \frac{D}{\|D\|}, \tag{22.14}$$

with $0 < \ell_0 \leq \lambda \leq \ell_2$ and D a *shifting vector*.
The following is required.

R_1. Each new link of the chain (a segment $[P_i, P_{i+1}]$) must have a length ℓ with $\ell_1 \leq \ell \leq \ell_2$.
R_2. The thresholds ℓ_1, ℓ_2 ought to decrease slowly with increasing iterations.
R_3. Avoid a complicated path with "inefficient meanders."

Concerning R_2, with i being the number of the iterations,

$$\ell_j(\text{new}) = \ell_j(\text{old}) \frac{i}{i+1}, \qquad \text{for } j = 1, 2,$$

while for R_3, an analysis of the distance between the new point P^* with radius $r \in [0, \ell_2]$. If necessary, a new point is inserted in the middle of the new link to ensure R_1.

Some Choices for the Direction D

The choice of D drastically influences the performances of the algorithm.

- A natural choice of D is the opposite gradient $G = -\nabla\Phi(P_H)$ at P_H. This is what was done in the algorithm of Choi and McKenna.
- Numerical experiments suggested a more efficient direction provided by the *projected gradient* G^*,

$$G^* = G - (G, T).T, \tag{22.15}$$

with T the tangent to the path at P_H defined as the unit vector parallel to

$$\frac{P_H - P_{H-1}}{\|P_H - P_{H-1}\|} + \frac{P_{H+1} - P_H}{\|P_{H+1} - P_H\|}$$

and $(., .)$ the scalar product of $\mathbb{R}^N$.

- Noting that each iteration requires the evaluation of $-G_i = \nabla\Phi(P_H)$, the gradient vector of Φ at P_H, and using the fact that at successive points surrounding the saddle point S,

$$G_i - G_{i+1} = H(P_{i+1} - P_i),$$

where P_i is the coordinate vector, we can estimate and update the Hessian H in S. Nevertheless, successive approximations are not used, to avoid the errors accumulating from rounding off. The Hessian is estimated directly using the last $(N + 1)$ gradient evaluations in N independent directions ($N + 1$ points for N directions). Diagonalization of the symmetric matrix H gives its index i_h and the *quadratic direction* Q,

$$Q = H^{-1}G. \tag{22.16}$$

The third possible choice for D is the direction Q when it is *operative*. This choice is the best and should be used as soon as possible.

Definition 22.1. Let q be the Euclidean norm of Q. We say that the direction Q is *operative* if

    i. $i_H = 1$ (characterizing a saddle point),
    ii. $q \leq \ell_2$ (to avoid quasi-singular Hessian),
    iii. $|(Q, G)| > \varepsilon q\|G\|$, where $\varepsilon > 0$ is a fixed threshold (for numerical stability).

So to summarize, the *outline* of the algorithm is the following (see the flow chart in Figure 22.5).

    Step 1. Set the thresholds to be used and consider a trial chain with endpoints M_1 and M_2 and such that the length of each link is less than ℓ_2.
    Step 2. Find P_H evaluate $G = -\nabla\Phi(P_H)$ (update H and compute Q (use (22.16))).

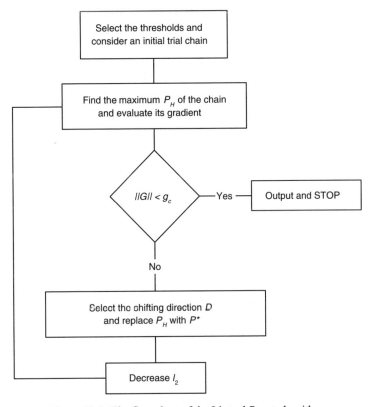

Figure 22.5. The flow chart of the Liotard-Penot algorithm.

Step 3. If $\|G\| < g_c$ then stop.

Step 4. Select the direction D (increasing interpolation or decreasing shifting).

Step 5. Search optimum $\lambda \in [\ell_0, \ell_2]$. Get new P^* (use (22.14)).

Step 6. Replace P_H with P^*. If necessary, bypass meanders and insert new points (requests R_1 and R_3).

Step 7. Decrease ℓ_2 ($\ell_2(\text{new}) := \ell_2(\text{old})\dfrac{i}{i+1}$) and related thresholds. Return to Step 2.

No convergence of this numerical procedure has been established. However, a large number of chemical applications have used it and no divergence case seem to be known.

◇ 22.II Numerical PDE Solver

Each iterative algorithm (mountain pass algorithm, monotone iteration scheme, and so on) requires a *numerical PDE solver*. The three basic ones, the finite difference method (FDM), finite element method (FEM), and boundary element Method (BEM), are in principle all usable. The FDM is used in Korman's algorithm, while the FEM is used in the Choi-McKenna mountain pass algorithm. The mountain pass algorithm has also been implemented with the BEM by Chen et al. [220].

◇ 22.III Monotone Iteration Methods

Consider, for example, the problem

$$-\Delta u = f(x, u) \qquad \text{in } \Omega$$
$$u = 0 \qquad \text{on } \partial\Omega. \tag{22.17}$$

Let $\underline{u}, \bar{u}$ be, respectively, super- and subsolutions of (22.17), in the sense

$$-\Delta\bar{u} = f(x, \bar{u}), \qquad \text{in } \Omega \quad \bar{u} \geq 0 \qquad \text{on } \partial\Omega$$
$$-\Delta\underline{u} = f(x, \underline{u}), \qquad \text{in } \Omega \quad \underline{u} \leq 0 \qquad \text{on } \partial\Omega.$$

Choose a number $\lambda > 0$ such that (suppose that $F(x, u) \in C^1(\bar{\Omega} \times \mathbb{R}; \mathbb{R}))$

$$\lambda + \frac{\partial f(x, u)}{\partial u} > 0, \qquad \forall(x, u) \in \bar{\Omega} \times [\underline{u}(x), \bar{u}(x)]$$

and such that the operator $(\Delta - \lambda)$ with Dirichlet boundary data has its spectrum strictly contained in the open half complex plane. Then, setting $f_\lambda(x, u) = \lambda u + f(x, u)$, the iterations

$$\begin{cases} u_0(x) & = \bar{u}(x) \\ (\Delta - \lambda)u_{n+1} = -f_\lambda(x, u_n(x)) & \text{in } \Omega \\ u_{n+1} & = 0 & \text{on } \partial\Omega \end{cases} \tag{22.18}$$

and

$$\begin{cases} v_0(x) & = \underline{u}(x) \\ (\Delta - \lambda)v_{n+1} = -f_\lambda(x, v_n(x)) & \text{in } \Omega \\ v_{n+1} & = 0 & \text{on } \partial\Omega \end{cases} \tag{22.19}$$

yield iterates u_n and v_n satisfying

$$\underline{u} = v_0 \leq v_1 \leq \cdots \leq v_n \leq \cdots \leq u_n \leq \cdots \leq u_1 \leq u_0 = \bar{u}.$$

The *monotone iteration scheme* tries to use this fact with the hope that it converges to the *stable solution* (i.e., local minima) located between the sub- and supersolution. (See, e.g., [33, 34, 220, 318, 791]).

◇ 22.IV The Mountain Pass Algorithm and Symmetry

An important feature of the numerical mountain pass algorithm reported in [228] is that it is shown to leave invariant many subspaces associated with the symmetries of the domain Ω. Thus, for a symmetric domain, if the algorithm is begun in a symmetric subspace, the solution is found in that space.

◇ 22.V The Mountain Pass Algorithm in Various Applications

The mountain pass algorithm appears now in many papers devoted to the study of boundary value problems.

The original paper [228] contains several semilinear elliptic problems.

[231] A semilinear wave equation. The authors use a dual variational formulation and the fast Fourier transform to take advantage of the time periodicity of the solutions.

[398] Periodic solutions of a boundary value problem that model a suspension bridge introduced by Lazer and McKenna [553].

[487] Humphreys finds large-amplitude oscillations in a nonlinear model to explain multiple periodic solutions and the traveling-wave behavior in suspension bridges. An improvement there is that the point e is found *automatically*, by iterating with the steepest descent until an e with $\Phi(e) < \Phi(0)$ is found. Moreover, some solutions are obtained where there are no proofs of existence.

[473] A coupled system of partial differential equations with Dirichlet boundary conditions.

[223, 224] Traveling wave solutions to the nonlinear beam equation, $u_{tt} + u_{xxxx} + f(u) = 0$ in R^{1+1}, which include the models for a nonlinearly suspended bridge in the case when $f(u) = \max(u, 0) - 1$.

[225] Traveling waves in a nonlinearly suspended beam using the concentration compactness method of Lions and the MPT.

[477] Some nonlinear boundary value problems are considered, related to some examples of anisotropic Sobolev spaces. Some regularity results for solutions are obtained. For numerical computation the mountain pass algorithm is used. A numerical algorithm based on multiscale discretization generated by biorthogonal wavelets is also described. The author discusses a priori and a posteriori error estimates.

[339] A *high-linking algorithm* based on the mountain pass algorithm to find sign-changing solutions of minimax type of semilinear elliptic Dirichlet problems without requiring any symmetry assumption on the domain or oddness on the nonlinearity. The algorithm is motivated by the high-linking theorem of Wang [949] for the existence of a third solution, with Morse index 2, proved by constructing a local linking at a mountain pass solution. The sign-changing solution obtained by Ding et al. [339] has a Morse index 2, while the solutions obtained by the original mountain pass algorithm of Choi and McKenna converges to solutions of Morse index 1. Moreover, in this paper, the idea of "local linking" is used for the first time to find such solutions with Morse index 2. The method uses constrained maximizations followed by a local minimization. Assuming a mountain pass solution w has been found, and by using an *ascent direction* and a *descent direction*[1] at w, we can form a triangle as a "local linking." Then we look for a maximum at the interior of this triangle. If the maximum is not in the interior, we deform the triangle so that the maximum becomes an interior point.

[573] To tentatively to justify the mountain pass algorithm of Choi and McKenna and the high-linking algorithm of Ding et al., Li and Zhou developed a *local*

---

[1] A point v is called a *descent* direction at a critical point u if there exists $\delta > 0$ such that
$$\Phi(u + tv) < \Phi(u), \qquad \text{for all } |t| < \delta.$$
An *ascent* direction is defined the same way by using the sign $>$ instead of $<$.

minimax method for finding multiple saddle points. This contrast with the MPT characterizes a critical value as a (constrained) *global* maximization on compacts sets at the first level and then a global minimization at the second level, and from a numerical point of view, this is not good.

As reported in [573], the solution obtained using the mountain pass algorithm may not be mathematically justified by the MPT. This is due to global characterization of the critical level in the MPT (the minimum is taken over all paths joining 0 to the minimum *e*), while the algorithm will tend to the nearest solution to the maximum of the energy functional to the maximum of Φ on $[0, e]$. Another point that should also be said is that until the moment of this writing, to the best of our knowledge, *no error estimates* have been given yet for the mountain pass algorithm in any of these papers. Moreover, in the case of boundary value problems that possess many solutions, the algorithm should not necessarily converge to the one characterized by the MPT. It should only converge to the one nearest to the maximum of the path used in the initialization (with a level greater than $c = \inf_{\gamma \in \Gamma} \max_{\gamma([0,1])} \Phi$).

Comparing to [228], Korman's algorithm does not require computation of the entire path. In [527], Korman and Ouyang proposed a more *direct approach* – the ingenious idea to introduce a fourth-order evolution equation such that the unstable solution of the original problem obtained by the MPT is a stable one (a stable steady state) and, hence, can be computed by using a direct method.

In [271], the authors investigate sign-changing solutions of a superlinear elliptic Dirichlet problem $-\Delta w = f(w)$ in Ω, whose existence reflects the symmetry of Ω. *Two numerical algorithms are proposed*. Numerical examples on several typical symmetric domains are presented (equilateral triangle, square, and disc). The examples indicate some advantages and disadvantages of each one of these algorithms. The FEM with linear splines is used to approximate the solution of the Dirichlet problem.

In [220], for a particular class of functionals, a simple scheme called a *scaling iterative method* is designed to find a solution of Morse index 1. A partial justification of that algorithm is given by the authors.

◇ 22.VI On the Algorithm of Liotard and Penot

In a heavy report of 126 pages [523], Kliesch deals with the mathematical description of adiabatic chemical reactions by reaction paths to whom the third chapter is consecrated. These are continuous curves that join two minimizers and pass through a saddle point on the way, which are used to describe reaction mechanisms. They are considered by the author as the heart of string models whose existence between two stable configurations of a molecular system would have a particular interest. The tool used there is, of course, the MPT, which permits a "mathematically correct" characterization of the transition structures of such systems. A numerical procedure for following the reaction path is proposed in the fifth chapter and the results of some path tracings are reported in the sixth chapter.

23

Perturbation from Symmetry and the MPT

In this chapter, we are interested in some methods used to find multiplicity results, as in the symmetric MPT, when the symmetry is broken by a "little" perturbation added to the functional. We focus on two methods closely related to the MPT.

As we saw in Chapter 11, when the functional under study is equivariant, corresponding MPTs yield multiplicity results. These have important applications in partial differential equations and Hamiltonian systems that are invariant under a group action.

A natural question that stems then is whether it is really the *symmetry* that is responsible for these spectacular multiplicity results. Would the results persist if some perturbation is injected? This *stability problem* is quite old, and mathematicians were interested in the study of the effects of breaking the symmetry by introducing a *small perturbation* since the appearance of the Ljusternik-Schnirelman theory.

> ...we shall develop methods, employing ideas contained in some of L.A. Ljusternik's work, which allow us to establish the existence of a denumerable number of stable critical values of an even functional – they do not disappear under small perturbations by odd functionals.
>
> Krasnosel'skii [534]

In this chapter, some methods used in this context are presented and developed in connection with the MPT. Before beginning, it is worth noting that mathematicians are unanimous in the belief that the results known until now are still partial and far from being satisfactory.

> In the last two chapters, several examples have been given of the existence of multiple critical points for functionals invariant under a group of symmetries. A natural question to ask is: What happens when such a functional is subjected to a perturbation which destroys the symmetry? Some special cases of this question have been studied and while progress has been made, they are not yet satisfactory general answers.
>
> Rabinowitz [748, p. 61], 1986

A natural question which even today is not adequately settled is whether the symmetry of the functional is important for results like Theorem 6.5 (*A symmetric MPT*) to hold.

Struwe [882, p. 110], 1990

However, there are no satisfactory general answers yet to the case where the group symmetry is broken by some non-equivariant – and even linear – perturbation.

Bolle et al. [135], 2000

We shall begin with the results of Rabinowitz, in which he formulated variationally and extended the principles behind some successful, but as said, still partial answers given to the aforementioned problem in the early 1980s by Bahri and Berestycki [81,83] and Struwe [868, 872].

23.1 "Standard" Method of Rabinowitz

The idea in the formulation of Rabinowitz is to think of the nonsymmetric functional Φ under study as a perturbation of its symmetrization,

$$\tilde{\Phi} = \frac{1}{2}(\Phi(u) + \Phi(-u)),$$

and then to estimate the growth rate at which the critical values of Φ move away from the perturbation from symmetry $\tilde{\Phi} - \Phi$.

Theorem 23.1 (Rabinowitz). *Suppose that $\Phi \in C^1(E; \mathbb{R})$ satisfies* (PS). *Let $W \subset E$ be a finite dimensional subspace of E, $w^* \in E \setminus W$, and let $W^* = W \oplus \mathbb{R}w^*$. Denote the upper "half-space" in W^* by*

$$W_+^* = \left\{ w + tw^*; \ w \in W, t \geq 0 \right\}.$$

Suppose that

1. $\Phi(0) = 0$.
2. *There exists $R > 0$ such that for any $u \in W$, $\|u\| \geq R$ implies that $\Phi(u) \leq 0$.*
3. *There exists $R^* \geq R$ such that for any $u \in W^*$, $\|u\| \geq R^*$ implies that $\Phi(u) \leq 0$.*

And let

$$\Gamma = \left\{ \gamma \in C(E; E); \ \gamma \text{ is odd and } \gamma(u) = u \text{ if } \max\{\Phi(u), \Phi(-u)\} \leq 0 \right\}$$

(in particular, if $u \in W$, and $\|u\| \geq R$ or if $u \in W^*$ and $\|u\| \geq R^*$).
Then, if

$$\beta^* = \inf_{\gamma \in \Gamma} \sup_{u^* \in W_+^*} \Phi(\gamma(u^*)) > \inf_{\gamma \in \Gamma} \sup_{u \in W} \Phi(\gamma(u)) = \beta \geq 0,$$

the functional Φ possesses a critical value greater or equal to β^*.

Proof. For $c \in]\beta, \beta^*[$, let

$$\Lambda = \left\{ \gamma \in \Gamma; \ \Phi(\gamma(u)) \leq c \text{ for all } u \in W \right\}.$$

By definition of β, the set Λ is nonempty. Hence,

$$c^* = \inf_{\gamma \in \Gamma} \sup_{u^* \in W^*_+} \Phi(\gamma(u^*)) \geq \beta^*$$

is well defined.

Claim 23.1. c^* is a critical value of Φ.

Indeed, suppose by contradiction that c^* is regular; we can apply the deformation lemma for $\bar{\varepsilon} = c^* - c > 0$ and $\varepsilon \in]0, \bar{\varepsilon}[$. Consider a functional $\gamma \in \Gamma$ such that

$$\sup_{u^* \in W^*_+} \Phi(\gamma(u^*)) \leq c^* + \varepsilon.$$

Define the odd mapping $\gamma' \colon W^* \to V$ by

$$\gamma'(u^*) = \begin{cases} \eta(\gamma(u^*), 1) & \text{if } u \in W^*_+, \\ -\eta(\gamma(-u^*), 1) & \text{if } -u \in W^*_+. \end{cases}$$

By the choice of $\bar{\varepsilon}$ and since $\gamma \in \Lambda$, we have that

$$\eta(\gamma(-u), 1) = \gamma(-u) = -\gamma(u) = -\eta(\gamma(u), 1) \qquad \text{for } u \in W.$$

Hence, γ' is well defined, odd, and continuous.

By the Tietze extension theorem, γ' may be extended to an odd mapping $\gamma'(E; E)$. Moreover, since $0 \leq \beta < c < c - c^* - \bar{\varepsilon}$, the mapping $\eta(., 1)$ keeps fixed any point U that satisfies $\Phi(u) \leq 0$ and $\Phi(-u) \leq 0$; by definition of Λ, so does γ and $\eta(., 1) \circ \gamma$. It follows that $\gamma' \in \Lambda$. So, we get

$$\sup_{u^* \in W^*_+} \Phi(\gamma'(u^*)) = \sup_{u^* \in W^*_+} \Phi(\eta(\gamma(u^*), 1)) < c^* + \varepsilon.$$

$\square$

It is crucial to understand how this abstract result is used in practice. A brief description is given here for illustration. Consider the problem

$$(\mathcal{P}) \quad \begin{cases} -\Delta u = f(x, u) & \text{in } \Omega, \\ u = 0 & \text{on } \partial\Omega, \end{cases}$$

where $f(x, s) \in C(\overline{\Omega} \times \mathbb{R}; \mathbb{R})$, $\Omega \subset \mathbb{R}^N$ and f satisfy the following:

f_1. There are $a_1, a_2 \geq 0$ such that

$$|f(x, s)| \leq a_1 + a_2|s|^{b-1}, \text{ where } b < 2^* = \frac{2N}{N-2} \text{ if } N > 2.$$

f_2. There exist $\mu > 2$ and $r \geq 0$ such that for $|s| \geq r$,

$$0 < \mu F(x, s) \leq sf(x, s).$$

f_3. $f(x, s)$ is odd in s.

By the symmetric MPT, the energy functional associated with $(\mathcal{P})$ has an unbounded sequence of critical values and $(\mathcal{P})$ an unbounded sequence of weak solutions.

Perturb the problem $(\mathcal{P})$ by a term $h \in L^2(\Omega)$,

$$(\mathcal{P}_p) \quad \begin{cases} -\Delta u = f(x, u) + h(x) & \text{in } \Omega, \\ u = 0 & \text{on } \partial\Omega, \end{cases}$$

and the corresponding functional

$$\Phi(u) = \frac{1}{2} \int_\Omega |\nabla u(x)|^2 \, dx - \int_\Omega [F(x, s) - h(x)u(x)] \, dx$$

is no longer even. Nevertheless, we have the following result.

Theorem 23.2 (Rabinowitz). *If f satisfies conditions f_1, f_2, and f_3 and $h \in L^2(\Omega)$, then the perturbed problem $(\mathcal{P}_p)$ possesses an unbounded sequence of weak solutions provided that b in f_1 satisfies*

$$\beta = \frac{2b}{N(b-2)} - 1 > \frac{\mu}{\mu - 1}. \tag{23.1}$$

Remark 23.1.

i. The inequality (23.1) is equivalent to

$$b < \frac{\mu N + (\mu - 1)(N + 2)}{\mu N + (\mu - 1)(N - 2)}. \tag{23.2}$$

Since $b = 1$ and $\mu = 2$ satisfy (23.2), (23.1) is nonvacuous.

ii. This result improves earlier ones by Bahri and Berestycki [81, 83] and Struwe [868]. It was then extended by Bahri and Lions [86] who used Morse theory and could improve (23.1) to

$$\frac{2b}{N(b-2)} > \frac{\mu}{\mu - 1}.$$

First, a priori bounds for critical points of Φ in terms of the critical values are obtained. By condition f_2, there are $a_4, a_5 > 0$ such that

$$F(x, s) \geq a_5 |s|^\mu - a_4, \quad \text{for all } s \in \mathbb{R}.$$

Therefore, there exists $a_3 > 0$ such that

$$\frac{1}{\mu}(sf(x, s) + a_3) \geq F(x, s) + a_4 \geq a_5 |s|^\mu, \quad \text{for all } s \in \mathbb{R}.$$

Under the hypotheses of Theorem 23.2, there exists a constant A depending on $\|f\|_{L^2(\Omega)}$ such that if u is a critical point of Φ,

$$\int_\Omega (F(x, u) + a_4) \, dx \leq A(\Phi(u)^2 + 1)^{1/2}. \tag{23.3}$$

So, if u is a critical point of Φ, then condition f_2 and (23.3) easily yield a bound for u in E in terms of $\Phi(u)$. Consider then a function $\chi \in C^\infty(\mathbb{R}; \mathbb{R})$ such that

$$\begin{cases} \chi(\xi) \equiv 1 & \text{in } B(0, 1), \\ \chi(\xi) \equiv 0 & \text{for } \xi \geq 2, \\ \chi'(\xi) \in]1, 2[& \text{for } \xi \in]1, 2[. \end{cases}$$

We state now the perturbed problem by using Rabinowitz' notations:

$$Q(u) = 2A(\Phi(u)^2 + 1)^{1/2},$$

and

$$\Psi(u) = \chi\left(Q(u)^{-1} \cdot \int_\Omega (F(x, u) + a_4)\, dx \right).$$

Then, we have the following.

Proposition 23.3. *Under the hypotheses of Theorem 23.2,*

1. *$\Psi \in C^1(E; \mathbb{R})$.*
2. *There exists a constant β_1 depending on $\|f\|_{L^2(\Omega)}$ such that for all $u \in E$,*

$$|\Psi(u) - \Psi(-u)| \leq \beta_1(|\Psi(u)|^{1/\mu} + 1). \tag{23.4}$$

3. *There exists M_0 such that if $\Psi(u) \geq M_0$ and $\Psi'(u) = 0$, then $\Psi(u) = \Phi(u)$ and $\Psi'(u) = 0$.*
4. *There exists $M_1 \geq M_0$ such that for all $c > M_1$, Ψ satisfies (PS)$_c$.*

So, by condition 3, to solve $(\mathcal{P}_p)$ it suffices to show that Ψ has an unbounded sequence of critical points. This is done using Theorem 23.2 through a series of steps. We require an estimate on the "perturbation from symmetry" of Φ of the form

$$|\Phi(u) - \Phi(-u)| \leq \beta_1(|\Phi(u)|^{1/\mu} + 1), \qquad \text{for any } u \in E. \tag{23.5}$$

Unfortunately, Φ does not satisfy (23.5); however, it can be modified in such a way that a new functional Ψ satisfies (23.5) and large critical values and points of Ψ are critical values and points of Φ.

Let $(\varphi_j)_j$ be the eigenfunctions of $-\Delta$ with Dirichlet conditions on the boundary. Let $E_j = \text{span}\{\varphi_1, \ldots, \varphi_j\}$. Then, there is an $R_j > 0$ such that

$$\Psi(u) \leq 0 \text{ if } u \in E_j \setminus B(0, R_j).$$

Set $D_j = E_j \cap B(0, R_j)$,

$$\Gamma_j = \{\gamma \in C(D_j; E);\ \gamma_j \text{ is odd and } \gamma|_{\partial B(0, R_j)} = I_d\},$$

and

$$b_j = \inf_{\gamma \in \Gamma_j} \max_{u \in D_j} \Psi(\gamma(u)), \qquad j \in \mathbb{N}.$$

These minimax values are not generally critical values unless $h \equiv 0$. They are used as a part of a comparison argument to prove that Ψ has an unbounded sequence of critical values. Indeed, we have the following estimate on the rate of divergence of the sequence $(\beta_j)_j$.

Proposition 23.4. *There is a constant $\beta_2 > 0$ and $\tilde{k} \in \mathbb{N}$ such that, for all $k \geq \tilde{k}$,*

$$b_k \geq \beta_2 k^\beta, \tag{23.6}$$

where β is defined in (23.1).

To get critical values of Ψ from the sequence $(b_k)_k$, another set of minimax values is introduced. Define

$$U_k = \left\{ u = w + t\varphi_{k+1};\ t \in [0, R_{k+1}],\ w \in B(0, R_{k+1}) \cap E_k,\ \|u\| \leq R_{k+1} \right\}$$

and

$$\Lambda_k = \left\{ \mathcal{H} \in C(U_k; E);\ \mathcal{H} \text{ is odd},\ \mathcal{H}|_{D_k} \in \Gamma_k \text{ and } \mathcal{H} = I_d \right.$$
$$\left. \text{for } u \in Q \equiv (\partial B(0, R_{k+1}) \cap E_{k+1}) \cup [(B(0, R_{k+1}) \setminus B(0, R_k)) \cap E_k] \right\}.$$

Set

$$c_k = \inf_{\mathcal{H} \in \Lambda_k}\ \max_{u \in U_k} \Psi(\mathcal{H}(u)).$$

We can show that $c_k \geq b_k$. If $c_k > b_k$ for a sequence of indices that tend to infinity, then Ψ has an unbounded sequence of critical values. Indeed, this follows from Theorem 23.2 and Proposition 23.4. The last step consists of showing that $c_k = b_k$ for large k (say $k \geq k^*$) is impossible because otherwise, we get that for $k \geq \hat{k}$ for some $\hat{k} \geq k^*$, and for a constant $w > 0$,

$$b_k \leq wk^{\mu/(\mu-1)},$$

and combining this with (23.6) yields a contradiction with (23.1).

23.2 Homotopy-like MPT

We pass now to a more recent method initiated by Bolle [134] to deal with problems with broken symmetry. It is based on a phenomenon of preservation of critical values along a path of functionals. Bolle's abstract result is a *homotopy-like* variant of the MPT, which deserves interest on its own, regardless of its application to the stability problem we are concerned with in this chapter.

He considers a continuous path of functionals $(\Phi_\theta)_{\theta \in [0,1]}$ such that at its endpoints, at $\theta = 0$, Φ_0 is the symmetric functional and at $\theta = 1$, Φ_1 is the functional we want to treat. We will see that the preservation of minimax critical levels along the path $(\Phi_\theta)_{\theta \in [0,1]}$ depends only on the variation $(\partial/\partial\theta)\Phi_\theta(x)$ at the critical points of Φ_θ.

Let E be a Hilbert space and consider a C^2-functional $\Phi = \Phi(\theta, x) \colon [0, 1] \times E \to \mathbb{R}$. For $\theta \in [0, 1]$, we shall use the notation Φ_θ for $\Phi(\theta, .)$. Suppose in all the sequels that Φ satisfies the (PS) condition, which means there that for every sequence $(\theta_n, x_n)_n$ with $\theta_n \in [0, 1]$, $x_n \in E$ such that

$$\begin{cases} \Phi'(\theta_n, x_n) \to 0 \text{ as } n \to \infty, \text{ and} \\ (\Phi(\theta_n, x_n))_n \text{ is bounded,} \end{cases}$$

there is a subsequence converging in $[0, 1] \times E$ to some pair (θ, x).

Consider the following conditions.

H_1. For all $b > 0$, there is a constant $C_1(b)$ such that

$$|\Phi_\theta(x)| < b \quad \text{implies that} \quad \left| \frac{\partial}{\partial \theta} \Phi(\theta, x) \right| \le C_1(b) \left(\|\Phi'_\theta(x)\| + 1 \right) (\|x\| + 1).$$

H_2. There exist two continuous functionals f_1 and $f_2 \colon [0, 1] \times \mathbb{R} \to \mathbb{R}$, with $f_1 \le f_2$, that are Lipschitz continuous relative to the second variable and such that, for all critical points x of Φ_θ,

$$f_1(\theta, \Phi_\theta(x)) \le \frac{\partial}{\partial \theta} \Phi(\theta, x) \le f_2(\theta, \Phi_\theta(x)).$$

H_3. There are two closed subsets A and B of E such that
  i. Φ_0 has an upper bound on A and $\displaystyle\lim_{\substack{\|x\| \to \infty \\ x \in A}} \left(\sup_{\theta \in [0,1]} \Phi_\theta(x) \right) = -\infty$.
  ii. And for

$$\mathcal{D}_{A,B} = \big\{ \gamma \in C(E; E); \ \gamma(x)|_B = I_d \text{ and } \exists R > 0 \text{ such that}$$
$$\gamma(x) = x \text{ for } x \in E \text{ with } \|x\| \ge R \big\}.$$

we have that

$$c_{A,B} = \inf_{\gamma \in \mathcal{D}_{A,B}} \sup_{x \in A} \Phi_0(\gamma(x)) > c_B = \sup_B \Phi_0.$$

In connection with this last condition, denote by Ψ_i $(i = 1, 2)$ the functions defined on $[0, 1] \times \mathbb{R}$ by

$$\begin{cases} \frac{\partial}{\partial \theta} \Psi_i(0, s) = f_i(\theta, \Psi_i(\theta, s)), \\ \Psi_i(0, s) = s. \end{cases}$$

They are continuous and for all $\theta \in [0, 1]$, $\Psi_1(\theta, .)$ and $\Psi_2(\theta, .)$ are nondecreasing on $\mathbb{R}$. Moreover, since $f_1 \le f_2$, we have $\Psi_1 \le \Psi_2$. In the sequel, we will denote $\overline{f}_i(s) = |f_i(\theta, s)|$, $i = 1, 2$.

Theorem 23.5 (Homotopy-like MPT, Bolle). *Assume that $\Phi \in C^2(E; \mathbb{R})$ and satisfies the Palais-Smale condition and conditions H_1–H_3. If the inequality*

$$\Psi_2(1, c_B) < \Psi_1(1, c_{A,B})$$

holds (i.e., for $\theta = 1$), then Φ_1 has a critical value at a level $\overline{c}$ such that

$$\Psi_1(1, c_B) \le \overline{c} \le \Psi_2(1, c_{A,B}).$$

Suppose in the sequel that Φ satisfies the (PS) condition, H_1–H_3, and $\Psi_2(1, c_A) < \Psi_1(1, c_{A,B})$.

Lemma 23.6. *Let* $\eta \in C([0, 1] \times E; E)$ *be such that*

> i. $\eta(0, .) = I_d$, *and*
> ii. *there is* $R > 0$ *such that, for any* $\theta \in [0, 1]$,
>
> $$x \in B \text{ and } \|x\| > R \qquad implies \qquad \eta(\theta, x) = x.$$
>
> iii. *For any* $\theta \in [0, 1]$, $\eta(\theta, A) \cap D = \varnothing$.
> *Then* $\eta(1, B) \cap D \neq \varnothing$.

Proof. Since η is continuous, A and D are closed subsets of E, and $\eta([0, 1] \times A) \cap D = \varnothing$, there is an open neighborhood U of A such that $\eta([0, 1] \times U) \cap D = \varnothing$. Let $V \subset E$ be open and satisfy $A \subset \overline{V} \subset U$. Let $\ell \in C(E; [0, 1])$ be such that $\ell|_V = 0$ and $\ell|_{E \setminus U} = 1$.

Let $g(x) = \eta(\ell(x), x)$. It is clear that $g \in \mathcal{D}_{A,B}$. By definition of $c_{B,A}$, there is $x \in B$ such that $g(x) \in D$. Since x cannot belong to U because $\eta([0, 1] \times U) \cap D = \varnothing$, $\ell(x) = 1$ and $\eta(1, x) \in D$. So, $\eta(1, B) \cap D \neq \varnothing$. $\square$

Corollary 23.7. *Let* $\eta \in C([0, 1] \times E; E)$ *and* $\kappa \in C([0, 1] \times E; E)$ *satisfy conditions* i *and* ii *of Lemma 23.6. Moreover, assume that*

> iv. *for all* $\theta \in [0, 1]$, $\kappa(\theta, .)$ *is a homeomorphism and* $\kappa_{-1} : [0, 1] \times E \to E$ *defined by* $\kappa_{-1}(\theta, x) = (\kappa(\theta, .))^{-1}(x)$ *is continuous.*
> v. *For all* $\theta \in [0, 1]$, $\eta(\theta, A) \cap \kappa(\theta, D) = \varnothing$.
> *Then* $\eta(1, B) \cap \kappa(1, D) \neq \varnothing$.

Proof. Set $\overline{\eta}(\theta, x) = (\kappa(\theta, .))^{-1}(\eta(\theta, x))$. We see that $\overline{\eta}$ satisfies the assumptions of Lemma 23.6, from which the result follows. $\square$

For two functions η and κ satisfying the assumptions of Corollary 23.7, the following *linking relation* holds. Let

$$A' = \eta(1, A), \qquad B' = \eta(1, B), \qquad \text{and} \qquad D' = \kappa(1, D).$$

Then, for all $\gamma \in \mathcal{D}_{A',B'}$,

$$\gamma(B') \cap D' \neq \varnothing. \tag{23.7}$$

Apply Corollary 23.7 to the functionals

$$\overline{\eta}(\theta, x) = \eta(2\theta, x) \qquad\qquad\qquad\qquad \text{if } 0 \leq \theta \leq 1/2,$$
$$\overline{\eta}(\theta, x) = (2 - 2\theta)\eta(1, x) + (2\theta - 1)\kappa(\eta(1, x)) \quad \text{if } 1/2 \leq \theta \leq 1,$$

and

$$\tilde{\kappa}(\theta, x) = \kappa(2\theta, x) \qquad \text{if } 0 \leq \theta \leq 1/2,$$
$$\tilde{\kappa}(\theta, x) = \kappa(1, x) \qquad \text{if } 1/2 \leq \theta \leq 1.$$

By the (PS) condition and condition H_2, we get the following result.

Lemma 23.8. *For any* $\delta > 0$ *and* $b > 0$, *there exists* $\zeta > 0$ *such that*

$$|\Phi_\theta| \le b \text{ and } \|\Phi'_\theta\| < \zeta$$

implies that

$$f_1(\theta, \Phi_\theta(x)) - \delta < \frac{\partial}{\partial\theta}\Phi(\theta, x) < f_2(\theta, \Phi_\theta(x)) + \delta.$$

Now, we can prove the homotopy-like MPT version of Bolle.

Proof of Theorem 23.5. As in [134], we will denote by $J_\theta(x)$ the derivative $(\partial/\partial\theta)\Phi(\theta, x)$. For $\delta > 0$, consider two functions $\overline{\Psi_1}$ and $\overline{\Psi_2} \colon [0, 1] \times \mathbb{R} \to \mathbb{R}$ defined by

$$\begin{cases} \overline{\Psi_1}(0, s) = s, \\ \dfrac{\partial}{\partial\theta}\overline{\Psi_1}(\theta, s) = f_1(\theta, \overline{\Psi_1}(\theta, s)) - \delta, \end{cases}$$

and

$$\begin{cases} \overline{\Psi_2}(0, s) = s, \\ \dfrac{\partial}{\partial\theta}\overline{\Psi_2}(\theta, s) = f_2(\theta, \overline{\Psi_2}(\theta, s)) + \delta. \end{cases}$$

Since $\overline{\Psi_2}(1, c_A) < \overline{\Psi_1}(1, c_{A,B} - \eta)$, for δ small enough we can assume that

$$\overline{\Psi_2}(1, c_A) < \overline{\Psi_1}(1, c_{A,B} - \eta). \tag{23.8}$$

Set $\phi_1(\theta) = \overline{\Psi_1}(\theta, c_{A,B} - \eta)$ and $\phi_2(\theta) = \overline{\Psi_2}(\theta, c_A)$. Since $f_1 - \delta \le f_2 + \delta$, it is easy to see that (23.8) implies

$$\phi_2(\theta) < \phi_1(\theta), \qquad \text{for all } \theta \in [0, 1]. \tag{23.9}$$

Denote

$$\alpha = \inf_{\theta\in[0,1]} \phi_2(\theta), \qquad \text{and} \qquad \beta = \sup_{\theta\in[0,1]} \phi_1(\theta).$$

Consider also a function $u \in C^\infty(\mathbb{R}; [0, 1])$ such that

$$\begin{cases} u(t) = 0 & \text{if } t \in (-\infty, \alpha - 2] \cup [\beta + 2, +\infty), \\ u(t) = 1 & \text{if } t \in [\alpha - 1, \beta + 1]. \end{cases}$$

By Lemma 23.8, there is $\zeta > 0$ such that

$$\alpha - 2 < \Phi_\theta(x) < \beta + 2$$

and

$$\|\Phi'_\theta(x)\|\zeta$$

implies that

$$f_1(\theta, \Phi_\theta(x)) - \delta < \frac{\partial}{\partial\theta}\Phi(\theta, x) < f_2(\theta, \Phi_\theta(x)) + \delta.$$

Let $v \in C^{\infty}(\mathbb{R}; [0, 1])$ such that

$$\begin{cases} v(t) = 0 & \text{if } |t| \leq \zeta/2, \\ v(t) = 1 & \text{if } |t| \geq \zeta. \end{cases}$$

Set

$$X_1(\theta, x) = \left(-\left(\frac{\partial}{\partial \theta} \Phi(\theta, x) \right)^{-} (x) + 1 + f_1^+(\theta, \phi_1(\theta)) \right) u(\Phi_\theta(x)) v(\|\Phi_\theta'(x)\|) \frac{\Phi_\theta'(x)}{\|\Phi_\theta'\|^2},$$

and

$$X_2(\theta, x) = \left(-\left(\frac{\partial}{\partial \theta} \Phi(\theta, x) \right)^{+} (x) - 1 - f_2^-(\theta, \phi_1(\theta)) \right) u(\Phi_\theta(x)) v(\|\Phi_\theta'(x)\|) \frac{\Phi_\theta'(x)}{\|\Phi_\theta'\|^2},$$

where $a^+ = \sup(a, 0)$ and $a^- = \inf(-a, 0)$.

Now let $\rho : [0, 1] \times E \to E$ be the flow of X_i defined by

$$\begin{cases} \rho_i(0, x) = x, \\ \dfrac{\partial}{\partial \theta} \rho_i(\theta, x) = X_i(\theta, \rho_i(\theta, x)). \end{cases}$$

It is well defined on $[0, 1] \times E$.

Using the definition of ρ_i, $\rho_i(\theta, .)$ is a homeomorphism for all $\theta \in [0, 1]$ and the function $(\theta, x) \mapsto \rho_i(\theta, .)^{-1}(x)$ is continuous on $[0, 1] \times E$.

Moreover, since $\lim_{|x| \to \infty} \sup_{\theta \in [0, 1] \atop x \in B} \Phi_\theta(x) = -\infty$, there is $R > 0$ such that, for all $x \in B$,

$$|x| > R \qquad \text{implies that} \qquad X_i(\theta, x) = 0, \qquad \text{for any } \theta \in [0, 1].$$

Hence, for all $\theta \in [0, 1]$ and for all $x \in B$ such that $|x| > R$, $\rho_i(\theta, x) = x$.

We want to apply Corollary 23.7, so we have to check that

$$\rho_2(\theta, A) \cap \rho_1(\theta, D) = \varnothing \qquad \text{for all } \theta \in [0, 1]. \tag{23.10}$$

We can prove similarly that

$$\Phi_0(x) \leq c_A \qquad \text{implies that} \qquad \forall \theta \in [0, 1], \ \Phi(\rho_2(\theta, x)) \leq \varphi_2(\theta) \tag{23.11}$$

and

$$\Phi_0(x) \geq c_{A, B} - \eta \quad \text{implies that} \quad \forall \theta \in [0, 1], \ \Phi_\theta(\rho_1(\theta, x)) \geq \varphi_1(\theta), \tag{23.12}$$

and we get (23.10) from (23.9), (23.11), and (23.12).

By Corollary 23.7, $\rho(1, B) \cap \rho(1, D) \neq \varnothing$. So, for $A' = \rho(1, A)$, $B' = \rho(1, B)$, and $D' = \rho(1, D)$, we get by the linking relation (23.7); that is, for all $\gamma \in \mathcal{D}_{A', B'}$,

$$\gamma(B') \cap D' \neq \varnothing.$$

Therefore,

$$\inf_{\gamma \in \mathcal{D}_{A', B'}} \sup_{\gamma(B')} \Phi_1 \geq \inf_{D'} \Phi_1 \geq \varphi_1(1) > \varphi_2(1) \geq \sup_{A'} \Phi_1,$$

and since Φ_1 satisfies (PS), it admits at least a critical level greater than or equal to $\varphi_1(1)$.

Since δ and η can be chosen arbitrarily small,

$$\varphi_1 \xrightarrow[\substack{\delta\to 0 \\ \eta\to 0}]{} \Psi_1(1, c_{A,B}).$$

Using (PS) again, we get that Φ_1 has at least one critical point of level greater than or equal to $\Psi_1(1, c_{A,B})$. $\square$

We have proved only that Φ_1 has at least one critical point of level greater than or equal to $\Psi_1(1, c_{A,B})$, but in fact there is a critical level in $[\Psi_1(1, c_B), \Psi_2(1, c_{A,B})]$.

This homotopy-like version of the MPT was derived to treat the boundary value problem

$$\ddot{x} + \nabla V(x) = 0, \qquad x(0) = x_0, \qquad x(T) = x_1,$$

where x_0 and x_1 are two fixed points of $\mathbb{R}^N$ and V is a superquadratic potential defined on $\mathbb{R}^N$. Bolle proved that this problem has a sequence of solutions x_n, whose energies $1/2|\dot{x}_n|^2 + V(x_n)$ tend to infinity. He extended the results of Ekeland et al. [371], who proposed more restrictive growth conditions on the potential V.

This result has been extended by Bolle et al. [135] to a form that gave better estimates compared to Rabinowitz' approach (cf. the previous section) in some situations (second-order partial differential equations and Hamiltonian systems) where the critical points of Φ_θ obey certain *conservation laws*. You may compare the papers [135] and [371]. When these conservation laws fail, for example, in a nonlinear wave equation, this is no longer the case.

A newer variant of the homotopy-like MPT due to Bolle et al. [135] has recently appeared. In the presence of a splitting $E = E_- \oplus E_+$, and $(E_n)_n$ is an increasing sequence of subspaces of E such that $E_0 = E_-$ and $E_{n+1} = E_n \oplus \mathbb{R}e_{n+1}$. If E_- is finite dimensional, set

$$\Gamma = \left\{\gamma \in \mathcal{C}(E; E); \; \gamma \text{ is odd and } \gamma(x) = x \text{ for } x \in E \text{ and } \|x\| \text{ large}\right\}$$

and

$$c_k = \inf_{\gamma \in \Gamma} \sup_{x \in E_k} \Phi(\gamma(x)).$$

Theorem 23.9 (Homotopy-like MPT, Bolle et al.). *Assume* Φ *satisfies conditions* H_1–H_3. *In addition, suppose*

H'_4. Φ_0 *is even and for any finite dimensional space* W *of* E, *we have* $\sup_{\theta\in[0,1]} \Phi(\theta, y) \to -\infty$ *as* $y \in W$ *and* $\|y\| \to \infty$.

Then, there is $C > 0$ *such that, for every* k,

1. *either* Φ_1 *has a critical level* $\overline{c_k}$ *with* $\Psi_2(1, c_k) < \Psi_1(1, c_{k+1}) \leq \overline{c_k}$, *or*
2. $c_{k+1} - c_k \leq C(\overline{f_1}(c_{k+1}) + \overline{f_2}(c_k) + 1)$.

In the proof of this theorem, two cases are distinguished:

1. When $\Psi_2(1, c_k) < \Psi_1(1, c_{k+1})$, the authors show that we are in the context of Bolle's theorem, so we get the first alternative.
2. Otherwise $\Psi_2(1, c_k) \geq \Psi_1(1, c_{k+1})$, and the local Lipschitz continuity of f_i with respect to the second argument yields the second alternative.

Comments and Additional Notes

◇ 23.I On a Perturbed Superlinear Problem

For the problem

$$(\mathcal{P}_p) \begin{cases} -\Delta u = |u|^{p-1}u + h & \text{in } \Omega \\ \quad u = 0 & \text{on } \partial\Omega \end{cases}$$

when $h \in C(\overline{W}; \mathbb{R})$,

1. Using the "standard method," we get infinitely many solutions for $1 < p < \dfrac{n+2}{n}$.
2. Using Bolle's result, we also get the same result for $1 < p < \dfrac{n}{n-2}$. Notice that this is the estimate obtained by Bahri and Lions [86].
3. However, with the aforementioned result of Bolle et al., we can go to $1 < p < \dfrac{n+1}{n+2}$.
4. The problem rests *still open* whether we can go to the Sobolev exponent, that is, $1 < p < \dfrac{n+2}{n-2}$.

The proof of Bolle et al. is a direct verification of the assumptions of the theorem, through a certain number of lemmata, applied with

$$\begin{cases} f_1(\theta, s) = -a(s^2 + 1)^{1/4} \\ f_2(\theta, s) = b(s^2 + 1)^{1/4}. \end{cases}$$

◇ 23.II A $\mathbb{Z}_2$-Equivariant Ljusternik-Schnirelman Theory for Noneven Functionals

Consider the nonlinear elliptic Dirichlet problem,

$$(\mathcal{P}) \begin{cases} -\Delta u = f(x, u) & \text{in } \Omega \subset \mathbb{R}^N \\ \quad u = 0 & \text{on } \partial\Omega, \end{cases}$$

where $f(x, .)$ is *odd* for each $x \in \Omega$.

We know that the oddness ensures that the energy functional Φ associated with this problem is even. So, the Ljusternik-Schnirelman procedure may be used to produce what Ekeland and Ghoussoub call *virtual critical points* in [361], that is, either the usual critical points or $\mathbb{Z}_2$-*resonant points*:

$$\begin{cases} \Phi(x) = \Phi(-x) \\ \Phi'(x) = \lambda\Phi'(-x). \end{cases}$$

They developed in [361] a $\mathbb{Z}$-equivariant Ljusternik and Schnirelman theory for noneven functionals and extended the classical existence and multiplicity theory of Ljusternik and Schnirelman for critical points of even functionals to virtual critical points of noneven functionals.

Their proof relies on the fact that the equivariant Ljusternik-Schnirelman min-max levels for the original functional Φ and for the even functional $\psi(x) = \max\{\Phi(x), \Phi(-x)\}$ are the same.

◇ 23.III A Critical Point Theory for Nonsymmetric Perturbations of G-Invariant Functionals

Let G be a compact Lie group and M a complete Finsler manifold. In [238], Clapp obtains some abstract critical point theorems for functionals of the form $\Phi = I + \Psi$, where I is G-invariant and Ψ is a nonsymmetric perturbation. In particular, she considers the case when M is a G-Hilbert space and Φ has the mountain pass geometry.

24

Applying the MPT in Bifurcation Problems

Bifurcation theory is a very far-reaching discipline in the midst of stormy development.

> E. Zeidler, *Nonlinear functional analysis and its applications.*
> *I. Fixed-point theorems.* Springer-Verlag, 1984

Rabinowitz successfully investigated bifurcation phenomena in some variational problems by an
ingenious application of the ideas used in the proof of the MPT.

The solution set of a nonlinear problem may be very complicated and may change
dramatically under a small perturbation, when the qualitative behavior of the natural
phenomenon it models lacks stability. In particular, some branching and bifurcating
solutions may appear. The object of bifurcation theory is precisely to study when
there is a "branching" of new solutions. Bifurcation is a very important subject in
modern nonlinear analysis that dates from the nineteenth century and to which many
books and a multitude of papers have been devoted. It has found applications in many
areas including elasticity theory, fluid dynamics, geophysics, astrophysics, meteorol-
ogy, statistical mechanics, chemical kinetics, and so on. The mathematical side is
also very rich. Bifurcation problems have been treated using many approaches: methods of function theory, algebra, algebraic geometry, critical point theory (both Morse
and Ljusternik-Schnirelman approaches), algebraic topology, and differential topology. Nevertheless, the simplest cases may be solved using only the implicit function
theorem.

In this chapter, we do not intend to go far in detail into bifurcation theory, which
is far behind the scope of this book. We will focus our attention on the way the MPT
has been applied successfully by Rabinowitz to investigate bifurcation phenomena in
some variational problems.

24.1 Preliminaries

We begin with fixing the sense and the notations used in the sequel. To easily understand
bifurcation phenomena in nonlinear equations, let us consider the simple situation of a

real equation

$$F(\lambda, u) = 0, \tag{24.1}$$

where F is a function of class C^1 in the neighborhood of a pair (λ_0, u_0). If $F(\lambda_0, u_0) = 0$ and $F_u(\lambda_0, u_0) \neq 0$, then by the *implicit function theorem* there is exactly one solution curve in the neighborhood of (λ_0, u_0) that passes through (λ_0, u_0).

If, on the other hand, the derivative $F_u(\lambda_0, u_0) = 0$, then it is possible that there is a *bifurcation* at (λ_0, u_0), that is, a branching of the solution curve.

Example 24.1. Consider again a function we saw many times before, $F(\lambda, u) = (\lambda - \lambda_0)^2 - (u - u_0)^2$. Equation (24.1) has two solution curves that pass through (λ_0, u_0) whose equations are $\lambda - \lambda_0 = u - u_0$ and $\lambda - \lambda_0 = u_0 - u$.

Let X and Y be two Banach spaces and $F: \mathbb{R} \times X \to Y$ be a functional depending on a real parameter λ. We are interested in studying equations of the type of (24.1). Suppose that $F(\lambda, 0) = 0$ for all $\lambda \in \mathbb{R}$; that is, the points of the curve $u = 0$ form trivial solutions. The parameter λ often has a physical interpretation and its introduction proved to be a convenient way to investigate when new solutions are generated.

Definition 24.1. We say that λ^* is a *bifurcation point for F from the trivial solution* $u = 0$ if there is a sequence $(\lambda_n, u_n) \in \mathbb{R} \times X$ with $u \neq 0$ and $F(\lambda_n, u_n) = 0$ such that

$$(\lambda_n, u_n) \to (\lambda^*, 0).$$

This is equivalent to requiring $(\lambda^*, 0)$ to belong to the closure in $\mathbb{R} \times X$ of the set of nontrivial solutions

$$\mathcal{S} = \{(\lambda, u) \in \mathbb{R} \times X; \ u \neq 0, \ F(\lambda, u) = 0\}.$$

Using the implicit function theorem, provided F is sufficiently regular, we get that a necessary condition for λ^* to be a bifurcation point of F is that the partial derivative $F_u(\lambda^*, 0)$ is not invertible. The fact that $F(\lambda, 0) = 0$ for all $\lambda \in \mathbb{R}$ means that the solution curve $u = 0$ would be the only one in the neighborhood of $(\lambda^*, 0)$ if $F_u(\lambda^*, 0)$ was invertible.

When $X = Y$ and f has the particular form $F(\lambda, u) = \lambda u - G(u)$, a necessary condition for λ^* to be a bifurcation point from the origin is that it belongs to the spectrum $\sigma(G'(0))$ of $G'(0)$. This necessary condition is not sufficient, however, as we may see in the case of a particular linear function G in the following example (see, for example [49, 747, 983]).

Example 24.2. Let $X = Y = \mathbb{R}^2$ and consider $G: X \to Y$ defined by

$$G(x, y) = (x + y^3, y - x^3).$$

The value $\lambda^* = 1$ is an eigenvalue of $G'(0) = I_d$ but is not a bifurcation point from the origin for $F(\lambda, (x, y)) = \lambda(x - y) - G(x, y)$. Indeed, consider a solution (x, y) of

$F(\lambda, (x, y)) = 0$; it follows that

$$\begin{cases} \lambda x = x + y^3 \\ \lambda y = y - x^3, \end{cases}$$

and then $x^4 + y^4 = 0$; that is, $(x, y) = (0, 0)$. Therefore, $F(\lambda, (x, y)) = 0$ has only the trivial solution and there are no bifurcation points for F.

24.2 The Ljapunov-Schmidt Reduction

Dealing with bifurcation problems in infinite dimensional Banach spaces is done in general through the *projection method* known as the *Ljapunov-Schmidt reduction*. This method reduces the problem to a finite system of finite dimensional nonlinear scalar equations (i.e., with a finite number of unknowns). Indeed, consider a functional $F \in C^2(\mathbb{R} \times X; Y)$ such that we already know a solution branch $F(\lambda, u) = 0$ for all $\lambda \in \mathbb{R}$. Since the possible bifurcation points λ^* from the origin for F require that $F_u(\lambda^*, 0)$ not be invertible, set

$$\begin{cases} L = F_u(\lambda^*, 0), \\ V = \ker(L) & \text{the kernel of } L, \text{ and} \\ R = \mathcal{R}(L) & \text{the range of } L. \end{cases}$$

Suppose that

a. V has a topological complement W in X, so that

$$\begin{cases} V \cap W = \{0\}, \\ X = V \oplus W. \end{cases}$$

b. R is closed and has a topological complement Z in Y,

$$\begin{cases} Z \cap R = \{0\}, \\ Y = Z \oplus R. \end{cases}$$

This holds true in particular when $\dim V < \infty$ and $\operatorname{codim} R < \infty$; that is, L is a Fredholm operator. Let P and Q denote the projections of Y onto Z and R, respectively (see Figure 24.1), and set

$$F(\lambda, u) = Lu + \varphi(\lambda, u).$$

To reduce the solvability of a bifurcation problem in infinite dimensional Banach spaces to that of finitely many nonlinear equations with finitely many variables, we pass by the *Ljapunov-Schmidt equations*.
Writing $u = v + w \in V \oplus W$, $F(\lambda, u) = 0$ is equivalent to

$$PF(\lambda, v + w) = 0, \tag{24.2}$$

and

$$QF(\lambda, v + w) = 0. \tag{24.3}$$

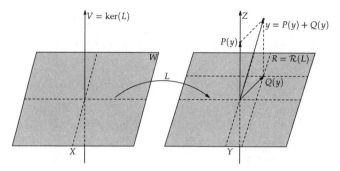

Figure 24.1. The Ljapunov-Schmidt reduction.

The idea is to solve (24.2) in a neighborhood of $(\lambda^*, 0)$ using the implicit function theorem. Since $Lv = 0$, we get that

$$F(\lambda, u) = Lw + \varphi(\lambda, v + w).$$

And since $Lw \in R$, we have that $Lw = QLw$, so (24.3) becomes

$$Lw + Q\varphi(\lambda, v + w) = 0.$$

Set $\Phi \colon \mathbb{R} \times V \times W \to R$, $(\lambda, v, w) \mapsto Lw + Q\varphi(\lambda, v + w)$. Since $F \in C^2(\mathbb{R} \times X; Y)$, then Φ is of class C^2. Moreover, $\Phi_w(\lambda^*, 0, 0) \colon w \to Lw + Q\varphi_u(\lambda^*, 0)w$. But since $\varphi(\lambda, u) = F_u(\lambda, u) - Lu$, we have that

$$\varphi_u(\lambda^*, 0) = F_u(\lambda^*, 0) - L \equiv 0 \text{ (Null mapping of } \mathcal{L}(X; Y)). \qquad (24.4)$$

Thus, $\Phi_w(\lambda^*, 0, 0) = L|_W$, which belongs to Isom $(W; R)$. Then, by the implicit function theorem, there exists

  i. a neighborhood Λ of λ^* in $\mathbb{R}$,
 ii. a neighborhood $\mathcal{V}$ of 0 in V,
iii. a neighborhood $\mathcal{W}$ of 0 in W, and
 iv. a function $\gamma \in C^2(\Lambda \times \mathcal{V}; \mathcal{W})$ such that the unique solutions of (24.4) are given by $(\lambda, v, \gamma(\lambda, v))$.

In particular $\gamma(\lambda, 0) = 0$ for all $\lambda \in \mathbb{R}$. So, substituting

$$w = \gamma(\lambda, v) \qquad (24.5)$$

in (24.2), we get

$$P(F(\lambda, v + \gamma(\lambda, v))) = 0. \qquad (24.6)$$

Equation (24.6) in the unknowns $(\lambda, v) \in \Lambda \times \mathcal{V}$ is called the *bifurcation equation* (and *Ljapunov's bifurcation equation*, and also the *branching equation*). Equations (24.5) and (24.6) together are equivalent in $\Lambda \times \mathcal{V} \times \mathcal{W}$ to the initial equation $F(\lambda, u) = 0$.

24.3 The MPT in Bifurcation Problems

So, how could Rabinowitz use the MPT in bifurcation problems? Consider the equation

$$Lu + H(u) = \lambda u, \tag{24.7}$$

where $L \colon E \to E$ is continuous linear $H \in \mathcal{C}(E; E)$ with $H(u) = o(\|u\|)$ at $u = 0$ and $\lambda \in \mathbb{R}$. A solution of (24.7) is a pair $(\lambda, u) \in \mathbb{R} \times E$ that satisfies the equation. In particular, for any $\lambda \in \mathbb{R}$, $(\lambda, 0)$ is a trivial solution of (24.7), and $(\mu, 0)$ is a bifurcation point for (24.7) if every neighborhood of $(\mu, 0)$ contains nontrivial solutions of (24.7).

In this situation, a necessary condition for $(\mu, 0)$ to be a bifurcation point is $\mu \in \sigma(L)$, the spectrum of L. We saw earlier that this necessary condition is not sufficient. However, when E is a Hilbert space and a variational structure exists, that is, $Lu + H(u) = \Phi'(u)$, Böhme [132] proved, roughly speaking, that $\mu \in \sigma(L)$ is also sufficient.

Theorem 24.1 (Böhme). *Let E be a Hilbert space and $\Phi \in \mathcal{C}^2(E; \mathbb{R})$, where $\Phi'(u) = Lu + H(u)$, L is linear, and $H(u) = o(\|u\|)$ at $u = 0$. If $\mu \in \sigma(L)$ is an isolated eigenvalue of finite multiplicity, then $(\mu, 0)$ is a bifurcation point.*

One can also study the behavior of the set of solutions of (24.7) as a function of λ rather than $\|u\|$. Indeed, in a **beautiful theorem**, to quote the reviewer of the paper [736], Rabinowitz obtained the following.

Theorem 24.2 (Rabinowitz). *Under the hypotheses of Theorem 24.1, either*

  i. *$(\mu, 0)$ is not an isolated solution of (24.7) in $\{u\} \times E$, or*
  ii. *there is a one-sided neighborhood Λ of μ in $\mathbb{R}$ such that for each $\lambda \in \Lambda \setminus \{u\}$, (24.7) has at least two nontrivial solutions, or*
  iii. *there is a neighborhood X of μ in $\mathbb{R}$ such that for each $\lambda \in X \setminus \{\mu\}$, (24.7) has at least one nontrivial solution.*

Case i occurs, for example, if the nonlinearity $H = 0$; that is, (24.7) is a linear eigenvalue problem.

The proofs of Theorem 24.2 and Proposition 24.4 (see the final notes) are based in part on ideas that go into the proof of the MPT. We will sketch them now.
Recall that we are interested in solving

$$\Phi'(u) = \lambda u \tag{24.8}$$

for each (λ, u) near $(\mu, 0)$ where μ is an isolated eigenvalue of L of finite multiplicity.

First, the problem is reduced to a finite dimensional one using the method of Ljapunov and Schmidt.

Write $E = \ker(L - \mu I_d) \oplus \ker^\perp$ where $\ker(A)$ is the null space of the linear operator A, and $\ker^\perp$ its orthogonal complement ($u \in E$ is written $v + w \in \ker \oplus \ker^\perp$). Let P and $P^\perp$ denote, respectively, the orthogonal projectors on $\ker$ and $\ker^\perp$; (24.8) is equivalent to

$$\begin{cases} \text{i. } \mu v + PH(v + w) = \lambda v, \\ \text{ii. } Lw + P^\perp H(v + w) = \lambda w, \end{cases} \tag{24.9}$$

since $Lv = \mu v$ for $v \in \ker$. Set

$$F(\lambda, v, w) = Lw - \lambda w + P^{\perp} H(v + w), \qquad (24.10)$$

so that $F: \mathbb{R} \times \ker \times \ker^{\perp} \to \ker$ is of class C^1 near $(\mu, 0, 0)$, $F(\mu, 0, 0) = 0$, and the derivative $F_w(\mu, 0, 0) = L - \mu I_d$ is an isomorphism of $\ker^{\perp}$ onto $\ker$. Hence, by the implicit function theorem, there is a neighborhood of $(\mu, 0)$ in $\mathbb{R} \times \ker$ in which the zeros of F and therefore solutions of (24.9ii) are given by $w = \varphi(\lambda, v)$ with φ of class C^1 near $(0, 0)$.

Also, since $L - \mu I_d \colon \ker^{\perp} \to \ker^{\perp}$ is an isomorphism, so is $L - \lambda I_d$ for all λ near μ. Thus,

$$\varphi(\lambda, v) = -(L - \lambda I_d)^{-1} P^{\perp} H(v + \varphi(\lambda, u)). \qquad (24.11)$$

Since $H(u) = o(||u||)$ at $u = 0$, (24.11) implies that $\varphi(\lambda, v) = o(||v||)$ at $v = 0$ uniformly for λ near μ.

Thus, to solve (24.8) near $(\mu, 0)$ in $\mathbb{R} \times E$, it suffices to solve the finite dimensional problem (24.9i) with $w = \varphi(\lambda, v)$ for (λ, v) near $(\mu, 0)$ in $\mathbb{R} \times \ker$. We can now convert (24.9i) to a finite dimensional variational problem. Set

$$f(\lambda, u) = \Phi(u) - \frac{\lambda}{2}||u|| \quad \text{and} \quad g(\lambda, v) = f(\lambda, \varphi(\lambda, v)),$$

since φ is of class C^1 and g is also of class C^1 near $(\mu, 0)$. By (24.9ii), it is easy to see that for a fixed λ near μ, critical points of $g(\lambda, .)$. near $v = 0$ are solutions of (24.9ii). Note that $v = 0$ is a critical point of $g(\lambda, .)$ corresponding to the solution $(\lambda, 0)$ of (24.8).

Either $v = 0$ is not an isolated critical point of $g(\mu, .)$ and the alternative i of Theorem 24.2 holds or it is isolated. In this latter case,

a. $v = 0$ is a strict local maximum or minimum of $g(\mu, .)$, or
b. $g(\mu, .)$ takes both positive and negative values in every neighborhood of 0.

Alternatives a and b imply, respectively, ii and iii of Theorem 24.2. While a and b are obtained respectively by a close study of the form of the functional g against an MPT background,

$$g(\lambda, v) = \frac{1}{2}(\mu - \lambda)||v||^2 + \frac{1}{2}((L - \lambda I_d)\varphi(\lambda, v), \Phi(\lambda, v)) + h(v, \Phi(\lambda, v)), \quad (24.12)$$

where $h'(u) = H(u)$ and $h(0) = 0$.

Since $\varphi(\lambda, v) = o(||v||)$ at $v = 0$ and $H(u) = o(||u||)$ at $u = 0$, it follows that $\lambda \neq \mu$. Supposing alternative a holds and, for example, $v = 0$ is a strict local minimum of $g(\mu, .)$, then for $r > 0$ small enough and λ near μ,

$$g(\lambda, v) \le \beta < 0, \qquad v \in \partial B_r \qquad \text{(in ker)}. \qquad (24.13)$$

So, for $\lambda < \mu$ and λ near μ, we get $\rho = \rho(\lambda) < r$ such that

$$g(\lambda, v) \ge \frac{1}{4}(\mu - \lambda)||v||^2 = \frac{1}{4}(\mu - \lambda)\rho^2 \equiv \alpha(\lambda), \qquad \text{for all } v \in \partial B_\rho. \qquad (24.14)$$

Considering $g(\lambda, v)$ in B_r, from (24.13) and (24.14), $g(\lambda, v)$ possesses a positive maximum $\bar{c}(\lambda)$ in B_r. Moreover, $g(\lambda, v)$ satisfies the geometry of the MPT. But we cannot use it directly since $g(\lambda, .)$ is only defined near 0. However, thanks to (24.13) and (24.14), a deformation η can be constructed, as in the proof of the usual deformation lemma, such that $\eta = I_d$ on ∂B_r, $\eta \in C([0, 1] \times B_r; B_r)$, and η satisfies the conclusions of the deformation lemma. The proof of the MPT then yields a second critical value $\underline{c}(\lambda) \geq \alpha(\lambda) > 0$ together with

$$\underline{c}(\lambda) = \inf_{\gamma \in \Gamma} \max_{\gamma} \Phi(\gamma),$$

where $\Gamma = \{\gamma : [0, 1] \to B_r \text{ continuous} ; \gamma(0) = 0, \gamma(1) \in \partial B_r\}$ is the collection of all curves in B_r that join 0 to ∂B_r. If $\underline{c}(\lambda) < \bar{c}(\lambda)$, we have two distinct nontrivial critical points of $g(\lambda, .)$, while if

$$\underline{c}(\lambda) = \bar{c}(\lambda) = \max_{B_r} g(\lambda, .),$$

then $\bar{c}(\lambda)$ is the maximum of $g(\lambda, .)$ over every curve in B_r joining 0 to B_r. Therefore, the sets of points at which $g(\lambda, .)$ achieves its maximum separates 0 and B_r in ker. In either event, ii of Theorem 24.2 holds. If 0 is a local maximum of $g(\mu, .)$, taking $\lambda > \mu$ and replacing g by $-g$, the same proof as that mentioned earlier works.

   Concerning case b, which is more delicate than case a, according to Rabinowitz [744], the main ideas are the following. Consider the negative gradient flow corresponding to $g(\mu, .)$ in a small neighborhood $\mathcal{B}$ of 0 in ker

$$\begin{cases} \dfrac{d\Psi}{dt} = -g_v(\mu, .), \\ \psi(0, v) = v. \end{cases} \qquad (24.15)$$

Computing $g_v(\lambda, v)$ gives $g_v(\lambda, v) = (\mu - \lambda)v + PH(v + \varphi(\lambda, v))$, which is also C^1 (so, g is of class C^2 with respect to v). Therefore, there is a unique solution to (24.15). Since by (b), $g(\mu, v)$ takes both positive and negative values near 0, one shows that the sets

$$S^+ = \{v \in \mathcal{B}; \ \Psi(t, v) \in \mathcal{B} \text{ for all } t \geq 0\}$$

and

$$S^- = \{v \in \mathcal{B}; \ \Psi(t, v) \in \mathcal{B} \text{ for all } t \leq 0\}$$

are both nonempty.

   So, we can find a neighborhood $Q \subset \mathcal{B}$ of 0 in ker and constants $c^+ > 0 > c^-$ such that the fact that $v \in \partial Q$ implies that

i. $g(\mu, v) = c^+$ or c^-, or
ii. $\psi(t, v) \in \partial Q$ for t close to 0.

Set $T^\pm = g(\mu, .)^{-1}(c^\pm) \cap S^\pm$ and

$$\mathcal{G} = \{\chi \in C([0, 1]; Q); \ \chi(0) = 0 \text{ and } \chi(1) \in T^-\}.$$

Define

$$c(\lambda) = \inf_{\chi \in \mathcal{G}} \max_{t \in [0,1]} g(\lambda, \chi(t)).$$

With ρ as in (24.14), we can assume that $B_{\rho(\lambda)} \subset Q$ and (24.14) holds if $\lambda < \mu$ and λ is near to μ. It follows that $c(\lambda) \geq \alpha(\lambda) > 0$. Moreover, $c(\lambda)$ is a critical value of $g(\lambda, .)$ by modifying the proof of the deformation lemma so that there exists $\eta \in \mathcal{C}([0, 1] \times Q; Q)$ satisfying the results of the deformation lemma. Then, the proof of the MPT shows that $c(\lambda, .)$ is a critical value of $g(\lambda, .)$.

We get the case $\lambda > \mu$ by replacing g by $-g$. Thus condition iii of Theorem 24.2 follows.

The proof of Corollary 24.4 parallels the above. Theorem 24.2 has interesting applications to bifurcation problems for partial differential equations.

Comments and Additional Notes

In the problem just seen, when Φ is *even*, we have a stronger result than Theorem 24.1.

Proposition 24.3. *If μ is of multiplicity n, then for each $r \in (0, r_0)$, (24.7) has at least n distinct pairs of solutions $(\lambda_j(r), u_j(r)) \to_{r \to 0} (\mu, 0)$ for each j.*

When Φ is even, we also have a stronger result than Theorem 24.2, as Rabinowitz conjectured in [736], holds.

Proposition 24.4 (Rabinowitz and Fadell). *Under the hypotheses of Corollary 24.3, either condition i holds or there exist integers $k, m \geq 0$ such that $k + m \geq n$ and left and right neighborhoods Λ_l, Λ_r of μ such that for each $\lambda \in \Lambda_l$ (resp. Λ_r), (24.7) possesses at least k (resp. m) distinct pairs of nontrivial solutions.*

In this chapter, we followed the nice expository given by Rabinowitz [744]. The Ljapunov-Schmidt procedure was introduced by Ljapunov [589, 590] and Schmidt [822].

Concerning bifurcation and the MPT, the reader may be interested in the large expository paper by Stuart [884], whose basic tools to study bifurcation into spectral gaps of the linear part are appropriate choices of the interpolation spaces, a variant of the Lyapunov-Schmidt reduction, and the MPT. You may also consult [508] by Jeanjean, who combined the MPT and an idea of Struwe.

25

More Climbs

I think there will always be some new slant that will keep us going.

Interview with Raoul Bott, *Notices of the AMS*,
48, no. 4, 2001

You may think that almost all of what could be said about the MPT has been said in the different chapters you read and their numerous notes. This chapter will show you that this is very far from the case. There are so many versions of the MPT and there are so many directions where it has been extended and used!

In this chapter, we will review briefly (or give only pointers to) some special versions or some special applications of the MPT that have not been covered in the text of the previous chapters.

◇ 25.I Directionally Constrained MPTs

In the following two notes, we are concerned with directionally constrained MPTs. We will present in detail a version due to Duc [349], and describe another one by Arcoya and Boccardo. The directionally constrained critical point theory is motivated by applications to variational problems for functionals lacking regularity. Let $\Omega \subset \mathbb{R}^N$ be an open subset and $f: \Omega \times \mathbb{R} \to \mathbb{R}$ a function. Consider the Dirichlet problem

$$(\mathcal{P}) \quad \begin{cases} -\Delta u = f(t, u) & \text{in } \Omega, \\ u = 0 & \text{on } \partial\Omega. \end{cases}$$

To find a weak solution of $(\mathcal{P})$ by variational methods, in general, we consider the energy functional

$$\Phi(u) = \frac{1}{2} \int_\Omega |\nabla u|^2 \, dx - \int_\Omega F(x, u) \, dx,$$

where $F(x, t) = \int_0^t f(x, s) \, ds$, in a *space of admissible functions* X, where $X = H_0^1(\Omega)$. If Φ is differentiable on X and u is a critical point of Φ, then we have

$$\langle \Phi'(u), h \rangle = \int_\Omega \{\Delta u + f(x, u)\} \, h(x) \, dt = 0$$

for every h in a *space of test functions* $Y \subset X$. Usually, for problems like $(\mathcal{P})$, $Y = C_0^\infty(\Omega)$. The second equality is exactly what characterizes a *weak solution* of $(\mathcal{P})$. This variational approach fails in the case *when Φ is not smooth*.

To get a weak solution, we see that $\langle \Phi'(u).h \rangle$ need not be defined at any h outside Y. Then, we can weaken, in this sense, the differentiability condition and the definition of a critical point and still be able to deal with such problems by using a variational approach. It is expected, naturally, that the growth condition required on $f(x, t)$ might be improved and that the regularity on f may be weakened.

Three forms of *constrained* versions of the MPT have appeared, where a weak notion of regularity (the one needed earlier) is used. Chronologically, the first one was introduced by Struwe in [873], it may be found in Section 10 of his book [882]. It is developed for functionals which may take infinite values and are only Gâteaux differentiable on their effective domains in directions of a dense space of testing functions, to quote Struwe.

The second one was given by Duc [349] and will be the main result in this note. We found no trace of Duc's version of the MPT elsewhere than in the preprint [349].

The third one is attributed to Arcoya and Boccardo [61]. This result is the object of the next note.

Duc Directionally Constrained MPT

Let X be a Banach space and $f : U \subset X \to \mathbb{R}$ a real functional where U is an open subset of X. Let $x \in U$ and Δ be *a linear subspace* of X.

Definition 25.1. A function f is Δ-differentiable in x if there exists a *linear mapping* $f_\Delta'(x) : \Delta \to \mathbb{R}$ such that

$$\lim_{t \to 0} \frac{f(x + th) - f(x)}{t} = \langle f_\Delta'(x), h \rangle, \qquad \text{for all directions } h \in \Delta,$$

and such that for each $h \in \Delta$, the mapping $x \mapsto f_\Delta'(x).h$ is continuous.

A function f is said to be Δ-differentiable on U if it is Δ-differentiable at any point $x \in U$.

Remark 25.1.

1. The subspace Δ is supposed to be *linear* because $f_\Delta'(x)$ is required to be linear.
2. The functional f_Δ' may be not continuous on the whole Δ.
3. If $\Delta = X$ and $f_\Delta'(x)$ is continuous on X for every $x \in X$, then f is Gâteaux differentiable.

We have the following proposition that expresses a property of *finite dimensional smoothness* on Δ-differentiable functions.

Proposition 25.1. *Let f be Δ-differentiable on an open subset $U \subset X$. Let Δ_1 be a finite dimensional subspace of Δ and let $h: \,]a, b[\subset \mathbb{R} \to X$ be a C^1-function such that*

$$h(]a, b[) \subset U \cap (x + \Delta_1).$$

Then $f \circ h$ is of class C^1 on $]a, b[$.

Banach Surfaces

Let E be a Banach space and $S \subset E$ be a subset.

Definition 25.2. We call S an E-surface if, for any $x \in S$, there is a hyperplane F, $e \in E$, and a neighborhood $\mathcal{N}$ of x in S such that:

  i. $E = F \oplus \operatorname{span}\{e\}$, and
  ii. there are two open subsets $W \subset F$, $V \subset \operatorname{span}\{e\}$ and $\varphi: W \to V$ of class C^1 such that

$$\mathcal{N} = \phi(W), \qquad \text{where} \qquad \phi(w) = (w, \varphi(w)), \qquad \text{for } w \in W.$$

An E-surface S is said to be *symmetric* if $x \in S$ implies that $-w \in S$, and it is *closed* if it is closed in E.

Example 25.1. Let $U \subset E$ be an open subset and $f: U \to \mathbb{R}$ a Fréchet continuously differentiable function. Let $r \in \mathbb{R}$ such that $f'(x) \neq 0$ whenever $f(x) = r$. Denote $\partial f_r = \{x \in U;\ f(x) = r\}$. Let $x \in \partial f_r$, and set $F = \{f'(x)\}^{-1}(0)$. Let $e \in E \setminus F$. Then we can consider $E = F \oplus \operatorname{span}\{e\}$.

By the implicit function theorem, we can find W, V, φ, ϕ as in Definition 25.2, and $\mathcal{N} = (W \times V) \cap \partial f_r$. Therefore, ∂f_r is an E-surface.

Consider an E-surface S and let $x \in S$. Let F, e, V, W, and φ be as in Definition 25.2. Let ℓ be a continuous linear functional on E such that $\ell(e) \neq 0$. Then, $\ell(v) \neq 0$ for every $v \in \operatorname{span}\{e\} \setminus 0$. Set

$$f(w, v) = \ell(v - \varphi(w)) \qquad \text{for } (w, v) \in W \times V.$$

It is easy to prove that $S \cap (W \times V)$ is a ∂f_0. Thus, we can and shall consider in the sequel an E-surface as a ∂f_0 in a local study.

Hereafter, the notations $x, S, \mathcal{N}, F, e, V, W, \varphi, \phi$, and ∂f_0 are as in the preceding discussion. For any $y \in S$, we set

$$T_y = \{\alpha'(0);\ \alpha \text{ is a curve on } S \text{ such that } \alpha(0) = y\}.$$

Constrained Critical Point Theory

A notion of *constrained critical point* of a function f defined on S is defined as follows.

Definition 25.3. Let f be a Δ-differentiable function on an open set $U \subset E$ and let S be an E-surface contained in U. Let $x \in S$; it is a constrained Δ-critical point of f on S if

$$\langle f'(x), h \rangle = 0 \qquad \text{for all } h \in \Delta \cap T_x,$$

and $c = f(x)$ is called a constrained Δ-critical value of f on S.

In the rest of this note, we will suppose that Δ is a *dense* linear subspace of E and that S is a *closed connected E-surface* contained in U where $U \subset E$ is open. Suppose also that φ' is Lipschitz continuous for every φ in Definition 25.2.

Let f be a Δ-differentiable function on U and set

$$\|f'(x)\|_S^\Delta = \sup\big\{|\langle f'(x), h \rangle|; \ h \in \Delta \cap T_x \text{ and } \|h\| = 1\big\},$$

where $\|f'(x)\|_S^\Delta$ may be infinite. The number $\|f'(x)\|_S^\Delta$ is the norm of the tangent part of $f'(x)$ corresponding to the manifold S.

To state an MPT related to this notion, we need of course to define a (PS) condition for this kind of differentiability.

Definition 25.4. A Δ-differentiable real function f on U satisfies the (PS) condition at the level c on S if

$$\big(\mathrm{PS}_S^\Delta\big)_c \ \begin{cases} \text{Any sequence } (x_n)_n \subset S \text{ such that} \\ f(x_n) \to c \quad \text{and} \quad \|f'(x)\|_S^\Delta \to 0 \\ \text{admits a converging subsequence.} \end{cases}$$

Let us adopt the following notation:

$$\mathbb{K} = \big\{x; \ x \text{ is a constrained } \Delta\text{-critical point of } f \text{ on } S\big\},$$

$$\mathbb{K}_c = \big\{x \in \mathbb{K}; \ f(x) = c\big\},$$

$$f_S^c = \big\{x \in S; \ f(x) \le c\big\},$$

$$N_\delta^S(\mathbb{K}_c) = \big\{x \in S; \ \|x - y\| < \delta, \forall y \in \mathbb{K}_c\big\},$$

and

$$B(c, \varepsilon, \delta) = \big\{f_S^{c+\varepsilon} \setminus f^{c-\varepsilon}\big\} \setminus N_\delta^S(\mathbb{K}_c).$$

Then, we have the following result.

Lemma 25.2. *Suppose that f satisfies condition* $(\mathrm{PS}_S^\Delta)_c$. *Then, $\mathbb{K}_c$ is compact.*

Now the *constrained critical points* and the constrained (PS) condition make sense, we can state an appropriate *deformation lemma* and directionally constrained version of the MPT.

Lemma 25.3 (Deformation Lemma, Duc). *Let S be an E-surface contained in an open subset U of E. Let c be a real number and f be a Δ-differentiable real function on U satisfying $(\mathrm{PS}_S^\Delta)_c$. Let $\mathcal{N}$ be a neighborhood of $\mathbb{K}_c$ and v be a positive real number. Then, there exists $\mu \in]0, \varepsilon[$ and a homeomorphism η from S into S such that*

    i. $\eta(x) = x$ *for all* $x \notin f_{c-\varepsilon}^{c+\varepsilon}$, *ii.* $\eta(f^{c+\mu} \setminus \mathcal{N}) \subset f^{c-\mu}$, *and*

    iii. *if* $\mathbb{K}_c = \varnothing$, *then* $\eta(f^{c+\mu}) \subset f^{c-\mu}$.

Theorem 25.4 (Directionally Constrained MPT, Duc). *Let S be a connected closed E-surface contained in an open subset U of E. Let f be a real Δ-differentiable function on U satisfying the condition $(PS_S^\Delta)_c$ for every real number c. Assume that there are two distinct points v and w in S such that*

$$\max\{f(v), f(w)\} < \inf\{f(x), \|x - v\| = r\},$$

where $r < \|v - w\|$.

If the set

$$\Gamma = \left\{ \gamma \in \mathcal{C}([0, 1]; S); \ \gamma(0) = v \ and \ \gamma(1) = w \right\}$$

is not empty, then

$$d = \inf_{\gamma \in \Gamma} \max_{u \in \gamma([0,1])} f(x)$$

is a constrained Δ-critical value of f on S.

The Directionally Constrained MPT of Arcoya and Boccardo

Another version of the directionally constrained MPT is attributed to Arcoya and Boccardo [61], which was announced already in [60]. Similarly to Duc's results, it deals with functionals that are not differentiable in all directions.

Consider a Banach space $(X, \| \cdot \|_X)$, and a subspace $Y \subset X$, which is a normed space with a norm $\| \cdot \|_Y$. Suppose that $\Phi \colon X \to \mathbb{R}$ is a functional on X that is continuous in $(Y, \| \cdot \|_X + \| \cdot \|_Y)$ and satisfies the following:

    a. Φ has a directional derivative at each $u \in X$ for any direction $v \in Y$.

    b. For fixed $u \in X$, the function $\langle \Phi'(u), v \rangle$ is linear in $v \in Y$ and for fixed $v \in Y$, the function $\langle \Phi'(u), v \rangle$ is continuous in $u \in X$.

A *critical point* $u \in X$ of Φ is a point such that $\langle \Phi'(u), v \rangle = 0$ for all $v \in Y$.

Theorem 25.5 (Directionally Constrained MPT, Arcoya and Boccardo). *Suppose that Φ has the preceding form, satisfies conditions* a *and* b *and is such that, for some* $e \in Y$,

$$\max\{\Phi(0), \Phi(e)\} < \inf_{\gamma \in \Gamma} \max_{t \in [0,1]} \Phi(\gamma(t)),$$

where $\Gamma = \left\{ \gamma \in \mathcal{C}((Y, \| \cdot \|_X + \| \cdot \|_Y); \mathbb{R}); \ \gamma(0) = 0 \ and \ \gamma(1) = e \right\}$. *Then there exists a sequence* $(u_n)_n$ *in Y satisfying for some* $(K_n)_n \subset \mathbb{R}^+$ *and* $(\varepsilon_n)_n$, *such that* $\varepsilon_n \to 0$,

$$(\Phi(u_n))_n \quad \text{is bounded,}$$

$$\|u_n\|_Y \leq 2K_n, \quad \text{for all } n \in \mathbb{N}, \tag{25.1}$$

$$|\langle \Phi'(u_n), v \rangle| \leq \varepsilon_n \left(\frac{\|v\|_Y}{K_n} + \|u\|_X \right), \quad \text{for all } v \in Y.$$

For the following form of the Palais-Smale condition,

> Any sequence $(u_n)_n$ in the Banach space Y satisfying for some $(K_n)_n \subset \mathbb{R}^+$ and $(\varepsilon_n)_n$ where $\varepsilon_n \to 0$ the condition (25.1) possesses a convergent subsequence in X;

there exists a subsequence of $(u_n)_n$ that converges to some critical point (in the sense of the preceding definition). In the regular case $X = Y$, we use, of course, the usual form of (PS).

Applications were given in [61,62] to the existence and multiplicity of non-negative critical points of functionals that may fail to be differentiable in $H_0^1(\Omega)$ but are indeed $\mathcal{C}^1$ on $H_0^1(\Omega) \cap L^\infty(\Omega)$.

In the same spirit, in [708] (Critical points for non differentiable functionals), Pellacci deals with the existence of critical points for integral functionals J defined in the Sobolev space $W_0^{1,p}(\Omega)$, not Fréchet-differentiable on $W_0^{1,p}(\Omega)$ but only Gâteaux-derivable along directions from $W_0^{1,p}(\Omega) \cap L_\infty(\Omega)$.

The basic tool is the variant of the metric critical point theory for continuous functionals seen in the previous chapter.

The Approach of Struwe

In [879], Struwe outlined a general critical point theory for functionals that are not Fréchet differentiable on their natural domains of definition (see [882, Section 10] for the details).

◇ 25.II Morse-Ekeland Index and the MPT

The main references for the Morse-Ekeland index of periodic solutions to convex Hamiltonian systems are the excellent book by Ekeland [360] and the papers [357,363, 364,369,433].

The index, as defined in [360], is an integer associated with a linear system on a time interval, provided the Hamiltonian is positive. Other definitions exist, valid for any quadratic Hamiltonian. We would like to signal here the combination of Morse-Ekeland index with the MPT in the treatment of Hamiltonian systems following Ekeland in [359]. Let

$$J = \begin{pmatrix} 0 & I_n \\ -I_n & 0 \end{pmatrix}$$

be the symplectic matrix on $\mathbb{R}^{2n}$, $H : \mathbb{R}^{2N} \to \mathbb{R}$ a $\mathcal{C}^2$-function. Consider the system

$$(H) \begin{cases} \dfrac{dx}{dt} = JH'(x) \\ x(0) = x(T), \end{cases}$$

where the period T is prescribed. Rabinowitz proved [741] the following existence result for (H).

Theorem 25.6 (Rabinowitz). *Assuming that*

   i. $H(x) > H(0)$ for all $x \neq 0$,
   ii. $H(x)||x||^{-2} \to 0$, and
   iii. There exist $r > 0$ and $\beta > 2$ such that $H'(x)x \geq \beta H(x)$ for all $||x|| \geq r$,

then for any positive T, the system (H) *has a nontrivial solution.*

Rabinowitz did not claim that the solution he finds has minimal period T because T/k-solutions are also T-periodic solutions.

Suppose now that instead of assumption ii we have

ii′. $H''(0) = 0$.

Consider the following extra conditions:

1. H is strictly convex.
2. $H'(x)x \geq \beta H(x)$.
3. There is a k such that $H(x) \leq \frac{k}{\beta}||x||$.

The Fenchel transform of H,

$$G(y) = H^*(y) = \sup_{x \in \mathbb{R}^{2n}} [(x, y) - H(x)]$$
$$= [(x, y) - H(x); \ y = H'(x)],$$

is well defined by condition 1 and by conditions 2 and 3,

$$G(x) \geq \frac{k^*}{\alpha}||\alpha||, \qquad \text{where } \frac{1}{\alpha} + \frac{1}{\beta} = 1.$$

Set

$$\Phi(u) = \int_0^T \left[\frac{1}{2}(J\dot{u}, u) + G(-J\dot{u}) \right] dt$$

on $E = \{u : [0, T] \to \mathbb{R}; \ \dot{u} \in L^\alpha \text{ and } u(0) = u(T)\}$. The mapping Φ is C^1 but not C^2 because $1 < \alpha < 2$, for $k > \alpha$, there are essentially no C^k-maps on L^α, and a C^2-map on L^α is exactly quadratic.

Moreover, $\Phi'(u) = 0$ if and only if there is $\xi \in \mathbb{R}^{2n}$ such that $\bar{u}(t) = u(t) + \xi$ is a solution of (H).

Since $\Phi(u + \xi) = \Phi(x)$ for all $\xi \in \mathbb{R}^{2N}$, we may consider $v = \dot{u}$ as the *true variable*. For $u \in L_0^\alpha = \{u \in L^\alpha(0, T; \mathbb{R}^{2n}; \int_0^T u \, dt\}$, we define the primitive of u in L_0^α, $\Pi \colon L_0^\alpha \to L_0^\alpha$ by $\Pi v = u$ if and only if $\dot{u} = v$, $\int_0^T u \, dt = 0$, and $\Psi(v) = \Phi(\Pi v)$.

Since Π is an isomorphism, we are reduced to finding a nonzero critical point of Ψ. And the method is, as you may have guessed, the MPT.

Indeed, Ψ is of class C^1 and satisfies (PS). Moreover, if $v(t) = \exp(2\pi Jt/T\xi)$ for some $\xi \in \mathbb{R}^{2n}$ such that $||\xi|| = 1$, then for λ large enough $\Psi(\lambda v)$ is negative.

Morse-Ekeland Index

Fix a nonzero critical point u of Ψ and consider the associated quadratic form

$$Q(u)(w, w) = \int_0^T (Jw, \Pi w) + G''(-Ju)(\dot{J}w, Jw)dt.$$

As remarked before, Ψ is not twice differentiable; however, $Q(u)$ plays the role of the second derivative of Ψ. Assuming H'' to be positive for $x \neq 0$, G'' is positive definite for $y \neq 0$. Thus $\int_0^T G''(-Ju)(Jw, Jw)$ is positive definite and, since Π is compact, $Q(u)$ has a well-defined Morse index and nullity.

Definition 25.5. The Morse-Ekeland "Index (u)" is the index of $Q(u)$. And the notation "Nullity (u)" stands for the nullity of $Q(u)$.

> This new index can be easily computed by counting *conjugate points*, whereas the Morse index of the strongly indefinite functional Φ is infinite because of the spectral properties of $u \mapsto J\dot{u}$, and therefore would be useless there.

Definition 25.6. For $s \in [0, T]$, we say that s is *conjugate to zero* if, u being the solution of (H) such that $\dot{u} = v$, the system

$$\begin{cases} \dot{y} = JH''(u(t))y \\ y(0) = y(s) \end{cases} \qquad (H_s'')$$

has a nontrivial solution.

Theorem 25.7. Index (u) *is the number of points in* $]0, T[$ *conjugate to zero counted with their multiplicity.*

Corollary 25.8. *If u is T/k-periodic, then* Index $(u) \geq k - 1$.

Minimal Period Problem

Theorem 25.9 (Hofer [481]). *Suppose $H \in C^2(\mathbb{R}^{2n}; \mathbb{R})$ satisfies*

a. $H''(x) > 0$ *for* $x \neq 0$,
b. $H''(0) = 0$,
c. *there exist $r > 0$ and $\beta > 2$ such that $H'(x).x \geq \beta H(x)$ for $\|x\| \geq r$.*

Then for all T, the system (H) has at least one solution with minimal period T.

In the same spirit, as reported earlier, Rabinowitz [741] proved without supposing the convexity of H the existence (only) of a nontrivial solution. Ambrosetti and Mancini [47] proved the theorem stated earlier (both existence and minimality) with the extra condition that the Fenchel conjugate satisfies

$$(G'(y), y) \geq \lambda(G''(y)y, y) \qquad \text{for some suitable } \lambda > 1.$$

More recently, Girardi and Matzeu [431], still in the convex case, proved this result for a broader class of Hamiltonians.

A solution $\overline{u}$ of (H) is obtained as a mountain pass critical point of the functional Ψ. We will sketch the way Ekeland proves that $\overline{u}$ is not T/k-periodic for $k \geq 2$; that is, T is the minimal period of $\overline{u}$.

First, by a result from Hofer concerning the structure of the critical set in the MPT, $\overline{u}$ is either a local minimum or is of mountain-pass type (see Chapter 12).

Case 1. $\overline{u}$ is a local minimum. Then, $\overline{u}$ has index 0. By Corollary 25.8, if $\overline{u}$ is T/k-periodic, we must have $k - 1 \leq 0$, that is, $k \leq 1$.

Case 2. $\overline{u}$ is of mountain-pass type, the proof is delicate but beautiful. If Ψ was C^2, this would be easy to show; we would use an extension of the Morse lemma by Gromoll and Meyer [455] improved by Hofer [482].

◇ 25.III Homological and Homotopical Linking

We describe the concept of linking from the points of view of the homotopy (Liu) and the homology (Chang). Some definitions were already introduced in Chapter 12. They will not be recalled there.

Let M be a C^2-Finsler manifold and $\Phi \in C^1(M; \mathbb{R})$.

Definition 25.7. Let D be a k-topological ball in M and let $S \subset M$ be a subset. We say that ∂D and S *homotopically link* if $\partial D \cap S = \varnothing$ and $\gamma(D) \cap S \neq \varnothing$ for each $\gamma \in C(D, M)$ such that $\gamma|_{\partial D} = I_d$.

This notion is used in the result.

Theorem 25.10. *Assume that ∂D and S homotopically link. If $\Phi \in C(M; \mathbb{R})$ satisfies*

$$\Phi(x) > \alpha, \forall x \in S \tag{25.2}$$

and

$$\Phi(x) \leq \alpha, \forall x \in \partial D, \tag{25.3}$$

then $\pi_k(\Phi^b, \Phi^a) \neq 0$, where $b > \max_{x \in \overline{D}} \Phi(x)$.

A similar notion of homological linking was introduced by Liu [585].

Definition 25.8. Let D be a k-topological ball in M and S be a subset in M. We say that ∂D and S *homologically link* if $\partial D \cap S = \varnothing$ and $|\tau| \cap S \neq \varnothing$ for each singular k-chain τ with $\partial \tau = \partial D$ where $|\tau|$ is the support of τ.

The same way as before, we get the following.

Theorem 25.11. *Assume that ∂D and S homologically link. If $\Phi \in C(M; \mathbb{R})$ satisfies (25.2) and (25.3), then $H_k(\Phi^b, \Phi^a) \neq 0$.*

The Particular Situation of the MPT. Let Ω be a neighborhood of 0 in a Banach space X. Let $e \notin \overline{\Omega}$, $S = \partial D = \{0, e\}$, and let $D = [z_0, z_1]$ be the segment joining z_0 and z_1.

The Situation of the Generalized MPT. Let $X = X_1 \oplus X_2$ be a Banach space where $\dim X_1 < +\infty$. Let $e \in X_2$, $||e|| = 1$ and let R_1, R_2, $\rho > 0$ with $\rho < R_1$. Set $S = X_2 \cap \partial B$ and $D = \{x + se; \ x \in X_1 \cap B_{R_2}, s \in [0, R_1]\}$.

In the two cases, it is easily shown that ∂D and S homotopically link exactly as for the linking theorem (see Theorem 19.2 on page 229). The detailed proofs may be found, for example, in [114, 205, 628, 882].

Actually, ∂D and S homologically link, too.

Theorem 25.12. *Suppose the boundary ∂D of the k-topological ball D and S homotopically link. Assume that*

1. $S \cap D =$ *single point,*
2. S *is a path-connected orientable submanifold with codimension k, and*
3. *there exists a tubular neighborhood N of S that $N \cap D$ is homeomorphic to D.*

Then ∂D and S homologically link.

Theorem 25.13. *Let $a < c$ be regular values of Φ. Set*

$$c = \inf_{Z \in \alpha} \sup_{x \in Z} \Phi(x) \qquad with \qquad \alpha \in \pi_k(\Phi^b, \Phi^a) \ nontrivial,$$

and

$$c^* = \inf_{\tau \in \alpha} \sup_{x \in |\tau|} \Phi(x) \qquad with \qquad \alpha \in H_k(\Phi^b, \Phi^a) \ nontrivial.$$

Assume $c > a$ (resp. $c^* > a$) and that Φ satisfies $(PS)_c$ (resp. $(PS)_c^*$). Then, c (resp. c^*) is a critical value of Φ. Moreover, we have $c^* \le c$.

Proof. It suffices to check that the families $\mathcal{F} = \{|Z|; \ Z \in \alpha\}$ (and $\mathcal{F}^* = \{|\tau|; \ \tau \in \alpha\}$) are invariant with respect to $\Phi_{\varepsilon_0} = \{\gamma \in C(M, M); \ \gamma \sim Id, \gamma|_{\Phi^{c-\varepsilon_0}} = Id\}$. This follows by the homotopy invariance of the homology classes. The inequality $c* \le c$ follows from $\mathcal{F} \subset \mathcal{F}^*$. $\square$

The cases of the MPT and generalized MPT are contained in the particular case $\alpha = \pi_k(\Phi^b, \Phi^a)$, where

$$c = \inf_{\gamma \in \Gamma} \sup_{x \in \gamma(D)} \Phi(x)$$

and $\Gamma = \{\gamma \in C(D, M); \ \gamma|_{\partial D} = Id\}$.
The assumption

$$\Phi|_{\partial D} \ge d > a, \qquad \forall x \in S \tag{25.4}$$

guarantees that $c \ge d > a$, so c is a critical value.

If, instead of $\Phi|_{\partial D} \le a < d \le \Phi|_S$, we suppose only $\Phi|_{\partial D} \le a \le \Phi|_S$, we still get the result (the key to the problem being that ∂D is compact). The following interesting proof is given in [205].

Claim 25.1. If $c = a$, then $\mathbb{K}_c \neq \emptyset$.

Otherwise, since ∂D is compact and $\partial D \cap S = \emptyset$, there exist two open neighborhoods $U_2 \supset \overline{U_1} \supset U_1$ of ∂D such that $U_2 \cap S = \emptyset$. According to the first deformation lemma, there exist $\varepsilon > 0$ and a deformation $\eta: (-\Phi)^{-c+\varepsilon} \setminus U_2 \to (-\Phi)^{-c-\varepsilon}$ and $\eta|_{U_1} = I_d$ on U_1; that is, $\eta: \Phi_{c-\varepsilon} \setminus U_2 \to \Phi_{c+\varepsilon}$ and $\eta|_{U_1} = I_d$ on U_2. Therefore, $\eta(S) \subset \Phi_{c+\varepsilon}$. Let $S_1 = \eta(S)$, then we have

$$\begin{cases} S_1 \cap \partial D = \emptyset, \\ S_1 \cap \gamma(D) = \eta(S \cap \eta^{-1}\gamma(D)) \neq \emptyset, \forall \gamma \in \Gamma, \end{cases}$$

that is, S and ∂D link (homotopically).

Since $\Phi(x) \geq c + \varepsilon$ for all $x \in S_1$, we get the contradiction

$$c = \inf_{\gamma \in \Gamma} \sup_{x \in D} \Phi(\gamma(x)) \geq \inf_{x \in S_1} \Phi(x) \geq c + \varepsilon.$$

$\square$

The strength of Morse theory comes from the extra information we get on critical points. Using the homological link, Liu [585] proved the following result.

Theorem 25.14. *Assume that $\alpha \in H_k(\Phi^b, \Phi^a)$ is nontrivial and*

$$c^* = \inf_{Z \in \alpha} \sup_{x \in |Z|} \Phi(x). \tag{25.5}$$

Suppose that Φ satisfies $(PS)_{c^*}$ and that $\mathbb{K}_{c^*}$ is isolated. Then there exists $x_0 \in \mathbb{K}_{c^*}$ such that $C_k(\Phi, x_0) \neq 0$.

This theorem also has a symmetric parent with a similar proof when Φ is G-invariant and m is a G-manifold for some compact Lie group G. It suffices to consider instead of α, a nontrivial set $[Z] \in H_G^k(\Phi^b, \Phi^a)$ and as a result $C_G^k(g, \mathcal{O}) \neq 0$ for some critical orbit $\mathcal{O} \subset \mathbb{K}_c$ where

$$c = \inf_{Z \in [Z]} \sup_{x \in [Z]} \Phi(x).$$

In the case of the MPT that interests us,

$$c = c^* = \inf_{\gamma \in \Gamma} \sup_{t \in [0,1]} \Phi(\gamma(t)),$$

where $\Gamma = \{\gamma \in C[0, 1]; X); \gamma(0) = 0, \gamma(1) = e\}$. By this result, there is $x_0 \in \mathbb{K}_c$ such that

$$C_1(\Phi, x_0) \neq 0. \tag{25.6}$$

A point satisfying (25.6) is called a *mountain pass point*. We have some precise information on such points.

Theorem 25.15. *Suppose that $\Phi \in C^2(M; \mathbb{R})$ and that x_0 is a mountain pass point. Assume that $\Phi''(x_0)$ is a Fredholm operator such that*

$$\dim \ker(\Phi''(x_0)) = 1 \text{ if } 0 \in \sigma(\Phi''(x_0)); \tag{25.7}$$

then

$$\text{rank } C_k(\Phi, x_0) = \delta_{q1}.$$

◇ 25.IV Topological Indices and the MPT

The main idea in both Morse and Ljusternik-Schnirelman theories is that a difference in the topological nature of the level sets of a functional indicates the presence of a critical level. The "detection" of these topological changes through the analytical behavior of the functional is generally done using some of the many existing topological indices. We describe some of these indices in relation to the MPT.

Information on the critical groups and Morse index at a critical point of mountain-pass type are known since the work of Ambrosetti [40] and Hofer [482]. In addition to the Morse index, other indices were used to investigate the specific situation of the MPT. We cite the following:

Morse-Conley-Benci Index. Very interesting results in connection with the MPT using a generalization of Morse-Conley index [250] were obtained by Benci in a series of papers [106–109]. The interested reader may also consult the monograph by Bartsch [93].

Relative Category and Strong Relative Category. These two indices (resp. cat and Cat) were introduced by Fournier and Willem [405] to prove a *mountain circle theorem* [406]. Other definitions of the relative category may be found in [380, 895]. This theorem is *similar* to the MPT, but gives existence of two critical points instead of one.

Theorem 25.16 (Mountain Circle Theorem, Fournier and Willem). *Let* $\Phi \in C^1(B \times \mathbb{R}^N; \mathbb{R})$ *satisfy the Palais-Smale condition on* $\Phi^{-1}([a, \infty[)$ *where* B *is a Banach space and* $a, b \in \mathbb{R}$.
If there exists $x \in \mathbb{R}^N$ *and* $r, s, t \colon B \to \mathbb{R}$ *continuous and such that* $r(u) > s(u) > t(u) \geq 0$, *for all* $u \in B$, *and*

$$\Phi(u, y) > b > a \geq \Phi(u, z), \qquad \forall u \in B, \ \forall \|x - y\| = s(u) \text{ and}$$
$$\|x - z\| = r(u), t(u),$$

then Φ *has two critical points in* $(B \times \mathbb{R}^N) \setminus \Phi^a$.

◇ 25.V The MPT for Upper Semicontinuous Compact-Valued Mappings

Recall that a multivalued mapping $F \colon X \to 2^Y$, where X and Y are topological spaces, is u.s.c. at a point $x_0 \in X$ if, for every open set V in Y such that $F(x_0) \subset V$, there exists an open neighborhood W of x_0 such that $F(x) \subset V$ for every $x \in W$, and F is u.s.c. if it is u.s.c. at every point $x \in X$. The mapping F is *usco* if it is u.s.c. and $F(x)$ is compact for every $x \in X$.

A notion of *weak slope* was introduced by Ribarska et al. [768] as an analog of the weak slope for "sup F" to extend the metric critical point theory to l.s.c. (single-valued) functionals.

Definition 25.9. Let (X, dist) be a metric space and $F: X \to 2^{\mathbb{R}}$ be an usco mapping. The *weak slope* of F at a point $x \in X$, denoted by $|dF|(x)$, is the supremum of the reals $\sigma \in [0, +\infty]$ such that there exist $\delta > 0$ and a continuous deformation $\mathcal{H}: B(x, \delta) \to [0, \delta] \to X$ such that for each $y \in B(x, \delta)$ and each $t \in [0, \delta]$, we have

   i. $\text{dist}(\mathcal{H}(y, t), y) \leq t$, and
   ii. $\sup F(\mathcal{H}(y, t)) \leq \sup F(y) - \sigma t$.

Then, Ribarska et al. establish an extension to usco mappings of the deformation lemma of Corvellec et al. [261].

Lemma 25.17 (Deformation Lemma for usco Mappings, Ribarska et al.). *Let (X, dist) be a metric space and $F: X \to 2^{\mathbb{R}}$ be an usco mapping. Let $\varepsilon, \delta > 0$ and $S \subset X$. If Q is an open neighborhood of $F^{-1}([c - \varepsilon, c + \varepsilon]) \cap S_\delta$ such that for every $y \in Q$, the weak slope $|dF|(y) > 2\varepsilon/\delta$, then there exists a continuous mapping $\eta: X \times [0, +\infty) \to X$ with the following properties:*

   i.   *$\eta(x, 0) = x$ for every $x \in X$,*
   ii. *$\eta(x, t) = x$ for every $x \in X \setminus Q$ and $t \in [0, +\infty)$,*
   iii. *if $x \in \overline{S}$ and $\sup F(x) \leq c + \varepsilon$, then $\sup F(\eta(x, \delta)) \leq c - \varepsilon$,*
   iv. *$\text{dist}(x, \eta(x, t)) \leq t$ whenever $x \in X$ and $t \in [0, +\infty)$.*

This provides an alternative approach to the lower semicontinuous case since $F(x) = \{t \in \mathbb{R};\ (x, t) \in \overline{\text{graph}(f)}\}$ defines an usco mapping whenever f is a *locally bounded lower semicontinuous* function. The following MPT for usco mappings is then deduced.

Let (X, dist) be a complete metric space and $F: X \to 2^{\mathbb{R}}$ be an usco mapping. Let M be a subset of X and $\mathcal{M} \subset 2^X$. Denoting

$$c(F, \mathcal{M}) = \inf\left\{ \sup\left(\bigcup_{x \in A \cap M} F(x) \right);\ A \in \mathcal{M} \right\},$$

we have the following result.

Theorem 25.18 (MPT for usco Mappings, Ribarska et al.). *Let D be a closed subset of X and u, v be two points of X from two disjoint components of $X \setminus D$. Assume*

$$c(X, F, \Gamma) = c(D, F, \Gamma) = c,$$

where Γ is the set of continuous paths joining u and v. If F satisfies (PS)$_c$, *that is,*

   If there is a sequence $(x_n)_n$ such that there is $y_n \in F(x_n)$,

$$y_n \to c \quad \text{and} \quad |dF|(x_n) \to 0,$$

   then c is a critical value; that is, there exists $x_0 \in X$ such that $c \in F(x_0)$ and $|dF|(x_n) = 0$.

   Then c is a critical value of F.

◇ 25.VI A Critical Point Theory for Multivalued Mappings with Closed Graph

Let (X, dist) be a complete metric space and $F: X \to 2^{\mathbb{R} \cup \{+\infty\}}$ be a multivalued mapping with closed graph and nonempty values. The graph of F,

$$\text{graph } F = \{(u, c) \in X \times \mathbb{R};\ c \in F(u)\},$$

is a metric space for the metric $\text{dist}\,((u, c), (v, b)) = \left(\text{dist}\,(u, v)^2 + |b - c|^2\right)^{1/2}$. Frigon extends in [409] the notion of weak slope and the corresponding critical point theory to multivalued mappings of this form.

Definition 25.10. Let $(u, c) \in \text{graph } F$. The weak slope of F at (u, c), denoted $|dF|(u, c)$, is the supremum of all $\sigma \in [0, +\infty]$ such that there exist $\delta > 0$ and a continuous function

$$H = (H_1, H_2): B((u, c); \delta) \times [0, \delta] \to \text{graph } F,$$

where $B((u, c); \delta)$ is the open ball of center (u, c) and radius δ in graph F, such that

  i. $\text{dist}\,((H(v, b), t), (v, b)) \leq t\sqrt{1 + \sigma^2}$, and
  ii. $H_2((v, b), t) \leq b - \sigma t$.

A point $u \in X$ is said to be a critical point of F at level $c \in \mathbb{R}$ if $c \in F(u)$ and $|dF|(u, c) = 0$.

In the case where $F(u) = \{f(u)\}$ is a continuous single-valued function, then $|dF|(u, f(x)) = |df|(u)$. Frigon defines then a (PS) condition, obtains a deformation lemma, and applies it to prove a corresponding version of the MPT.

More recently, this approach was investigated more deeply in two papers by Kristaly and Varga, who extended Frigon results to a form requiring a weaker compactness condition (like that of Cerami for single-valued mappings) in [540] and investigated some multiplicity results in [541] using an index theory as in Chapter 11. We present now the formulation of the MPT of Kristaly and Varga.

Theorem 25.19 (The MPT for Multivalued Mappings with Closed Graph, Kristaly and Varga's formulation). *Let (X, dist) be a complete metric space and $F: X \to 2^{\mathbb{R} \cup \{+\infty\}}$ be a multivalued mapping with closed graph and nonempty values, and let (u_0, c_0) and (u_1, c_1) be two points in the same path-connected component of graph F. Let*

$$\Gamma = \left\{\gamma \in C([0, 1]; \text{graph } F);\ \gamma(0) = (u_0, c_0) \text{ and } \gamma(1) = (u_1, c_1)\right\}$$

and set

$$c = \inf_{\gamma \in \Gamma} \max_{t \in [0, 1]} \pi_{\mathbb{R}}(\gamma(t)),$$

where the functional $\pi_{\mathbb{R}}: \text{graph } F \to \mathbb{R}$ is defined by $\pi_{\mathbb{R}}(u, c) = c$.

Suppose that the set

$$D_c = \{(x, b) \in \text{graph } F; \ b \geq c\}$$

separates (u_0, c_0) *and* (u_1, c_1).
Then, there exists a sequence $(x_n, b_n) \in \text{graph } F$ *such that*

    a. $b_n \to c$,
    b. $(1 + \|x_n\|)|dF|(x_n, b_n) \to 0$, *and*
    c. $\text{dist}((x_n, b_n), D_c) \to 0$.

The approach of Frigon also provides a new approach to use with single-valued lower semicontinuous functionals and refines the extension by Ribarska et al.

Frigon applied his version to study the differential inclusion $-\Delta u \in G(u)$ in $\Omega, u = 0$ on $\partial\Omega$, where Ω is a bounded domain of $\mathbb{R}^n$ with smooth boundary and $G: \mathbb{R} \to \mathbb{R}$ is upper semicontinuous with nonempty, compact, convex values. See also the note by Ribarska et al. [772].

◇ 25.VII An MPT on Closed Convex Sets for Functionals Satisfying the Schauder Condition

Suppose H is a Hilbert space, $f: H \to \mathbb{R}$ is a C^1-functional that satisfies the (PS) condition, and M is a closed convex set in H. Let $f(x) = \|x\|^2 - g(x)$, $Ax = g'(x)$. If the condition $AM \subset M$ is satisfied, then we say that $f(x)$ satisfies the *Schauder condition* on M. In [887] (The Schauder condition in the critical point theory), Sun first proves that if M is a closed convex set in H, $f(x)$ satisfies the Schauder condition on M, and

$$c = \inf_{x \in M} f(x) > -\infty,$$

then c is a critical value of $f(x)$. Then, he proves an MPT on a closed convex set.

Theorem 25.20 (MPT on Closed Convex Sets for Functionals Satisfying the Schauder Condition, Sun). *Suppose that M is a closed convex set in H, and $f(x)$ satisfies the Schauder condition on M. Let D be an open subset in M, $x_0 \in D$, $x_1 \in M \setminus \overline{D}$ satisfying*

$$\inf_{x \in \partial D} f(x) > \max\{f(x_0), f(x_1)\},$$

where ∂D is the boundary of D with respect to M, then $c = \inf_{h \in \Phi} \max_{t \in [0,1]} f(h(t))$ is a critical value of $f(x)$, where $\Phi = \{h(t): [0, 1] \to M$ is continuous; $h(0) = x_0$, $h(1) = x_1\}$.

◇ 25.VIII A Relative MPT

Suppose that $\Phi\colon H_0^1 \to \mathbb{R}$ is a C^1-functional and let (t, u_0), $0 \le t < \eta(u_0)$ be the right-direction saturation of the initial value problem

$$\begin{cases} u' = f(u) \\ u(0) = u_0 \in H_0^1. \end{cases}$$

Definition 25.11. A set N is an invariant set of descent flow of Φ if

$$\{u(t, u_0); \; 0 \le t \le \eta(u_0), u_0 \in N \setminus \mathbb{K}\},$$

where $\mathbb{K}$ is the critical set of Φ.

Let $c \in \mathbb{R}$, N is an invariant set of descent flow of Φ. We say that Φ has the *retracting property* for c in N if, for any $b > c$, $\Phi^{-1}(c, b] \cap N \cap \mathbb{K} = \varnothing$, then $\Phi^c \cap N$ is a retract of $\Phi^b \cap N$; that is, there exists a continuous deformation $\eta\colon \Phi^b \cap N \to \Phi^c \cap N$ such that $\eta(\Phi \cap N) \subset \Phi^c \cap N$ and $\eta|_{\Phi^c \cap N} = I_d$.

In [570], Li and Zhang proved the following form of the MPT.

Theorem 25.21 (Relative MPT, Li and Zhang). *Consider a C^1-functional Φ on $H_0^1(\Omega)$ that has a retracting property for all $m \in \mathbb{R}$ on a connected invariant set N of the descent flow of Φ in $C_0^1(\Omega)$ that satisfies the Palais-Smale condition and $\inf_{x \in \partial B(x_0, \rho)} \Phi(x) > \max\{\Phi(x_0), \Phi(x_1)\}$, where $x_0, x_1 \in N$, $B(x_0, \rho) \subset N$, and $x_1 \notin B(x_0, \rho)$. Then, either*

  i. *Φ has infinitely many critical points in N, or*
  ii. *Φ has a mountain pass point relative to N. Moreover,*

$$c = \inf_{\gamma \in \Gamma_N} \sup_{t \in [0,1]} \Phi(\gamma(t))$$

is a critical value where $B(x_0, \rho) = \{x \in H_0^1; \; \|x\| < \rho\} \cap N$ and Γ_N is the set of all continuous paths joining x_0 and x_1 and taking their values in N ($\gamma([0, 1]) \subset N$).

This theorem was used to prove that, under some conditions, an elliptic semilinear problem admits at least six solutions: two positive, two negative, and two sign-changing, three of them being relative mountain pass points.

◇ 25.IX An MPT with a Retractable Property Instead of (PS)

The Palais-Smale condition depends on the topologies of spaces. The following *retractable property* is defined in [988] (A variant mountain pass lemma) by Zhang to deal with a wider class of spaces.

Definition 25.12. Let X be a Banach space, and let $f\colon X \to \mathbb{R}$ be a C^1-functional, $c \in \mathbb{R}$. We say that f possesses the retractable property with respect to c if, for each $b > c$ such that $f^{-1}(c, b] \cap \mathbb{K} = \varnothing$, f^c is a retraction of f^b.

He extends the MPT in two directions:

1. Replace the (PS) condition by the retractable property.
2. Permit the limiting case.

Theorem 25.22 (The MPT for Functionals Satisfying the Retractable Property, Zhang). *Suppose that $f \in C^1(X; \mathbb{R})$ possesses the retractable property with respect to $c = f(\theta)$. Assume that θ is a local minimum of f, and that there is an $x_0 \in X$ such that $f(x_0) = f(\theta)$; then f has at least one nontrivial critical point.*

(A more general formulation was announced in the *Notices*, American Mathematical Society, **26**, 805-49-21, 1983). The retractable property was verified on the space C_0^1 for certain functionals occurring in second-order elliptic boundary value problems. A general theorem which combines the sub- and supersolution method with the preceding MPT was given. Further applications were obtained in this paper.

◇ 25.X An MPT for Locally Lipschitz Functionals on Locally Convex Closed Subsets of Banach Spaces

In [942, 944], Wang proves a minimax principle that includes the MPT for a locally Lipschitz functional Φ on *locally convex closed subsets* S of a Banach space. He introduces appropriate notions of critical points and (PS) conditions. The proof uses a deformation lemma.

◇ 25.XI An MPT for a Convex Functional on the Set of All Rearrangements of a Given Functional in L^p

Let (Ω, μ) and (Ω', μ') be two positive measure spaces such that $\mu(\Omega) = \mu(\Omega') < \infty$.

Definition 25.13. Two measurable functions $f: \Omega \to \mathbb{R}$ and $g: \Omega' \to \mathbb{R}$ are called *rearrangements* of one another if $\mu(f^{-1}[\beta, \infty)) = \mu'(g^{-1}[\beta, \infty))$ for every real β.

Burton [162] considers the class $\mathcal{F} \subset L^p$ of all rearrangements of a given function f_0 in an L^p space and studies the existence of critical points of a functional Φ relative to the class $\mathcal{F}$. He establishes a version of the MPT for a convex functional Φ relative to $\mathcal{F}$.

This result is applied to investigate multiple configurations for steady vortices in the study of the flow of an ideal fluid confined by a solid boundary. The functions in $\mathcal{F}$ represent possible configurations of a specified distribution of vorticity in the fluid, while Φ represents the kinetic energy.

◇ 25.XII An MPT for Continuous Convex Functionals

In [757] (Mountain-pass type theorems for nondifferentiable convex functions), Radulescu extends the MPT to *continuous convex functions* using *subdifferentiability*. He uses a pseudo-gradient lemma and Ekeland's variational principle in the proof of his variant of the MPT.

◇ 25.XIII An MPT in Variational Inequalities

In [995] (The mountain pass lemma in variational inequalities, in Chinese), Zhong generalizes the MPT to the case of *variational inequalities*. He uses Ekeland's variational principle.

◇ 25.XIV A Product MPT

A very nice idea was introduced by Sere [827] to deal with multiplicity of homoclinic orbits of first-order Hamiltonian systems. It seems to have attracted the attention of many mathematicians working on Hamiltonian systems (see [13, 14, 235, 273, 274, 903], for example).

◇ 25.XV Jacobian Conjecture

Jacobian conjectures are concerned with some global properties concerning vectorial functions $F \colon \mathbb{R}^n \to \mathbb{R}^n$ based on the information on their (local) Jacobian matrices.

We cite the celebrated *Keller Jacobian conjecture on* $\mathbb{R}^n$ (1939),

Every polynomial map $F \colon \mathbb{R}^n \to \mathbb{R}^n$ such that $\det F'(\mathbf{x}) \equiv 1$ is *injective*;

the *Markus-Yamabe conjecture* (1960), the *weak polynomial Markus-Yamabe conjecture* (1963), the *real Jacobian conjecture on* $\mathbb{R}^n$,

Every polynomial map $F \colon \mathbb{R}^n \to \mathbb{R}^n$ such that $\det F'(\mathbf{x}) \neq 0$ is *injective*;

and what interests us here, the *Chamberland conjecture*,

Every C^1 map $F \colon \mathbb{R}^n \to \mathbb{R}^n$ such that the eigenvalues of $F'(\mathbf{x}) \neq 0$ are uniformly bounded away from zero is *injective*;

which remains still open even for the case $n = 2$.

In [201] (A mountain pass to the Jacobian conjecture), Chamberland and Meisters prove, using the MPT, that if the eigenvalues of $F'(\mathbf{x})F'(\mathbf{x})^\perp$ are uniformly bounded away from zero for $\mathbf{x} \in \mathbb{R}^n$, where $F \colon \mathbb{R}^n \to \mathbb{R}^n$ is a class C^1 map, then F is injective.

This was discovered in a attempt by the authors to prove the Chamberland conjecture. If true, it would imply (via *reduction of degree*) the Keller Jacobian conjecture, which is also still open even for $n = 2$.

For a similar approach, also consult the paper by Rabier [732].

◇ 25.XVI The MPT in Quantum Chemistry

Besides what has been said in Chapter 22 about the existence of an algorithm based on the MPT, we would like to bring attention to a recent report (of 126 pages) [523] where the author treats a *mechanical* model that describes adiabatic chemical reactions by the geometrical method of reaction paths [469] using an advanced discussion of the results of Pucci and Serrin [727] on the structure of the critical set in the MPT. The reaction paths are used to study the dynamics of the chemical reactions. We report the following quotation from [523]:

> In the case of unimolecular reactions, educt and product are associated with a minimizer of the energy function E. The transition structure is associated with a saddle of the energy function .... The course of adiabatic chemical reaction is frequently likened to a walk on the potential energy surface. Such a walk starts in an educt valley, passes over a col, and ends in a product valley. In the case of unimolecular reactions this situation can be described by a reaction path, i.e., a continuous curve configuration space which joins two minimizers of the energy function via a saddle point. A reaction path describes a 'reaction walk' only if it passes through the saddle point that is associated with the lowest energy barrier between educt valley and product valley. In Chapter 3, the reaction path concept is discussed against the background of the Mountain-Pass Theorem.

◇ 25.XVII The MPT Proving Uniqueness in a Nonlinear Problem

Alama et al. [11] considered a Ginzburg-Landau system of the form

$$\begin{cases} -f'' - \frac{1}{r}f' + \frac{(d-S)^2}{r^2}f = x^2(1-f^2)f, \\ -S'' + \frac{1}{r}S' = (d-S)f^2, \end{cases} \tag{$\mathcal{GL}$}$$

where $f \geq 0$, $f(r)$, $S(r) \to 0$ as $r \to 0$, and $f(r) \to 1$, $S(r) \to d$ as $r \to \infty$. A solution (f, S) of this problem is said to be "admissible" if $f \geq 0$ for all $r \geq 0$, $S(0) = 0$, and if the Ginzburg-Landau free energy is finite.

The authors of [11] show that any such solution is a nondegenerate relative minimum point of the free energy functional constrained to a convex set. Then they use a version of the MPT to derive a contradiction if there should be more than one solution.

Theorem 25.23. *Let $d \in \mathbb{Z} \setminus \{0\}$ and $x \in \mathbb{R}$ with $x^2 \geq 2d^2$. Then there exists a unique admissible solution to the Ginzburg-Landau equations ($\mathcal{GL}$).*

Recall that, thanks to the MPT, a functional that satisfies (PS) and admits two local minima should possess a third critical point that is not a local minimum.

◇ 25.XVIII Proof of GOD

We finish with a *divine Note*.

In [937], (On mountain passes to a proof of God?, in Dutch) van Groesen gives an elementary introduction to minimax methods (with a very nice figure).

He quotes Maupertuis on the fact that the economy of laws of motion or actions assumes the existence of a *Supreme Being*:

> ...des lois du mouvement où l'action est toujours employée avec la plus grande économie démontrent l'existence de l'Etre suprème ;
> > in *Examen philosophique de la preuve de l'existence de Dieu*,
> > Maupertuis, 1757

A

Background Material

It is convenient to call these spaces Sobolev spaces; in addition to the brevity of this designation, it is appropriate since he proved some important results concerning these functions [851] and popularized them in his book [852].

C.B. Morrey Jr., *Multiple integrals in the calculus of variations.*
Springer-Verlag, 1966

For the convenience of the reader, we briefly recall in this appendix the definitions and some basic properties of Sobolev spaces and relate some details regarding Nemytskii operators. Mastering these notions is essential to treat nonlinear differential problems by critical point theorems in general and by the MPT in particular.

The material presented here is for people who are interested in applications to nonlinear boundary value problems or for those that are still beginning with variational methods.

A.1 Sobolev Spaces

Some standard references exclusively devoted to Sobolev spaces include [1, 120, 630]; you may also consult [145, 411]. The reader is supposed to be acquainted with L^p spaces and their basic properties. These can be found in standard textbooks on functional analysis like [145, 784, 977].

Let $\Omega \subset \mathbb{R}^N$ be an open set and m and p positive integers such that $1 \leq p \leq \infty$.

A.1.1 Definitions and Basic Properties

We begin by recalling the sense of a *weak* or *distributional derivative*.

Definition A.1. Let $u \in L^1_{loc}(\Omega)$ and $\alpha = (\alpha_1, \ldots, \alpha_N) \in \mathbb{N}^N$ be a *multi-index*. The weak derivative $D^\alpha u$ of u, when it exists, is a function $v_\alpha \in L^1_{loc}(\Omega)$ that satisfies

$$\int_\Omega D^\alpha \varphi(x).u(x)\,dx = (-1)^{|\alpha|} \int_\Omega v_\alpha(x)\varphi(x)\,dx, \qquad \text{for all } \varphi \in \mathcal{C}_c^\infty(\Omega),$$

where $|\alpha| = \alpha_1 + \cdots + \alpha_N$.

Remark A.1. Notice that when the weak derivative exists, it is unique.

Consider now

$$\|u\|_{m,p} = \left(\sum_{0 \le |\alpha| \le m} \|D^\alpha u\|_p^p \right)^{1/p} \qquad \text{if} \quad 1 \le p < \infty$$

and

$$\|u\|_{m,\infty} = \max_{0 \le |\alpha| \le m} \|D^\alpha u\|_\infty$$

when the right side makes sense. These are norms on any space of functions in which the right side takes finite values [1].

Definition A.2. We define the *Sobolev spaces*

$$W^{m,p}(\Omega) = \left\{ u \in L^p(\Omega); \ D^\alpha u \in L^p(\Omega) \text{ for } 0 \le |\alpha| \le m \right\},$$

where $D^\alpha u$ is the weak derivative defined earlier. The spaces $W^{m,p}(\Omega)$ are equipped with one of the aforementioned norms, and we denote by $W_0^{m,p}(\Omega)$ the closure of $C_c^\infty(\Omega)$ in $W^{m,p}(\Omega)$.

Definition A.3. Let X and Y be two Banach spaces.

- X is *continuously embedded* into Y and we write $X \hookrightarrow Y$ if

    i. X is a subspace of Y, and
    ii. every convergent sequence in X is still convergent in Y,

    or, in other words, the embedding $I: X \to Y$, $u \mapsto u$ is continuous. That is, there exists $C > 0$ such that $\|u\|_Y \le C\|u\|_X$, for every u in X.
- The space X is said to be *compactly embedded* into Y and we write $X \overset{c}{\hookrightarrow} Y$ if $X \hookrightarrow Y$ and I is compact.

The next properties of Sobolev spaces are easy to check:

- $W^{0,p}(\Omega) = L^p(\Omega)$.
- $W_0^{0,p}(\Omega) = L^p(\Omega)$ if $1 \le p < \infty$.
- $W_0^{m,p}(\Omega) \hookrightarrow W^{m,p}(\Omega) \hookrightarrow L^p$, for all p.

We also have some useful information on Sobolev spaces.

Theorem A.1. *The following properties hold:*

- $W^{m,p}(\Omega)$ *is a Banach space.*
- $W^{m,p}(\Omega)$ *is separable if* $1 \le p < \infty$.
- $W^{m,p}(\Omega)$ *is reflexive if* $1 < p < \infty$.
- $W^{m,2}(\Omega)$ *is a Hilbert space with inner product*

$$(u, v)_{m,2} = \sum_{0 \le |\alpha| \le m} (D^\alpha u, D^\alpha v)_2,$$

where $(.,.)_2$ denotes the inner product in $L^2(\Omega)$. The associated norm is denoted $\|.\|_{m,2}$.

$W^{m,2}(\Omega)$ is denoted sometimes as $H^m(\Omega)$.

Since $W_0^{m,p}(\Omega)$ is a closed subspace of $W^{m,k}(\Omega)$, it inherits the same properties seen earlier. Moreover,

$$\sum_{|\alpha|=m} (D^\alpha u, D^\alpha v)_2$$

is another inner product on $W_0^{m,2}(\Omega)$ that yields a norm equivalent to $\|.\|_{m,2}$ on $W_0^{m,2}(\Omega)$.

When $\Omega = \mathbb{R}^N$, then $W_0^{m,p}(\Omega) = W^{m,p}(\Omega)$.

A.1.2 Embedding Theorems

In addition to the *reflexivity* of Sobolev spaces, their embedding properties are responsible for their usefulness. The embedding results require some geometric regularity and smoothness on boundary of the domains like the *cone*, the *segment*, or the *local Lipschitz* properties (the interested reader may consult [1] for more details). The reader may also consult the nice book by Brézis [145] introducing functional analysis in a very accessible way and where proofs of the results given in this section can be found. The following important property holds, for the so-called C^1-sets (see [145], for example).

Proposition A.2. *Let $\Omega \subset \mathbb{R}^N$ be an open set of class C^1 with bounded boundary $\partial\Omega$; then there exists a* continuous linear extension operator

$$P: W^{1,p}(\Omega) \to W^{1,p}(\mathbb{R}^N).$$

This means that any functional $h \in W^{1,p}(\Omega)$ may be extended to a functional $\tilde{h}$ that is in $W^{1,p}(\mathbb{R}^N)$ and this correspondence is *continuous*. Many nice applications of this result are given in [145].

Proposition A.3. *Let $\Omega \subset \mathbb{R}^N$ be an open set of class C^1; then the subset of $W^{1,p}(\Omega)$ defined by*

$$\left\{ v = u\big|_\Omega; \ \text{where } u \in C_c^\infty(\mathbb{R}^N) \right\}$$

is dense in $W^{1,p}(\Omega)$.

The next two theorems contain results about continuous and compact embedding of Sobolev spaces into L^p spaces and spaces of continuous functionals, which are useful in applications.

Theorem A.4 (Sobolev). *Let $\Omega \subset \mathbb{R}^N$ be an open set of class C^1 with bounded boundary.*

- *If $1 \le p < N$, then $W^{1,p}(\Omega) \hookrightarrow L^{p^*}(\Omega)$, where $\dfrac{1}{p^*} = \dfrac{1}{p} - \dfrac{1}{N}$.*
- *If $p = N$, then $W^{1,p}(\Omega) \hookrightarrow L^q(\Omega)$ for $N \le q < \infty$.*

- *If $N < p < \infty$, then we have*

$$\begin{cases} W^{1,p}(\Omega) \hookrightarrow L^\infty(\Omega) \\ W^{1,p}(\Omega) \hookrightarrow C^{0,\alpha}(\overline{\Omega}), \quad \text{where } \alpha = 1 - \dfrac{N}{p}. \end{cases}$$

Theorem A.5 (Rellich and Kondrachov). *The following compact embedding holds for Sobolev spaces.*

- *If $p < N$, then $W^{1,p}(\Omega) \overset{c}{\hookrightarrow} L^q(\Omega)$ for $1 \le q < p^*$, where $\dfrac{1}{p^*} = \dfrac{1}{p} - \dfrac{1}{N}$.*
- *If $p = N$, then $W^{1,p}(\Omega) \overset{c}{\hookrightarrow} L^q(\Omega)$ for $1 \le q < \infty$.*
- *If $N \ge p$, then $W^{1,p}(\Omega) \overset{c}{\hookrightarrow} C(\overline{\Omega})$. In particular, $W^{1,p}(\Omega) \overset{c}{\hookrightarrow} L^q(\Omega)$ for all q.*

Remark A.2. The embedding of a Sobolev space (a space of *classes* of functions) into a space of continuous functions must be understood in the sense that *to each class of the considered Sobolev space belongs a continuous function*. In general, we say that some property holds for a function in a Sobolev space if it is true for some of its representatives.

A.1.3 Poincaré Inequality

In some cases, the norm of $W_0^{1,p}(\Omega)$ may be simplified using Poincaré inequality.

An open set $\Omega \subset \mathbb{R}^N$ is said to be *bounded in one direction* if there is $e \in \mathbb{R}^N$, $\|e\| = 1$, and two real numbers $a, b \in \mathbb{R}$ such that

$$(x, e) \in]a, b[\qquad \text{for any } x \in \mathbb{R}^N.$$

Proposition A.6 (Poincaré Inequality). *Suppose that Ω is an open subset of $\mathbb{R}^N$ bounded in one direction. Then there is a constant $C > 0$ such that, for any $u \in W_0^{1,p}(\Omega)$,*

$$\|u\|_p \le C \|\nabla u\|_p.$$

The Poincaré inequality holds true even if Ω is supposed to be of *finite measure*.

This result has the following consequence that has proven many times to be important in applications.

Corollary A.7. *If Ω is bounded in one direction or has a finite measure, then $\|\nabla u\|_p$ is an equivalent norm on $W_0^{1,p}(\Omega)$ to the norm induced by $W_0^{1,p}(\Omega)$.*

A.1.4 Generalized and Classical Solutions

Considering a boundary value problem, a *classical solution* (also called *strong solution*) is a smooth solution that satisfies the equations of the boundary value problem in the *usual sense*, using Fréchet derivatives. For example, in the situation of the Dirichlet problem (1.1) seen earlier, a classical solution is a C^2-functional that satisfies (1.1).

But in general, boundary value problems may possess no classical solution (examples exist, see [430]). This occurs mainly because the spaces C^k *are not reflexive.*

The variational approach consists of defining *weak solutions* that are specific to the problem and in proving their existence in some space (Sobolev space $W^{k,p}(\Omega)$ or a manifold) using critical point theorems. These solutions correspond to the critical points of an *energy functional* defined on that specific space.

The drawback is that, in general, these weak solutions have bad differential properties. But when both the functions appearing in the statement of the problem and the domain are smooth enough, one can hope to prove, using *regularity theory*, that the weak solutions are in fact strong. Unfortunately, the tools used in regularity theory are, for the time being, not as elegant as functional analysis results used in variational methods.

A.2 Nemytskii Operators

Nemytskii operators play an important role in variational methods. In this section, we present their continuity and differentiability properties with some details. Long and technical proofs are avoided, but we give the exact references where they can be found.

The main reference for Nemytskii operators remains the book by Vainberg [935]. Vainberg's book uses quite old notations. We tried to combine his results and the more modern notations and approach of de Figueiredo [296]. The notes in [296] and the book by Ambrosetti and Prodi [49] are also good references for a reader who is only interested in applications to nonlinear variational problems.

Let Ω be an open set in $\mathbb{R}^N$ where $N \geq 1$ and denote by $\mathcal{M}(\Omega)$ the set of all real-valued functions $u : \Omega \to \mathbb{R}$ that are measurable on Ω.

Definition A.4. Let $g : \Omega \times \mathbb{R}^M \to \mathbb{R}$. It is said to be a *Carathéodory function* if

    i. $s = (s_1, s_2, \ldots, s_M) \mapsto g(x, s_1, s_2, \ldots, s_M)$ is continuous for a.e. $x \in \Omega$, and

    ii. $x \mapsto g(x, s_1, s_2, \ldots, s_M)$ is measurable for all $s = (s_1, s_2, \ldots, s_M) \in \mathbb{R}^M$.

The notion of Carathéodory function is a kind of regularity that has a link with the *S-property* due to Lusin [600, p. 65]. Writing meas (A) for the Lebesgue measure of a set A, the definition of the *S*-property is the following.

Definition A.5. Let E be a measurable set in a finite dimensional Euclidean space. A function ϕ defined on E that is finite almost everywhere has the *S-property* if, for every $\varepsilon > 0$, there exists a *closed set* $F \subset E$ such that meas $(E \setminus F) \leq \varepsilon$ and ϕ is continuous on F.

The S-property was characterized by Lusin in the following well-known result.

Theorem A.8 (Lusin). *A function ϕ that is finite almost everywhere on a measurable set E of a finite dimensional Euclidean space is measurable on E if and only if it has the S-property.*

The relation between *S*-property and Carathéodory functions was described by Vainberg in [933,934] by the following result.

Theorem A.9 (Vainberg). *A function* $g\colon \Omega \times \mathbb{R}^M \to \mathbb{R}$ *is a Carathéodory function if and only if, for every* $\varepsilon > 0$, *there is a closed set* $F \subset \Omega$ *such that* $\mathrm{meas}\,(\Omega \setminus F) < \varepsilon$ *and* g *is continuous with respect to* $(x, s_1, s_2, \ldots, s_M)$ *in* $F \times \mathbb{R}^M$.

The necessary and sufficient condition appearing in this theorem is what Vainberg calls *strong S-property*. A proof can be found in [935, pp. 148–152].

Theorem A.10. *Let* $g\colon \Omega \times \mathbb{R}^M \to \mathbb{R}$ *be a Carathéodory function. Then the function*

$$x \mapsto g(x, u_1(x), u_2(x), \ldots, u_M(x))$$

is measurable for all functions $u_i \in \mathcal{M}(\Omega)$, $i = 1, 2, \ldots, M$.

Remark A.3. This result was first proved by Carathéodory [181, pp. 665–666], who introduced this notion to study differential equations.

Definition A.6. Let $g\colon \Omega \times \mathbb{R} \to \mathbb{R}$ be a Carathéodory function. Then g generates an operator $H_g \colon \mathcal{M}(\Omega) \to \mathcal{M}(\Omega)$ defined by

$$H_g(u) = g(x, u(x))$$

called the *Nemytskii operator* associated to g.

Remark A.4. The symbol H is used for historical reasons because the roman character H corresponds to N (like in Nemytskii) in the Russian alphabet as pointed out by Fučik in [411]. Nemytskii operators are also called *superposition operators*.

Before studying the continuity of H_g in L^p spaces, we will see that when $\mathrm{meas}\,(\Omega) < \infty$, as a mapping on the set of measurable functions on Ω, the Nemytskii operator is *asymptotically continuous*.

Theorem A.11. *Assume that* $\mathrm{meas}\,(\Omega) < \infty$, *then* H_g *is continuous in measure, if* $\left(u_n^{(1)}\right)_n, \left(u_n^{(2)}\right)_n, \ldots, \left(u_n^{(M)}\right)_n$ *are sequences of measurable functions on* Ω *that converge, respectively, to* $u^{(1)}, u^{(2)}, \ldots, u^{(M)}$. *Then the sequence* $\left(g(x, u_n^{(1)}(x), u_n^{(2)}(x), \ldots u_n^{(M)}(x))\right)_n$ *converges in measure to* $\left(g(x, u^{(1)}(x), u^{(2)}(x), \ldots u^{(M)}(x)\right)$, *that is, in* $\mathcal{M}(\Omega)$ *equipped with the topology of convergence in measure.*

A.2.1 Continuity of Nemytskii Operators

The proofs of the principal results in this section are attributed to Nemytskii. They can be found, for example, in [935]. When considering an equation of the type

$$Au = g(x, u(x)), \tag{A.1}$$

we are confronted with the problem of determining the "shape" to be required for the function $g(x, u(x))$ to get a solution $u(x)$ in some appropriate space. In other terms, when seeking a solution of (A.1) in a particular functions space, which kind of properties – smoothness, asymptotic behavior, etc. – must we require on $g(x, u(x))$?

For example, when continuous solutions of (A.1) are wanted and the operator A is continuous, it seems natural to suppose $(x, s) \mapsto g(x, s)$ to be continuous when one looks for a solution in the L^p space. Nemytskii has shown that Carathéodory functions are the right type of functions to deal with. More precisely, let $g: \Omega \times \mathbb{R} \to \mathbb{R}$ be a Carathéodory function and consider the *growth condition*

$$|g(x, s)| \leq a(x) + b|s|^{p/q}, \tag{A.2}$$

where $p, q \in [1, \infty[$, $b \geq 0$ is a constant and $a(x) \in L^q(\Omega)$.

Theorem A.12. *The following assertions hold:*

  i. *When (A.2) is satisfied, $H_g: L^p(\Omega) \to L^q(\Omega)$ is well defined, continuous, and bounded.*
  ii. *Conversely, if H_g maps $L^p(\Omega)$ into $L^q(\Omega)$, then it is continuous, bounded, and satisfies (A.2).*

Remark A.5. In fact, by the growth condition, assertion i shows immediately that $H_g(u) \in L^q(\Omega)$, for all $u \in L^p(\Omega)$ and then assertion i follows from assertion ii.

Theorem A.13. *Let $g: \Omega \times \mathbb{R} \to \mathbb{R}$ be a Carathéodory function and $p \in [1, \infty[$. If H_g maps $L^p(\Omega)$ into $L^\infty(\Omega)$, then there is some $M > 0$ such that*

$$|g(x, s)| \leq M, \qquad \text{for all } s \in \mathbb{R} \text{ and almost all } x \in \Omega.$$

Remark A.6. When considering Nemytskii operators generated by Carathéodory functions on spaces of vector functions,

$$g_i: \Omega \times \mathbb{R}^M \to \mathbb{R}, \qquad i = 1, \dots, L \text{ and } \Omega \subset \mathbb{R}^N,$$

defined by a family of L functions $g_i(x, s_1, \dots, s_M)$ for $i = 1, \dots, L$. The Nemytskii operator is then

$$H_g = (H_{1,g}, \dots, H_{M,g}),$$

where

$$H_{i,g}(u) = g_i(x, u_1(x), \dots, u_L(x)).$$

In this case, the results concerning the continuity and boundedness of Nemytskii operators obtained earlier remain still true. That is, for each $i \in \{1, \dots, L\}$, the operator $H_{i,g}$ takes $(L^p(\Omega))^M$ into $L^q(\Omega)$ with a growth condition of the form

$$|g_i(x, s_1, s_2, \dots, s_M)| \leq a_i(x) + b \sum_{j=1}^{M} |s_j|^{p/q},$$

where $a_i(x) \in L^q(\Omega)$ and $b > 0$.

A.2.2 Potential Nemytskii Operators

Now we will seek an operator Γ defined on an L^p-space and whose derivative yields the Nemytskii operator H_g. Let $p > 1$ and $g(x, s)$ be a Carathéodory function satisfying a growth condition that permits H_g to map $L^p(\Omega)$ into $L^{p'}(\Omega)$, where $\frac{1}{p} + \frac{1}{p'} = 1$. That is

$$|g(x,s)| \leq a(x) + b|s|^{p-1}, \tag{A.3}$$

where $a(x) \in L^{p'}(\Omega), (p/p' = p/[p/(p-1)] = p - 1)$.

Denoting the potential associated to g by

$$G(x,s) = \int_0^s g(x,s)\,dt,$$

we have the relation

$$|G(x,s)| \leq A(x) + B|s|^p,$$

where $A(x) \in L^1(\Omega)$ and $B > 0$.

Then $G(x,s)$ is a Carathéodory function,

$$\begin{cases} H_g \text{ maps } L^p(\Omega) \text{ into } L^{p'}(\Omega), \text{ and} \\ H_G \text{ maps } L^p(\Omega) \text{ into } L^1(\Omega). \end{cases}$$

Theorem A.14. *The functional*

$$\Gamma(u) = \int_\Omega G(x, u(x))\,dx$$

is a Fréchet-differentiable real-valued function defined on $L^p(\Omega)$ whose derivative $\Gamma' = H_g$, provided the growth condition (A.3) is satisfied.

Remark A.7. The operator Γ is a special case of the so-called *Hammerstein operator*

$$\int_\Omega K(x,y)G(x,u(x))\,dx$$

with a *kernel* $K(x,y)$ defined on $\Omega \times \Omega$.

Regularity properties of Γ are important when solving nonlinear problems by variational methods.

Remark A.8. According to Vainberg, Alexiewics and Orlicz were the first to give examples of operators that are Gâteaux differentiable everywhere but are nowhere Fréchet differentiable. In this context, Vainberg has showed that if a Carathéodory function $g(x,s)$ is nonlinear in s and generates a Nemytskii operator in $L^2(\Omega)$, this operator is Gâteaux differentiable everywhere, whereas it does not have a Fréchet derivative at any point.

Bibliography

[1] R.A. Adams, *Sobolev spaces*. Pure and Applied mathematics series, Academic Press, New York 1975.

[2] Adimurthi, Some remarks on the Dirichlet problem with critical growth for the n-Laplacian. *Houston J. Math.*, **17**, no. 2, 285–298 (1991).

[3] S. Adly and D. Goeleven, Homoclinic orbits for a class of hemivariational inequalities. *Appl. Anal.*, **58**, no. 3–4, 229–240 (1995).

[4] S. Adly, D. Goeleven, and D. Motreanu, Periodic and homoclinic solutions for a class of unilateral problems. *Discrete Contin. Dynam. Systems*, **3**, no. 4, 579–590 (1997).

[5] S. Agmon, The L^p approach to the Dirichlet problem. *Ann. Scuola Norm. Sup. Pisa*, **13**, 405–448 (1959).

[6] J. Aguirre and I. Peral, Existence of periodic solutions for a class of nonlinear equations. Contributions to nonlinear partial differential equations (Madrid, 1981). Res. Notes Math., **89**, Pitman, Boston, MA and London, 1–6, 1983.

[7] S. Ahmad, A.C. Lazer, and J.L. Paul, Elementary critical point theory and perturbations of elliptic boundary value problems at resonance. *Indiana Univ. Math. J.*, **25**, 933–944 (1976).

[8] J. Ai and X.P. Zhu, Positive solutions of inhomogeneous elliptic boundary value problems in the half space. *Commun. Part. Diff. Eq.*, **15**, no. 10, 1421–1446 (1990).

[9] S. Alama and M.A. Del Pino, Solutions of elliptic equations with indefinite nonlinearities via Morse theory and linking. *Ann. Inst. H. Poincare–An.*, **13**, no. 1, 95–115 (1996).

[10] S. Alama and Y.Y. Li, Existence of solutions for semilinear elliptic equations with indefinite linear part. *J. Diff. Eq.*, **96**, no. 1, 89–115 (1992).

[11] S. Alama, L. Bronsard, and T. Giorgi, Uniqueness of symmetric vortex solutions in the Ginzburg-Landau model of superconductivity. *J. Funct. Anal.*, **167**, no. 2, 399–424 (1999).

[12] F. Alessio and M. Calanchi, Homoclinic-type solutions for an almost periodic semilinear elliptic equation on $\mathbb{R}^n$. *Rend. Sem. Mat. Univ. Padova*, **97**, 89–111 (1997).

[13] F. Alessio and P. Montecchiari, Multibump solutions for a class of Lagrangian systems slowly oscillating at infinity *Ann. Inst. H. Poincaré–An.*, **16**, no. 1, 107–135 (1999).

[14] F. Alessio, M.L. Bertotti, and P. Montecchiari, Multibump solutions to possibly degenerate equilibria for almost periodic Lagrangian systems. *Z. Angew. Math. Phys.*, **50**, no. 6, 860–891 (1999).

[15] W. Allegretto and P.O. Odiobala, Nonpositone elliptic problems in $\mathbb{R}^n$. *Proc. Amer. Math. Soc.*, **123**, no. 2, 533–541 (1995).

[16] _____, Decaying solutions of elliptic systems in $\mathbb{R}^n$. *Rocky Mt. J. Math.*, **26**, no. 2, 419–437 (1996).

[17] W. Allegretto and L.S. Yu, Positive L^p-solutions of subcritical nonlinear problems. *J. Diff. Eq.*, **87**, no. 2, 340–352 (1990).

[18] _____, On the existence of multiple decaying positive solutions of elliptic equations. *Nonlinear Anal.*, **15**, no. 11, 1065–1076 (1990).

[19] _____, Decaying solutions of 2mth order elliptic problems. *Can. J. Math.*, **43**, no. 3, 449–460 (1991).

[20] L. Almeida, Topological sectors for Ginzburg-Landau energies. *Rev. Mat. Iberoam.*, **15**, no. 3, 487–545 (1999).

[21] F.J. Almgren Jr. and E.H. Lieb, The (non) continuity of symmetric decreasing rearrangement. Symposia mathematica, Vol. XXX (Cortona, 1988). Academic Press, London, 89–102, 1989.

[22] C.O. Alves, Multiple positive solutions for equations involving critical Sobolev exponent in $\mathbb{R}^N$. *Electron. J. Diff. Eq.*, no. 13, 10 pp. (1997) (http://edje.math.swt.edu/Volumes/1997/13/abstr.html).

[23] _____, Existence of positive solutions for an equation involving supercritical exponent in $\mathbb{R}^N$. *Nonlinear Anal.–Theor.*, **42A**, no. 4, 573–581 (2000).

[24] C.O. Alves and O.H. Miyagaki, Existence of positive solutions to a superlinear elliptic problem. *Electron. J. Diff. Eq.*, no.11, 12 pp. (2001) (http://edje.math.swt.edu/Volumes/2001/11/abstr.html).

[25] C.O. Alves, P.C. Carrião, and O.H. Miyagaki, Multiple solutions for a problem with resonance involving the p-Laplacian. *Abstr. Appl. Anal.*, **3**, no. 1–2, 191–201 (1998).

[26] _____, Existence and multiplicity results for a class of resonant quasi-linear elliptic problems on $\mathbb{R}^N$. *Nonlinear Anal.–Theor.*, **39A**, no. 1, 99–110 (2000).

[27] C.O. Alves, D.C. de Morais Filho, and M.A.S. Souto, Radially symmetric solutions for a class of critical exponent elliptic problems in $\mathbb{R}^N$. *Electron. J. Diff. Eq.*, no. 7, 12 pp. (1996) (http://edje.math.swt.edu/Volumes/1996/07/abstr.html).

[28] C.O. Alves, D.C. de Morais Filho, and M.A.S. Souto, On systems of elliptic equations involving subcritical or critical Sobolev exponents. *Nonlinear Anal.–Theor.*, **42A**, no. 5, 771–787 (2000).

[29] C.O. Alves, J.M. do, and M.A.S. Souto, Local mountain-pass for a class of elliptic problems in $\mathbb{R}^N$ involving critical growth. *Nonlinear Anal.–Theor.*, **46A**, no. 4, 495–510 (2001).

[30] C.O. Alves, J.V. Gonçalves, and O.H. Miyagaki, Remarks on multiplicity of positive solutions of nonlinear elliptic equations in $\mathbb{R}^N$ with critical growth. Dynamical systems and differential equations, Vol. I (Springfield, MO, 1996). *Discrete Contin. Dynam. Systems*, 51–57 (1998).

[31] _____, Multiple positive solutions for semilinear elliptic equations in $\mathbb{R}^N$ involving critical exponents. *Nonlinear Anal.*, **34**, no. 4, 593–615 (1998).

[32] _____, On elliptic equations in $\mathbb{R}^N$ with critical exponents. *Electron. J. Diff. Eq.*, no. 9, 11 pp. (1996) (http:edje.math.swt.edu/Volumes/1996/09/abstr.html).

[33] H. Amann, Fixed point equations and nonlinear eigenvalue problems in ordered Banach spaces. *SIAM Rev.*, **18**, no. 4, 620–709 (1976).

[34] _____, Supersolution, monotone iteration and stability. *J. Diff. Eq.*, **21**, 367–377 (1976).

[35] _____, Saddle points and multiple solutions of differential equations. *Math. Z.*, **169**, no. 2, 127–166 (1979).

[36] _____, A note on the degree of gradient mappings. *Proc. Amer. Math. Soc.*, **84**, 591–595 (1982).

[37] A. Ambrosetti, Teoria di Lusternik-Schnirelman su varietá con bordo negli spazi di Hilbert. *Rend. Sym. Mat. Univ. Padova*, **45**, 337–353 (1971).

[38] _____, A perturbation theorem for superlinear boundary value problems. *M.R.C., Univ. Wisconsin-Madison*, TSR **1446**, 1974.

[39] _____, A note on Ljusternik-Schnirelman theory for functionals which are not even. *Ric. Mat.*, **25**, 179–186 (1976).

[40] ———, Elliptic equations with jumping nonlinearities. *J. Math. Phys. Sci.*, **18**, no. 1, 1–12 (1984).

[41] ———, Nonlinear oscillations with minimal period. Nonlinear functional analysis and its applications, Part 1 (Berkeley, CA, 1983), *Proc. Sympos. Pure Math.*, **45**, Part 1, 29–35, American Mathematical Society, Providence, RI, 1986.

[42] ———, Variational problems in nonlinear analysis. *Bol. Unione Mat. Ital.*, **A(7) 2**, no. 2, 169–188 (1988).

[43] ———, *Critical points and nonlinear variational problems.* Mém. Soc. Math. France (N.S.), no. 49, Tome 120, Fascicule 2, Cours de la chaire Lagrange, 1992.

[44] ———, Variational methods and nonlinear problems: Classical results and recent advances. *Topological nonlinear analysis*, Progress in Nonlinear Differential Equations Application, **15**, 1–36, Birkhäuser Boston, Boston, MA, 1995.

[45] A. Ambrosetti and M. Badiale, The dual variational principle and elliptic problems with discontinuous nonlinearities. *J. Math. Anal. Appl.*, **140**, no. 2, 363–373 (1989).

[46] A. Ambrosetti and V. Coti Zelati, Solutions with minimal period for Hamiltonian systems in a potential well. *Ann. Inst. H. Poincaré–An.*, **4**, no. 3, 275–296 (1987).

[47] A. Ambrosetti and G. Mancini, Solutions of minimal period for a class of convex Hamiltonian systems. *Math. Ann.*, **255**, no. 3, 405–421 (1981).

[48] A. Ambrosetti and G. Prodi, On the inversion of some differentiable mappings with singularities between Banach spaces. *Ann. Mat. Pur. Appl.*, **93**, 231–246 (1972).

[49] ———, *A primer of nonlinear analysis.* Cambridge Studies in Advanced Mathematics, **34**, Cambridge University Press, 1993.

[50] A. Ambrosetti and P.H. Rabinowitz, Dual variational methods in critical point theory and applications. *J. Funct. Anal.*, **14**, 349–381 (1973).

[51] A. Ambrosetti and P.N. Srikanth, Superlinear elliptic problems and the dual principle in critical point theory. *J. Math. Phys. Sci.*, **18**, no. 5, 441–451 (1984).

[52] A. Ambrosetti and M. Struwe, A note on the problem $-\Delta u = \lambda u + u|u|^{2^*-2}$. *Manuscripta Math.*, **54**, no. 4, 373–379 (1986).

[53] ———, Existence of steady vortex rings in an ideal fluid. *Arch. Rat. Mech. Anal.*, **108**, no. 2, 97–109 (1989).

[54] A. Ambrosetti, H. Brézis, and G. Cerami, Combined effects of concave and convex nonlinearities in some elliptic problems. *J. Funct. Anal.*, **122**, no. 2, 519–543 (1994).

[55] A. Ambrosetti, V. Coti Zelati, and I. Ekeland, Symetry breaking in Hamiltonian systems. *J. Diff. Eq.*, **67**, 165–184 (1987).

[56] S. F. Aniţa, Solutions for elliptic variational inequalities. *An. Ştiinţ. Univ. "Al. I. Cuza" Iaşi Sect. I a Mat. (N.S.)*, **34**, no. 3, 247–256 (1988).

[57] F. Antonacci, Periodic and homoclinic solutions to a class of Hamiltonian systems with indefinite potential in sign. *Bol. Unione Mat. Ital.*, **B(7) 10**, no. 2, 303–324 (1996).

[58] J. Appel and P.P. Zabrejko, *Nonlinear superposition operators.* Cambridge Tracts in Mathematics, **95**, Cambridge University Press, 1990.

[59] D. Arcoya, Positive solutions for semilinear Dirichlet problems in an annulus. *J. Diff. Eq.*, **94**, no. 2, 217–227 (1991).

[60] D. Arcoya and L. Boccardo, A min-max theorem for multiple integrals of the calculus of variations and applications. *Atti Accad. Naz. Lin.*, (9), **6**, no. 1, 29–35 (1995).

[61] ———, Critical points for multiple integrals of the calculus of variations. *Arch. Rat. Mech. Anal.*, **134**, no. 3, 249–274 (1996).

[62] ———, Some remarks on critical point theory for nondifferentiable functionals. *NoDEA–Nonlinear Diff.*, **6**, 79–100 (1999).

[63] D. Arcoya and D.G. Costa, Nontrivial solutions for a strongly resonant problem. *Diff. Integral Eq.*, **8**, no. 1, 151–159 (1995).

[64] G. Arioli, A note on quasilinear elliptic equations at critical growth. *NoDEA–Nonlinear Diff.*, **15**, 83–97 (1998).

[65] G. Arioli and J. Chabrowski, Periodic motions of a dynamical system consisting of an infinite lattice of particles. *Dyn. Syst. Appl.*, **6**, no. 3, 387–395 (1997).

[66] G. Arioli and F. Gazzola, Periodic motions of an infinite lattice of particles with nearest neighbor interaction. *Nonlinear Anal.*, **26**, no. 6, 1103–1114 (1996).

[67] ———, Weak solutions of quasilinear elliptic PDE's at resonance. *Ann. Fac. Sci. Toulouse*, **6**, 573–589 (1997).

[68] ———, Quasilinear elliptic equations at critical growth. *NoDEA–Nonlinear Diff.*, **15**, 83–97 (1998).

[69] ———, Existence and multiplicity results for quasilinear elliptic differential systems. *Commun. Part. Diff. Eq.*, **25**, no. 1–2, 125–153 (2000).

[70] G. Arioli and B. Ruf, Periodic solutions for a system of forced and nonlinearly coupled oscillators with applications to electrical circuits. *Dyn. Syst. Appl.*, **4**, no. 1, 87–102 (1995).

[71] G. Arioli and A. Szulkin, Homoclinic solutions for a class of systems of second order differential equations. *Topol. Method Nonlinear Anal.*, **6**, no. 1, 189–197 (1995).

[72] C. Arzela, Sul principio di Dirichlet. *Rend. Ac. Sci. Bologna*, (N.S.) **1**, 71–84 (1896–1897).

[73] J.P. Aubin, *Applied abstract analysis*. Wiley, New York, 1977.

[74] J.P. Aubin and I. Ekeland, *Applied nonlinear analysis*. Pure and Applied Mathematics, Wiley Interscience, New York, 1984.

[75] M. Badiale, Critical exponent and discontinuous nonlinearities. *Diff. Integral Eq.*, **6**, no. 5, 1173–1185 (1993).

[76] M. Badiale and G. Citti, Concentration compactness principle and quasilinear elliptic equations in $\mathbb{R}^n$. *Commun. Part. Diff. Eq.*, **16**, no. 11, 1795–1818 (1991).

[77] A. Bahri, Topological results on a certain class of functionals and applications. *J. Funct. Anal.*, **41**, 397–427 (1981).

[78] ———, Critical points at infinity in some variational problems. Partial differential equations, Proc. Lat. Am. Sch. Math., ELAM-8 (Rio de Janeiro, Brazil, 1986) *Lect. Notes Math.*, **1324**, 1–29 (1988).

[79] ———, Critical points at infinity in the variational calculus. *Pitman Res. Notes Math.*, **182**, Halsted, New York, 1989.

[80] A. Bahri and H. Berestycki, A perturbation method in critical point theory and applications. *Trans. Am. Math. Soc.*, **267**, 1–32 (1981).

[81] ———, Forced vibrations of superquadratic Hamiltonian systems. *Acta Math.*, **152**, 143–197 (1984).

[82] ———, Existence of forced oscillations for some nonlinear differential equations. *Commun. Pure Appl. Math.*, **37**, 403–442 (1984).

[83] ———, Stability results for Ekeland ε-principle and cone extremal solutions. *Math. Oper. Res.*, **18**, 173–201 (1993).

[84] A. Bahri and J.M. Coron, On a nonlinear elliptic equation involving the critical Sobolev exponent: The effect of topology of the domain. *Commun. Pure Appl. Math.*, **41**, 253–294 (1988).

[85] A. Bahri and P.-L. Lions, Remarques sur la théorie variationnelle des points critiques et applications. (French) [Remarks on variational critical point theory and applications] *C. R. Acad. Sci. Paris Ser. I Math.*, **301**, no. 5, 145–147 (1985).

[86] ———, Morse index of some min-max critical points I. Applications to multiplicity results. *Commun. Pure Appl. Math.*, **41**, 1027–1037 (1988).

[87] ———, Remarks on the variational theory of critical points. *Manuscripta Math.*, **66**, 129–152 (1989).

[88] P. Bartolo, An extension of Krasnoselskii genus. *Boll. Unione Math. Ital.*, **1-C**, 347–356 (1982).

[89] T. Bartsch, Critical orbits of symmetric functionals. *Math. Oper. Res.*, **18**, 173–201 (1989).

[90] ———, On the genus of representation spheres. *Comment. Math. Helv.*, **65**, 85–95 (1990).

[91] ———, Infinitely many solutions of a symmetric Dirichlet problem. *Nonlinear Anal.*, **20**, no. 10, 1205–1216 (1993).

[92] ———, Simple proof of the degree formula for $\mathbb{Z}/p$-equivariant maps. *Math. Z.*, **212**, 285–292 (1993).

[93] ———, *Topological methods for variational problems with symmetries*. Lecture Notes in Mathematics, **1560**, Springer-Verlag, Berlin, 1993.

[94] T. Bartsch and M. Clapp, Bifurcation theory for symmetric potential operators and the equivariant cup-length. *Math. Z.*, **204**, no. 3, 341–356 (1990).

[95] ———, Critical point theory for indefinite functionals with symmetries. *J. Funct. Anal.*, **138**, no. 1, 107–136 (1996).

[96] T. Bartsch and Z.-Q. Wang, On the existence of sign changing solutions for semilinear Dirichlet problems. *Topol. Method. Nonlinear Anal.*, **7**, no. 1, 115–131 (1996).

[97] T. Bartsch and M. Willem, Infinitely many radial solutions of a semilinear elliptic problem on $\mathbb{R}^N$. *Arch. Rat. Mech. Anal.*, **124**, no. 3, 261–276 (1993).

[98] ———, On an elliptic equation with concave and convex nonlinearities. *Proc. Am. Math. Soc.*, **123**, no. 11, 3555–3561 (1995).

[99] T. Bartsch, K.-C. Chang, and Z.-Q. Wang, On the Morse indices of sign changing solutions of nonlinear elliptic problems. *Math. Z.*, **233**, no. 4, 655–677 (2000).

[100] T. Bartsch, M. Clapp, and D. Puppe, A mountain pass theorem for actions of compact Lie groups. *J. Reine Angew. Math.*, **419**, 55–66 (1991).

[101] P.W. Bates and I. Ekeland, A saddle point theorem. *Differential equations*, Proc. Eighth Fall Conf., Oklahoma State Univ. (Stillwater, OK, 1979), Academic Press, 123–126, 1980.

[102] M. Benabas, Orbite homocline dans un modèle à une infinité de particules. Une approche variationnnelle. (French) [A variational approach to homoclinic orbits in an infinite lattice of particles] *C.R. Acad. Sci. Paris, Sér. I Math.*, **326**, no. 7, 805–808 (1998).

[103] V. Benci, Some critical point theorems and applications. *Comm. Pure Appl. Math.*, **33**, 147–172 (1980).

[104] ———, A geometrical index for the group S^1 and some applications to the research of periodic solutions of O.D.E.'s. *Comm. Pure Appl. Math.*, **34**, 393–432 (1981).

[105] ———, On critical point theory for indefinite functionals in the presence of symmetries. *Trans. Am. Math. Soc.*, **274**, 533–572 (1982).

[106] ———, Some applications of the generalized Morse-Conley index. *Conf. Semin. Mat. Univ. Bari*, **218**, 1–32 (1987).

[107] ———, A new approach to Morse-Conley theory. *Recent advances in Hamiltonian systems*, Proc. Int. Conf., (L'Aquila, Italy, 1986), G. Dell'Antonio ed., World Scientific Publ, 1–52, 1987.

[108] ———, A new approach to the Morse-Conley theory and some applications. *Ann. Mat. Pur. Appl., IV. Ser.* **158**, 231–305 (1991).

[109] ———, Introduction to Morse theory: A new approach. *Topological nonlinear analysis*, Progress in Nonlinear Differential Equations Application, **15**, 37–177, Birkhäuser Boston, Boston, 1995.

[110] V. Benci and D. Fortunato, The dual method in critical point theory. Multiplicity results for indefinite functionals. *Ann. Mat. Pur. Appl.*, **132**, no. 4, 215–242 (1983).

[111] ———, A "Birkhoff-Lewis" type result for nonautonomous differential equations. *Partial differential equations* (Rio de Janeiro, 1986), Lecture Notes in Mathematics, **1324**, 85–96, Springer, Berlin and New York, 1988.

[112] V. Benci and F. Pacella, Morse theory for symmetric functionals on the sphere and an application to a bifurcation problem. *Nonlinear Anal.*, **9**, 763–773 (1985).

[113] V. Benci and P.H. Rabinowitz, Critical point theorems for indefinite functionals. *Invent. Math.*, **52**, 241–273 (1979).

[114] V. Benci, P. Bartolo, and D. Fortunato, Abstract critical point theory and applications to some nonlinear problems with "strong" resonance at infinity. *Nonlinear Anal.*, **7**, 981–1012 (1983).

[115] F. Benkert, On the perturbation of critical values of maximum-minimum type. *Z. Anal. Anwend.*, **11**, no. 2, 245–268 (1992).

[116] H. Berestycki and D.G. de Figueiredo, Double resonance in semilinear elliptic problems. *Commun. Part. Diff. Eq.*, **6**, 91–120 (1981).

[117] H. Berestycki and P.-L. Lions, Existence d'états multiples dans des équations de champs scalaires non linéaires dans le cas de masse nulle. (French) [Existence of multiple states in nonlinear field equations in the zero mass case] *C.R. Acad. Sci. Paris, Sér. I Math.*, **297**, no. 4, 267–270 (1983).

[118] H. Berestycki, J.-M. Lasry, G. Mancini, and B. Ruf, Existence of multiple periodic orbits on star-shaped hamiltonian systems. *Commun. Pure Appl. Math.*, **38**, 253–289 (1985).

[119] H. Berestycki, I. Capuzzo-Dolcetta, and L. Nirenberg, Variational methods for indefinite superlinear homogeneous elliptic problems. *Nonlinear Diff. Eq. Appl.*, **2**, no. 4, 553–572 (1995).

[120] J. Berg and J. Löfström, *Interpolation spaces*. Springer, 1976.

[121] M.S. Berger, *Nonlinearity and functional analysis*. Academic Press, 1977.

[122] M.S. Berger and M. Berger, *Perspectives in nonlinearity. An introduction to nonlinear analysis*. W. A. Benjamin, Inc., New York and Amsterdam, 1968.

[123] M.S. Berger and L. Zhang, Quasiperiodic solutions of saddle point type for the Duffing equations with small forcing. *Commun. Appl. Nonlinear Anal.*, **4**, no. 4, 123–145 (1997).

[124] D. Bertoloni Meli, *Equivalence and priority: Newton versus Leibnitz*. Clarendon Press, Oxford, 1977.

[125] F. Bethuel and O. Rey, Multiple solutions to the Plateau problem for nonconstant mean curvature. *Duke Math. J.*, **73**, no. 3, 593–646 (1994).

[126] F. Bethuel and J.-C. Saut, Travelling waves for the Gross-Pitaevskii equation. I. *Ann. Inst. H. Poincaré–An.*, **70**, no. 2, 147–238 (1999).

[127] G.D. Birkhoff, Dynamical systems with two degree of freedom. *Trans. Am. Math. Soc.*, **18**, 199–300 (1917).

[128] ———, *Dynamical systems*. American Mathematical Society, Providence, RI, 1927.

[129] G.D. Birkhoff and M.R. Hestenes, Generalized minimax principles in the calculus of variations. *Duke Math. J.*, **1**, 413–432 (1935).

[130] E. Bishop and R.R. Phelps, A proof that every Banach space is subreflexive. *Bull. Am. Math. Soc.*, **67**, 79–98 (1961).

[131] N.A. Bobylev, S.V. Emel'yanov, and S.K. Korovin, *Geometrical methods in variational problems*. Mathematics and its Applications, **485**, Kluwer Academic Publishers, Dordrecht, 1999.

[132] R. Böhme, Die Lösung der Verzweigungsgleichungen für nichtlineare Eigenwertprobleme. (German) *Math. Z.*, **127**, 105–126 (1972).

[133] P. Boleslaw, Steepest descent of locally Lipschitz functionals in superreflexive spaces. *Rend. Circ. Mat. Palermo II.*, **42**, no. 1, 29–34 (1993).

[134] P. Bolle, On the Bolza problem. *J. Diff. Eq.*, **152**, no. 2, 274–288 (1999).

[135] P. Bolle, N. Ghoussoub, and H. Tehrani, The multiplicity of solutions in non-homogenous boundary value problems. *Manuscripta Math.*, **101**, no. 3, 325–350 (2000).

[136] J.M. Borwein and D. Preiss, A smooth variational principle with applications to subdifferentiability and to differentiability of convex functions. *Trans. Am. Math. Soc.*, **303**, no. 2, 517–527 (1987).

[137] R. Bott, Lectures on Morse theory, old and new. Proceedings of the 1980 Beijing Symposium on Differential Geometry and Differential Equations, **1–3** (Beijing, 1980), 169–218, Science Press, Beijing, 1982.

[138] ———, Lectures on Morse theory, old and new. *Bull. Am. Math. Soc.*, (*N.S.*), **7**, no. 2, 331–358 (1982).

[139] ———, Morse theory indomitable. *Inst. Hautes Études Sci. Publ. Math.*, **68**, 99–114 (1989).

[140] M. Bouchekif, Some existence results for a class of quasilinear elliptic systems. *Ric. Mat.*, **46**, no. 1, 203–219 (1997).

[141] _____, On certain quasilinear elliptic equations with indefinite terms. systems. *Funkcial. Ekvac.*, **41**, no. 2, 309–316 (1998).

[142] Y. Bozhkov, On a quasilinear system involving K-Hessian operators. *Adv. Diff. Eq.*, **2**, no. 3, 403–426 (1997).

[143] D. Braess, *Nonlinear approximation theory*. Springer Series in Computational Mathematics, **7**, Springer-Verlag, Berlin and New York, 1986.

[144] A. Bressan, Dual variational methods in optimal control theory. *Nonlinear controllability and optimal control*, 219–235, Monographs Textbooks Pure Appl. Math., **133**, Dekker, New York, 1990.

[145] H. Brézis, *Analyse fonctionnelle, théorie et applications*. Masson, Paris 1987.

[146] _____, The Ambrosetti-Rabinowitz mountain pass lemma via Ekeland's minimization principle. (Manuscript)

[147] _____, Some variational problems with lack of compactness. *Proc. Symp. Pure Math.*, **45**, 165–201 (1986).

[148] _____, Points critiques dans les problèmes variationnels sans compacité. (French) [Critical points in variational problems without compactness.] *Séminaire Bourbaki*, Vol. **1987/88**. Astérisque no. 161–162, (1988), Exp. no. 698, 5, 239–256 (1989).

[149] H. Brézis and F. Browder, A general ordering principle in nonlinear functional analysis. *Adv. Math.*, **21**, 355–364 (1976).

[150] _____, Partial differential equations in the 20th century. *Adv. Math.*, **135**, 76–144 (1998).

[151] H. Brézis and E. Lieb, Minimum action solutions of some vector field equations. *Commun. Math. Phys.*, **96**, 97–113 (1984).

[152] H. Brézis and L. Nirenberg, Positive solutions of nonlinear elliptic equations involving critical Sobolev exponents. *Commun. Pure Appl. Math.*, **36**, 437–477 (1983).

[153] _____, Remarks on finding critical points. *Commun. Pure Appl. Math.*, **44**, no. 8–9, 939–963 (1991).

[154] _____, H^1 versus C^1 local minimizers. *C.R. Acad. Sci. Paris*, t.317, Série I, 465–472 (1993).

[155] H. Brézis, J.M. Coron, and L. Nirenberg, Free vibrations for a nonlinear wave equation and a theorem of P. Rabinowitz. *Commun. Pure Appl. Math.*, **33**, no. 5, 667–684 (1980).

[156] F. Browder, Infinite dimensional manifolds and nonlinear elliptic eigenvalue problems. *Ann. Math.*, **82**, 459–477 (1965).

[157] B. Buffoni and L. Jeanjean, Minimax characterization of solutions for a semilinear elliptic equation with lack of compactness. *Ann. Inst. H. Poincare–An.*, **10**, no. 4, 377–404 (1993).

[158] B. Buffoni and E. Séré, A global condition for quasi-random behavior in a class of conservative systems. *Commun. Pure Appl. Math.*, **49**, no. 3, 285–305 (1996).

[159] B. Buffoni, L. Jeanjean, and C.A. Stuart, Existence of a nontrivial solution to a strongly indefinite semilinear equation. *Proc. Am. Math. Soc.*, **119**, no. 1, 179–186 (1993).

[160] G.R. Burton, Multiple steady states for rotating rods. *Nonlinear Anal.*, **10**, no. 10, 1069–1076 (1986).

[161] _____, Rearrangement inequalities and duality theory for a semilinear elliptic variational problem. *J. Math. Anal. Appl.*, **121**, no. 1, 123–137 (1987).

[162] _____, Variational problems on classes of rearrangements and multiple configurations for steady vortices. *Ann. Inst. H. Poincaré–An.*, **6**, no. 4, 295–319 (1989).

[163] R. Cacciopoli, Un principio di inversione per le corrispondenze funzionali e sue applicazioni alle equazioni alle derivate parziali. *Atti. Accad. Naz. Lin.*, **16**, 392–400 (1932).

[164] L. Caklovic, S. Li, and M. Willem, A note on Palais-Smale condition and coercivity. *Diff. Integral. Eq.*, **3**, 799–800 (1990).

[165] P. Caldiroli, A new proof of the existence of homoclinic orbits for a class of autonomous second order Hamiltonian systems in $\mathbb{R}^N$. *Math. Nachr.*, **187**, 19–27 (1997).

[166] I. Campa and M. Degivanni, Subdifferential calculus and nonsmooth critical point theory. *SIAM J. Optimiz.*, **10**, 1020–1048 (2000).

[167]   A.M. Candela, F. Giannoni, and A. Masiello, Multiple critical points for indefinite func-
        tionals and applications. *J. Diff. Eq.*, **155**, no. 1, 203–230 (1999).
[168]   A. Canino, On p-convex sets and geodesics. *J. Diff. Eq.*, **75**, no. 1, 118–157 (1988).
[169]   _____, Multiplicity of solutions for quasilinear elliptic equations. *Topol. Method. Non-
        linear Anal.*, **6**, no. 2, 357–370 (1995).
[170]   _____, Multiplicity of solutions for quasilinear elliptic equations. *Topol. Method. Non-
        linear Anal.*, **6**, 357–370 (1995).
[171]   _____, On a variational approach to some quasilinear problems. Well posed problems
        and stability in optimization (Marseille, 1995), *Serdica Math. J.*, **22**, 399–426 (1996).
[172]   _____, On a jumping problem for quasilinear elliptic equations. *Math. Z.*, **226**, 193–210
        (1997).
[173]   A. Canino and M. Degiovanni, Nonsmooth critical point theory and quasilinear elliptic
        equations. Topological methods in differential equations and inclusions (Montreal, PQ,
        1994), 1–50, NATO Adv. Sci. Inst. Ser. C Math. Phys. Sci., **472**, Kluwer Academic
        Publishers, Dordrecht, 1995.
[174]   D.M. Cao, Nontrivial solution of semilinear elliptic equation with critical exponent in
        $\mathbb{R}^2$. *Commun. Part. Diff. Eq.*, **17**, no. 3–4, 407–435 (1992).
[175]   _____, Multiple positive solutions of inhomogeneous semilinear elliptic equations
        in unbounded domains in $\mathbb{R}^2$. *Acta Math. Sci.* (English ed.) **14**, no. 3, 297–312
        (1994).
[176]   D.M. Cao and H.S. Zhou, On the existence of multiple solutions of nonhomogeneous
        elliptic equations involving critical Sobolev exponents. *Z. Angew. Math. Phys.*, **47**, no. 1,
        89–96 (1996).
[177]   D.M. Cao, G. Li, and H.S. Zhou, The existence of two solutions to quasi-linear elliptic
        equations on $\mathbb{R}^N$. *Chinese J. Contemp. Math.*, **17**, no. 3, 277–285 (1996).
[178]   _____, Multiple solutions for nonhomogeneous elliptic equations involving critical
        Sobolev exponent. *Proc. Roy. Soc. Edin. A*, **124**, no. 6, 1177–1191 (1994).
[179]   H.T. Cao, The asymptotic and bifurcation results for potential systems involving Sobolev
        critical exponent. *Appl. Anal.*, **52**, no. 1–4, 91–101 (1994).
[180]   A. Capozzi, On subquadratic Hamiltonian systems. *Nonlinear Anal.*, **8**, no. 6, 553–562
        (1984).
[181]   C. Carathéodory, *Vorlesungen über reelle Funktionen*. Leipzig and Berlin, 1918.
[182]   J. Caristi, Fixed point theorems for mappings satisfying unwardness conditions. *Trans.
        Am. Math. Soc.*, **215**, 241–251 (1976).
[183]   P.C. Carrião and O.H. Miyagaki, Existence of homoclinic solutions for a class of time-
        dependent Hamiltonian systems. *J. Math. Anal. Appl.*, **230**, no. 1, 157–172 (1999).
[184]   P.C. Carrião, O.H. Miyagaki, and J.C. Pádua, Radial solutions of elliptic equations with
        critical exponents in $\mathbb{R}^N$. *Diff. Integral Eq.*, **11**, no. 1, 61–68 (1998).
[185]   A. Castro and J. Cossio, Multiple solutions for a nonlinear Dirichlet problem. *SIAM
        J. Math. Anal.*, **25**, no. 6, 1554–1561 (1994).
[186]   A. Castro and A.C. Lazer, Critical point theory and the number of solutions of a nonlinear
        Dirichlet problem. *Ann. Math. Pur. Appl.*, **70**, 113–137 (1979).
[187]   A. Castro, J. Cossio, and J.M. Neuberger, A sign-changing solution for a superlinear
        Dirichlet problem. *Rocky Mt. J. Math.*, **27**, 1041–1053 (1997).
[188]   A. Cauchy, Méthode générale pour la résolution des systèmes d'équations simultannées.
        C.R. Acad. Sci. Paris, **25**, 536–538 (1847).
[189]   G. Cerami, Un criterio di esistenza per i punti critici su varietà illimitate. *Rend. Accad.
        Sc. Lett. Inst. Lombardo*, **112**, 332–293 (1978).
[190]   J. Chabrowski, On pairs of decaying positive solutions of semilinear elliptic equations
        in unbounded domains. *Boll. Unione Mat. Ital. A* (7) **4**, no. 1, 21–30 (1990).
[191]   _____, On entire solutions of the p-Laplacian. Workshop on theoretical and numerical
        aspects of geometric variational problems (Canberra, 1990), 27–61, *Proc. Centre Math.
        Appl. Austral. Nat. Univ.*, **26**, Australia National University, Canberra, 1991.
[192]   _____, Concentration-compactness principle at infinity and semilinear elliptic

equations involving critical and subcritical Sobolev exponents. *Calc. Var. Part. Diff. Eq.*, **3**, no. 4, 493–512 (1995).

[193] _____, On multiple solutions for the nonhomogeneous p-Laplacian with a critical Sobolev exponent. *Diff. Integral Eq.*, **8**, no. 4, 705–716 (1995).

[194] _____, Introduction to the theory of critical points. The mountain pass theorem. Ekeland's variational principle. Instructional workshop on analysis and geometry, Part III (Canberra, 1995), 137–181, *Proc. Centre Math. Appl. Austral. Nat. Univ.*, **34**, Australia National University, Canberra, 1996.

[195] _____, On nodal radial solutions of an elliptic problem involving critical Sobolev exponent. *Comment. Math. Univ. Carolin.*, **37**, no. 1, 1–16 (1996).

[196] _____, On multiple solutions for nonhomogeneous system of elliptic equations. *Rev. Mat. Univ. Complut.*, **8**, no. 1, 207–234 (1996).

[197] _____, *Variational methods for potential operator equations. With applications to nonlinear elliptic equations*. de Gruyter Studies in Mathematics, **24**, Walter de Gruyter & Co., Berlin, 1997.

[198] J. Chabrowski and P. Drabek, On positive solutions of nonlinear elliptic equations involving concave and critical nonlinearities. *Stud. Math.*, **151**, no. 1, 67–85 (2002).

[199] J. Chabrowski and J. Yang, Existence theorems for elliptic equations involving supercritical Sobolev exponent. *Adv. Diff. Eq.*, **2**, no. 2, 231–256 (1997).

[200] _____, Nonnegative solutions for semilinear biharmonic equations in $\mathbb{R}^N$. *Analysis*, **17**, no. 1, 35–59 (1997).

[201] M. Chamberland and G. Meisters, A mountain pass to the Jacobian conjecture. *Can. Math. Bull.* **41**, no. 4, 442–451 (1998).

[202] K.C. Chang, Variational methods for non differentiable functions and their applications to PDE. *J. Math. Anal. Appl.*, **80**, 102–129 (1981).

[203] _____, A variant mountain pass lemma, *Sci. Sin. Ser. A*, **26**, 1241–1255 (1983).

[204] _____, Variational methods and sub- and super-solutions. *Sci. Sin. Ser. A*, **26**, 1256–1265 (1983).

[205] _____, *Infinite dimensional Morse theory and its applications*. Progress in Nonlinear Differential Equations and their Applications, **6**, Birkhäuser, 1985.

[206] _____, *Infinite-dimensional Morse theory and its applications*. Séminaire de Mathématiques Supérieures [Seminar on Higher Mathematics], **97**, Presses de l'Université de Montréal, Montreal, Quebec, 1985.

[207] _____, Morse theory on Banach spaces and its applications to partial differential equations. *Chinese Ann. Math.*, **4B**, 381–399 (1985).

[208] _____, On the mountain pass lemma. *Equadiff. 6* (Brno, 1985), Lecture Notes in Mathematics, Springer, Berlin, no. 1192, 203–208 (1986).

[209] _____, Critical groups, Morse theory and application to semilinear elliptic BVPs. *Chinese mathematics into the 21st century* (Tianjin, 1988), 41–65, Peking University Press, Beijing, 1991.

[210] _____, H^1 versus C^1 isolated critical points. *C.R. Acad. Sci. Paris Sér. I Math.*, **319**, no. 5, 441–446 (1994).

[211] _____, Morse theory in differential equations. Proceedings of the international congress of mathematicians, Vol. **1**, **2** (Zürich, 1994), 1065–1076, Birkhäuser, Basel, 1995.

[212] _____, Morse theory in nonlinear analysis. *Nonlinear functional analysis and applications to differential equations* (Trieste, 1997), 60–101, World Scientific Publishing, River Edge, NJ, 1998.

[213] K.C. Chang and J. Eells, Unstable minimal surface coboundaries. *Acta Math. Sin. (N.S.)*, **2**, no. 3, 233–247 (1986).

[214] K.C. Chang and C.W. Hong, Periodic solutions for the semilinear spherical wave equation. *Acta Math. Sin. (N.S.)*, **1**, no. 1, 87–97 (1985).

[215] K.C. Chang and S. Shi, A local minimax theorem without compactness. *Nonlinear and convex* (Santa Barbara, CA, 1985), 211–233, Lecture Notes in Pure and Applied Math., **107**, Dekker, New York, 1987.

[216] J.V. Chaparova, Nontrivial solutions of some semilinear sixth-order differential equations. *C.R. Acad. Bulg. Sci.*, **54**, no. 9, 29–32 (2001).

[217] C.N. Chen and S.-Y. Tzeng, Existence and multiplicity results for homoclinic orbits of Hamiltonian systems. *Electron. J. Diff. Eq.*, no. 7, 19 pp. (1997) (online).

[218] _____, Some properties of Palais-Smale sequences with applications to elliptic boundary-value problems. *Electron. J. Diff. Eq.*, no. 17, 29 pp. (1999) (online).

[219] _____, Some properties of Palais-Smale seqences with applications to elliptic boundary value problems. *Electron. J. Diff. Eq.*, no. 17, 29 pp. (1999) (online).

[220] G. Chen, J. Zhou, and W.-M. Ni, Algorithms and visualization for solutions of nonlinear elliptic equations. *Int. J. Bifurcat. Chaos*, **10**, no. 7, 1565–1612 (2000).

[221] J.Q. Chen and Y.Q. Li, A variant of the quantitative deformation lemma and its applications. (Chinese) *Fujian Shifan Daxue Xuebao Ziran Kexue Ban*, **15**, no. 3, 6–10, 25 (1999).

[222] W.X. Chen and W.Y. Ding, Scalar curvatures on S^2. *Trans. Am. Math. Soc.*, **303**, no. 1, 365–382 (1987).

[223] Y. Chen, A variational numerical method for finding the traveling waves of nonlinear suspension beam equations – The mountain pass algorithm. (Chinese. English summary) *J. Numer. Method. Comput. Appl.*, **21**, no. 1, 41–47 (2000).

[224] _____, Traveling wave solutions to beam equation with fast-increasing nonlinear restoring forces. *Appl. Math. J. Chinese Univ., Ser. B*, **15**, no. 2, 151–160 (2000).

[225] Y. Chen and P.J. McKenna, Traveling waves in a nonlinearly suspended beam: Theoretical results and numerical observations. *J. Diff. Eq.*, **136**, no. 2, 325–355 (1997).

[226] R. Chiappinelli, Nonlinear stability of eigenvalues of compact self-adjoint operators. *Recent trends in nonlinear analysis*. Appell, Juergen (ed.), Festschrift dedicated to Alfonso Vignoli on the occasion of his 60th birthday. Progress in Nonlinear Differential Equation Application, Birkhäuser, Basel, **40**, 93–103, 2000.

[227] Q.-H. Choi and T. Jung, Critical points theory and its application to a nonlinear wave equation. Proceedings of the fifth international colloquium on differential equations (Plovdiv, 1994), 73–82, VSP, Utrecht, 1995.

[228] Y.S. Choi and P.J. McKenna, A mountain pass method for the numerical solution of semilinear elliptic equations. *Nonlinear Anal.*, **20**, no. 4, 417–437 (1993).

[229] Y.S. Choi and A.W. Shaker, On symmetry in semi-linear elliptic systems with nonnegative solutions. *Appl. Anal.*, **53**, no. 1–2, 125–134 (1994).

[230] Y.S. Choi, K.C. Jen, and P.J. McKenna, The structures of the solution set for periodic oscillations in a suspension bridge model. *IMA J. Appl. Math.*, **47**, 283–306 (1991).

[231] Y.S. Choi, P.J. McKenna, and M. Romano, A mountain pass method for the numerical solution of semilinear wave equations. *Numer. Math.*, **64**, no. 4, 487–509 (1993).

[232] C.-C Chou, K.-F. Ng, and S. Shi, Multi-variational principle, minimax theorem, and applications. *J. Math. Anal. Appl.*, **237**, no. 1, 19–29 (1999).

[233] M. Choulli, R. Deville, and A. Rhandi, A general mountain pass principle for nondifferentiable functionals. *Rev. Mat. Apll.*, **13**, no. 2, 45–58 (1992).

[234] C.O. Christenson and W.L. Voxman, *Aspects of topology*. Dekker, New York, 1977.

[235] K. Cieliebak and E. Séré, Pseudoholomorphic curves and the shadowing lemma. *Duke Math. J.*, **99**, no. 1, 41–73 (1999).

[236] S. Cingolani and J.L. Gomez, Asymmetric positive solutions for a symmetric nonlinear problem in $\mathbb{R}^N$. *Calc. Var. Part. Diff. Eq.*, **11**, no. 1, 97–117 (2000).

[237] G. Citti, Existence of positive solutions of quasilinear degenerate elliptic equations on unbounded domains. *Ann. Mat. Pur. Appl., IV. Ser.* **158**, 315–330 (1991).

[238] M. Clapp, Critical point theory for perturbations of symmetric functionals. *Comment. Math. Helv.*, **71**, no. 4, 570–593 (1996).

[239] M. Clapp and D. Puppe, Invariants of the Lusternik-Schnirelman type and the topology of critical sets. *Trans. Am. Math. Soc.*, **298**, 603–620 (1991).

[240] _____, Critical point theory with symmetries. *J. Reine Angew. Math.*, **418**, 1–29 (1991).

[241] _____, Critical point theory of symmetric functions and closed geodesics. *Diff. Geom. Appl.*, **6**, no. 4, 367–396 (1996).

[242] D. Clark, A variant of the Ljusternik-Schnirelman theory. *Indiana J. Math.*, **22**, 65–74 (1973).

[243] F. Clarke, *Optimization and nonsmooth analysis.* Canadian Mathematical Series of Monographs and Advanced Texts, Wiley-Interscience, 1983. Second edition, SIAM, Philadelphia, 1990.

[244] F. Clarke and I. Ekeland, Hamiltonian trajectories with prescribed minimal period. *Commun. Pure Appl. Math.*, **33**, 103–116 (1980).

[245] F. Clarke, Yu S. Ledayev, R.J. Stern, and P.R. Wolenski, *Nonsmooth analysis and control theory.* Graduate Texts in Mathematics, **178**, Springer-Verlag, New York, 1998.

[246] P. Clément, P. Felmer, and E. Mitidieri, Solutions homoclines d'un système hamiltonien non-borné et superquadratique. (French) [Homoclinic orbits of an unbounded superquadratic Hamiltonian system.] *C.R. Acad. Sci. Paris Sér. I Math.*, **320**, no. 12, 1481–1484 (1995).

[247] _____, Homoclinic orbits for a class of infinite-dimensional Hamiltonian systems. *Ann. Scuola Norm. Sup. Pisa*, **24**, no. 2, 367–393 (1997).

[248] P. Clément, M. Garcia-Huidobro, R. Manasevich, and K. Schmitt, Mountain pass type solutions for quasilinear elliptic equations. *Calc. Var. Part. Diff. Eq.*, **11**, no. 1, 33–62 (2000).

[249] C.V. Coffman, Ljusternik-Schnirelman theory: Complementary principles and the Morse index. *Nonlinear Anal.–Theor.*, **12**, no. 5, 507–529 (1988).

[250] C. Conley, *Isolated invariant sets and the morse index.* C.B.M.S. Regional Conference Series American Mathematical Society, RI, **38**, 1978.

[251] M. Conti and F. Gazzola, Positive entire solutions of quasilinear elliptic problems via nonsmooth critical point theory. *Topol. Method. Nonlinear Anal.*, **8**, no. 2, 275–294 (1997).

[252] M. Conti and R. Lucchetti, The minimax approach to the critical point theory. Recent developments. *Well-posed variational problems*, 29–76, *Math. Appl.*, **331**, Kluwer Academic Publishers, Dordrecht, 1995.

[253] M.C. Conti, Walking around mountains. *Rend. Accad. Sci. Lett. Lombardo*, **A 128**, no. 1, 53–70 (1995).

[254] C. Cort zar, M. Elgueta, and P. Felmer, Existence of signed solutions for a semilinear elliptic boundary value problem. *Diff. Integral Eq.*, **7**, no. 1, 293–299 (1994).

[255] J.-N. Corvellec, Morse theory for continuous functionals. *J. Math. Anal. Appl.*, **195**, 1050–1072 (1995).

[256] _____, A note on coercivity of lower semicontinuous functions and nonsmooth critical point theory. *Serdica Math. J.*, **22**, no. 1, 57–68 (1996).

[257] _____, Quantitative deformation theorems and critical point theory. *Pacific J. Math.*, **187**, no. 2, 263–279 (1999).

[258] _____, On critical point theory with the (PS)* condition. Calculus of variations and differential equations (Haifa, 1998), 65–81, Chapman & Hall/CRC, Boca Raton, FL, 2000.

[259] _____, On the second deformation lemma. Prépublication no. 10, Laboratoire MANO, University of Perpignan, 2000.

[260] J.-N. Corvellec and M. Degiovanni, Nontrivial solutions of quasilinear equations via nonsmooth Morse theory. *J. Diff. Eq.*, **136**, 268–293 (1997).

[261] J.-N. Corvellec, M. Degiovanni, and M. Marzocchi, Deformation properties for continuous functionals and critical point theory. *Topol Method. Nonlinear Anal.*, **1**, 151–171 (1993).

[262] D.G. Costa, On a class of elliptic systems in $\mathbb{R}^N$. *Electron. J. Diff. Eq.*, 1994, no. 07 (online).

[263] D.G. Costa and O.H. Miyagaki, Nontrivial solutions for perturbations of the p-Laplacian on unbounded domains. *J. Math. Anal. Appl.*, **193**, no. 3, 737–755 (1995).

[264] D.G. Costa and A.S. Oliveira, Existence of solution for a class of semilinear elliptic problems at double resonance. *Bol. Soc. Bras. Mat.*, **19**, 21–37 (1988).

[265] D.G. Costa and E.A. de B. Silva, The Palais-Smale condition versus coercivity. *Nonlinear Anal.*, **16**, no. 4, 371–381 (1991).

[266] D.G. Costa and E.A. de B. Silva, A note on problems involving critical Sobolev exponents. *Diff. Integral Eq.*, **8**, no. 3, 673–679 (1995).

[267] D.G. Costa and H. Tehrani, On the sign of the mountain pass solution. *Nonlinear Anal.*, **44**, no. 1, 65–80 (2000).

[268] D.G. Costa and M. Willem, Ljusternik-Schnirelman theory and asymptotically linear Hamiltonian systems. Differential equations: Qualitative theory, Vol. I, II (Szeged, 1984), *Colloquia Methematica Societatis János Balayai*, **47**, 179–191, *Diff. Eq. Qual. Theory*, Szeged, Hungary, North-Holland, Amsterdam, 1987.

[269] _____, Points critiques multiples de fonctionnelles invariantes. *C.R. Acad. Sci. Paris*, **298**, 381–384 (1984).

[270] _____, Multiple critical points of invariant functionals and applications. *Nonlinear Anal.*, **10**, 843–852 (1986).

[271] D.G. Costa, Z. Ding, and J.M. Neuberger, A numerical investigation of sign-changing solutions to superlinear elliptic equations on symmetric domains. *J. Comput. Appl. Math.*, **131**, no. 1–2, 299–319 (2001).

[272] V. Coti Zelati, Solution of a BVP constrained in an infinitely deep potential well. *Rend. Istit. Mat. Univ. Trieste*, **18**, no. 1, 100–104 (1986).

[273] V. Coti Zelati and P.H. Rabinowitz, Homoclinic orbits for second order Hamiltonian systems possessing superquadratic potentials. *J. Am. Math. Soc.*, **4**, no. 4, 693–727 (1991).

[274] _____, Homoclinic type solutions for a semilinear elliptic PDE on $\mathbb{R}^n$. *Commun. Pure Appl. Math.*, **45**, no. 10, 1217–1269 (1992).

[275] V. Coti Zelati, I. Ekeland, and P.-L. Lions, Index estimates and critical points of functionals not satisfying Palais-Smale. *Ann. Scuola Norm. Sup. Pisa*, **17**, no. 4, 569–581 (1990).

[276] V. Coti Zelati, I. Ekeland, and E. Sere, A variational approach to homoclinic orbits in Hamiltonian systems. *Math. Ann.*, **288**, no. 1, 133–160 (1990).

[277] V. Coti Zelati, P. Montecchiari, and M. Nolasco, Multibump homoclinic solutions for a class of second order, almost periodic Hamiltonian systems. *NoDEA Nonlinear Diff. Eq. Appl.*, **4**, no. 1, 77–99 (1997).

[278] A. Crannell, The existence of many periodic non-travelling solutions to the Boussinesq equation. *J. Diff. Eq.*, **126**, no. 2, 169–183 (1996).

[279] R. Courant, *Dirichlet principle, conformal mappings and minimal surfaces*. Interscience, New York, 1950. Reprinted: Springer, New York, Heidelberg, and Berlin, 1977.

[280] R. Courant and D. Hilbert, *Methods of mathematical physics*, **1**, Wiley, New York, 1962.

[281] M. Cuesta, D.G. de Figueiredo, and J.-P. Gossez, Sur le spectre de Fučik du p-Laplacien. (French) [On the Fucik spectrum of the p-Laplacian.] *C.R. Acad. Sci. Paris Sér. I Math.*, **326**, no. 6, 681–684 (1998).

[282] T. D'Aprile, Existence and concentration of local mountain passes for a nonlinear elliptic field equation in the semi-classical limit. *Topol. Method. Nonlinear Anal.*, **17**, no. 2, 239–275 (2001).

[283] S. Dabuleanu and D. Motreanu, Existence results for a class of eigenvalue quasilinear problems with nonlinear boundary conditions. *Adv. Nonlinear Var. Inequal.*, **2**, no. 2, 41–54 (1999).

[284] B. Dacorogna, *Weak continuity and weak lower semi-continuity of nonlinear functionals*. Lecture Notes Math., Springer-Verlag, Berlin, 922, 1982.

[285] _____, *Direct methods in the calculus of variations*. Applied Mathematical Sciences, Springer-Verlag, Berlin, 78, 1989.

[286] S.H. Dai and Z.Q. Yan, Existence of nontrivial solutions of the equation $\Delta^2 u - a\Delta u + bu = f(x, u)$ – An application of the mountain pass lemma (Chinese). *Acta Sci. Natur. Univ. Jilin.*, no. 1, 21–28 (1984).

[287] R. Dalmasso, On singular superlinear elliptic problems of second and fourth orders. *Bull. Sci. Math. (2)*, **116**, no. 2, 247–263 (1992).

[288] E. Dancer, On the Dirichlet problem for weakly nonlinear elliptic partial differential equations. *Proc. Roy. Soc. Edin.*, **76**, 283–300 (1977).

[289] ———, The G-invariant implicit theorem in infinite dimensions. *Proc. Roy. Soc. Edin.*, **92**, 13–30 (1982).

[290] ———, Multiple solutions of asymptotically homogeneous problems. *Ann. Mat. Pur. Appl.*, **152**, no. 4, 63–78 (1988).

[291] E. N. Dancer and J. Wei, On the profile of solutions with two sharp layers to a singularly perturbed semilinear Dirichlet problem. *Proc. Roy. Soc. Edin. A*, **127**, no. 4, 691–701 (1997).

[292] J. Daneš, A geometric theorem useful in nonlinear functional analysis. *Boll. Unione. Mat. Ital.*, **6**, 369–375 (1972).

[293] ———, Equivalence of some geometric and related results of nonlinear functional analysis. *Comment. Math. Univ. Carolin.*, **26**, no. 3, 443–454 (1985).

[294] D.G. de Figueiredo, Semilinear elliptic equations at resonance: Higher eigenvalues and unbounded nonlinearities. *Recent advances in differential equations*, 88–99, Academic Press, New York and London, 1981.

[295] ———, On the existence of multiple ordered solutions of nonlinear eigenvalue problems. *Nonlinear Anal.*, **11**, no. 4, 481–492 (1987).

[296] ———, Lectures on the Ekeland variational principle with applications and detours. Tata Institute of Fundamental Research Lectures on Mathematics and Physics, **81**, Tata Institute of Fundamental Research, Bombay, Springer-Verlag, Berlin and New York, 1989.

[297] D.G. de Figueiredo and J.-P. Gossez, Un problème elliptique semilinéaire sans condition de croissance, *C.R. Acad. Sci. Paris*, t.308, Série I, 277–280 (1989).

[298] D.G. de Figueiredo and C.A. Magalhaes, On nonquadratic Hamiltonian elliptic systems. *Adv. Diff. Eq.*, **1**, no. 5, 881–898 (1996).

[299] D.G. de Figueiredo and I. Massabò, Semilinear elliptic equations with the primitive of the nonlinearity interacting with the first eigenvalue. *J. Math. Anal. Appl.*, **156**, no. 2, 381–394 (1991).

[300] D.G. de Figueiredo and B. Ruf, On a superlinear Sturm-Liouville equation and a related bouncing problem. *J. Reine Angew. Math.*, **421**, 1–22 (1991).

[301] ———, On the periodic Fučik spectrum and a superlinear Sturm-Liouville equation. *Proc. Roy. Soc. Edin. A*, **123**, no. 1, 95–107 (1993).

[302] D.G. de Figueiredo and S. Solimini, A variational approach to superlinear elliptic problems. *Commun. Part. Diff. Eq.*, **9**, no. 7, 699–717 (1984).

[303] D.G. de Figueiredo, O.H. Miyagaki, and B. Ruf, Corrigendum: "Elliptic equations in $\mathbb{R}^2$ with nonlinearities in the critical growth range." *Calc. Var. Part. Diff. Eq.*, **4**, no. 2, 203 (1996).

[304] D.G. de Figueiredo, J.V. Goncalves, and O.H. Miyagaki, On a class of quasilinear elliptic problems involving critical exponents. *Commun. Contemp. Math.*, **2**, no. 1, 47–59 (2000).

[305] E. De Giorgi, M. Degiovanni, A. Marino, and M. Tosques, Evolution equations for a class of nonlinear operators. *Atti. Accad. Naz. Lin.*, **75**, 1–8 (1984).

[306] M. Degiovanni, Homotopical properties of a class of nonsmooth functions. *Ann. Mat. Pur. Appl., IV. Ser.*, **156**, 37–71 (1990).

[307] ———, Nonsmooth critical point theory and applications. *Nonlinear Anal.*, **30**, 89–99 (1997).

[308] M. Degiovanni and S. Lancelotti, A note on nonsmooth functionals with infinitely many critical values. *Boll. Unione Mat. Ital., VII Ser.*, **A 7**, no. 2, 289–297 (1993).

[309]   _____, Perturbations of critical values in nonsmooth critical point theory. *Serdica Math. J.*, **22**, no. 3 (1996).

[310]   M. Degiovanni and M. Marzocchi, A critical point theory for nonsmooth functionals. *Ann. Mat. Pur. Appl.*, **567**, 73–100 (1994).

[311]   M. Degiovanni and V. Rădulescu, Perturbation of nonsmooth symmetric nonlinear eigenvalue problems. *C.R. Acad. Sci. Paris, Série I*, **239**, 281–286 (1999).

[312]   M. Degiovanni and F. Schuricht, Buckling of nonlinearly elastic rods in the presence of obstacles treated by nonsmooth critical point theory. *Math. Ann.*, **311**, 675–628 (1998).

[313]   M. Degiovanni and S. Zani, Euler equations involving nonlinearities without growth conditions. *Potential Anal.*, **5**, 505–512 (1996).

[314]   _____, Multiple solutions of semilinear elliptic equations with one-sided growth conditions. *Math. Comput. Model.*, **32**, no. 11–13, 1377–1393 (2000).

[315]   K. Deimling, *Nonlinear functional analysis*. Springer-Verlag, Berlin, 1985.

[316]   G. Dell'Antonio, Non-collision periodic solutions of the N-body system. *NODEA– Nonlinear Diff.*, **5**, no. 1, 117–136 (1998).

[317]   Y. Deng and G. Wang, On inhomogeneous biharmonic equations involving critical exponents. *Proc. Roy. Soc. Edin. A*, **129**, no. 5, 925–946 (1999).

[318]   Y. Deng, G. Chen, W.M. Ni, and J. Zhou, Boundary element monotone iteration scheme for semilinear elliptic partial differential equations. *Math. Comput.*, **65**, 943–982 (1996).

[319]   Y.B. Deng, The existence and nodal character of the solutions in $\mathbb{R}^n$ for semilinear elliptic equation involving critical Sobolev exponent. *Acta Math. Sci.* (English ed.), **9**, no. 4, 385–402 (1989).

[320]   _____, On the superlinear Ambrosetti-Prodi problem involving critical Sobolev exponents. *Nonlinear Anal.*, **17**, no. 12, 1111–1124 (1991).

[321]   _____, Existence of multiple positive solutions of inhomogeneous semi-linear elliptic problems involving critical exponents. *Commun. Part. Diff. Eq.*, **17**, no. 1–2, 33–53 (1992).

[322]   Y.-B. Deng, H.S. Zhou, and X.P. Zhu, On the existence and $L^p(\mathbb{R}^N)$ bifurcation for the semilinear elliptic equation. *J. Math. Anal. Appl.*, **154**, no. 1, 116–133 (1991).

[323]   Y.H. Deng, Y.T. Shen, and S.R. Li, On the existence of infinitely many solutions of the Dirichlet problem for semilinear elliptic equations. *Kexue Tongbao* (English ed.), **30**, no. 7, 853–857 (1985).

[324]   R. Deville and M. Choulli, Un théorème du col pour applications non différentiables. Séminaire d'Initiation à l'Analyse, Exp. no. 4, 7 pp., Publ. Math. Univ. Pierre et Marie Curie, 107, Univ. Paris VI, Paris.

[325]   R. Deville, G. Godefroy, and V. Zizler, Un principe variationnel utilisant des fonctions bosses. *C.R. Acad. Sci. Paris, Série I*, **312**, 281–286 (1991).

[326]   J. Dieudonné, *Foundations of modern analysis, Tomes 1 and 2*. Academic Press, New York, 1960.

[327]   J. Dieudonné, *History of functional analysis*. North-Holland, Amsterdam, 1981.

[328]   *Differential equations*. Proceedings of the First Latin American School, University of Sao Paulo, Sao Paulo, June 29–July 17, 1981. Edited by Djairo Guedes de Figueiredo and Chaim Samuel Hönig. Lecture Notes in Mathematics, **957**, Springer-Verlag, Berlin and New York, 1982.

[329]   G. Dinca and D. Pasca, Periodic solutions of superlinear convex autonomous Hamiltonian systems. *J. Global Optim.*, **17**, no. 1/4, 65–75 (2000).

[330]   G. Dinca, P. Jebelean, and J. Mawhin, A result of Ambrosetti-Rabinowitz type for p-Laplacian. *Qualitative problems for differential equations and control theory*, 231–242, World Scientific Publishing, River Edge, NJ, 1995.

[331]   W.Y. Ding, Ljusternik-Schnirelman theory for harmonic maps. *Acta Math. Sin.*, **2**, 105–122 (1986).

[332]   W.Y. Ding, Symmetric harmonic maps between spheres. *Commun. Math. Phys.*, **118**, no. 4, 641–649 (1988).

[333] Y. Ding, A remark on the linking theorem with applications. *Nonlinear Anal.*, **22**, 237–250 (1994).

[334] ⸻, Infinitely many entire solutions of an elliptic system with symmetry. *Topol. Method. Nonlinear Anal.*, **9**, no. 2, 313–323 (1997).

[335] ⸻, Solutions to a class of Schrödinger equations. *Proc. Am. Math. Soc.*, **130**, no. 3, 689–696 (2002).

[336] Y. Ding and M. Girardi, Periodic and homoclinic solutions to a class of Hamiltonian systems with the potentials changing sign. *Dyn. Syst. Appl.*, **2**, no. 1, 131–145 (1993).

[337] ⸻, Infinitely many homoclinic orbits of a Hamiltonian system with symmetry. *Nonlinear Anal.–Theor.*, **38**, no. 3, 391–415 (1999).

[338] Z. Ding, Nonlinear periodic oscillations in suspension bridges. *Control of nonlinear distributed parameter systems*. Proceedings of the conference advances in control of nonlinear distributed parameter systems, Texas A & M Univ., College Station, TX. Chen, G. et al. (eds.), dedicated to Prof. David L. Russell on the occasion of his 60th birthday. Dekker, New York, *Lect. Notes Pure Appl. Math.*, **218**, 69–84 (2001).

[339] Z. Ding, D. Costa, and G. Chen, A high-linking algorithm for sign-changing solutions of semilinear elliptic equations. *Nonlinear Anal.*, **38**, 151–172 (1999).

[340] J.M. do and M.A.S. Souto, On a class of nonlinear Schrödinger equations in $\mathbb{R}^2$ involving critical growth. *J. Diff. Eq.*, **174**, no. 2, 289–311 (2001).

[341] F. Dobarro and E. Lami Dozo, Variational solutions in solar flares with gravity. Partial differential equations (Han-sur-Lesse, 1993), 120–143, *Math. Res.*, **82**, Akademie-Verlag, Berlin, 1994.

[342] C.L. Dolph, Nonlinear integral equations of the Hammerstein type. *Trans. Am. Math. Soc.*, **66**, 289–307 (1949).

[343] C.G. Dong and S. Li, On the existence of infintely many solutions of the Dirichlet problem for some nonlinear elliptic equations. *Sci. Sin. Ser. A*, **25**, 468–475 (1982).

[344] H. Drygas, Spectral methods in linear minimax estimation. *Acta Appl. Math.*, **43**, no. 1, 17–42 (1996).

[345] ⸻, Reparametrization methods in linear minimax estimation. Matsusita, K. et al. (eds.), Statistical sciences and data analysis, Proceedings of the third Pacific Area statistical conference (Makuhari, Japan, December 11–13, 1991), 87–95, VSP, Utrecht, 1993.

[346] Y. Du, A deformation lemma and its applications, *Kexue Tongbao*, **36**, 103–106 (1991).

[347] ⸻, A deformation lemma and some critical point theorems. *Bull. Aust. Math. Soc.*, **43**, no. 1, 161–168 (1991).

[348] ⸻, Critical point theorems with relaxed boundary condition and applications. *Bull. Aust. Math. Soc.*, **47**, no. 1, 101–118 (1993).

[349] D.M. Duc, Critical values of functionals on Banach surfaces. Preprint, International Center of Theoretical Physics, Trieste, **IC/87/344**, November (1987), (http://www.ictp.trieste.it/~pub_off).

[350] M.D. Duong, Nonlinear singular elliptic equations. *J. London Math. Soc.*, **40**, no. 3, 420–440 (1989).

[351] J. Eels, A setting for global analysis. *Bull. Am. Math. Soc.*, **72**, 751–807 (1966).

[352] H. Egnell, Existence results for some quasilinear elliptic equations. *Variational methods* (Paris, 1988), 61–76, Progr. Nonlinear Differential Equations Appl., **4**, Birkhäuser Boston, Boston, 1990.

[353] I. Ekeland, Sur les problèmes variationnels. *C.R. Acad. Sci. Paris*, **275**, A1057–A1059 (1972).

[354] ⸻, On the variational principle. *J. Math. Anal. Appl.*, **47**, 324–353 (1974).

[355] ⸻, Nonconvex minimization problems. *Bull. Am. Math. Soc.*, **1**, 443–474 (1979).

[356] ⸻, Une théorie de Morse pour les systèmes Hamiltoniens convexes. *Ann. Inst. H. Poincaré–An.*, **1**, 143–197 (1984).

[357] _____, Index theory for periodic solutions of convex Hamiltonian systems. Proceedings AMS Summer Institute on Nonlinear Functional Analysis (Berkeley, CA, 1983), *Proc. Symp. Pure Math.*, **45**, 395–424 (1986).

[358] _____, The ε-variational principle revisited. *Methods of nonconvex analysis*, A. Cellina (ed.), Lecture Notes in Mathematics, Springer-Verlag, **1446**, 1–15, (1989).

[359] _____, The mountain pass theorem and applications. *Minimax results of Ljusternick-Schnirelman type and applications*, Sém. Math. Sup., 107, Presses Univ. Montréal, Montreal, PQ, **107**, 9–34, 1989.

[360] _____, *Convexity methods in Hamiltonian mechanics*. Ergebnisse der Mathematik und ihrer Grenzgebiete (3), **19**, Springer-Verlag, Berlin, 1990.

[361] I. Ekeland and N. Ghoussoub, $\mathbb{Z}_2$-equivariant Ljusternik-Schnirelman theory for non-even functionals. *Ann. Inst. H. Poincare–An.*, **15**, no. 3, 341–370 (1998).

[362] _____, New aspects of the calculus of variations in the large. *Bull. Am. Math. Soc.*, **39**, no. 2, 207–265 (2001).

[363] I. Ekeland and H. Hofer, Periodic solutions with prescribed minimal period for convex autonomous Hamiltonian systems. *Invent. Math.*, **81**, no. 1, 155–188 (1985).

[364] _____, Subharmonics for convex nonautonomous Hamiltonian systems. *Commun. Pure Appl. Math.*, **40**, no. 1, 1–36 (1987).

[365] _____, Symplectic topology and Hamiltonian dynamics I. *Math. Z.*, **200**, 355–378 (1989).

[366] _____, Symplectic topology and Hamiltonian dynamics II. *Math. Z.*, **203**, 553–567 (1990).

[367] I. Ekeland and J.M Lasry, Sur le nombre de points critiques de fonctions invariantes par des groupes. *C.R. Acad. Sci. Paris*, **282**, no. 11, A559–A562 (1976).

[368] _____, Sur le nombre de points critiques de fonctions invariantes par des groupes. *C.R. Acad. Sci. Paris*, **282**, no. 16, A841–A844 (1976).

[369] _____, On the number of periodic trajectories for a Hamiltonian flow on a convex energy surface. *Ann. Math.*, **112**, 283–319 (1980).

[370] I. Ekeland and R. Temam, *Convex analysis and variational methods*. North-Holland, 1976. French ed., Dunod, Paris, 1976.

[371] I. Ekeland, N. Ghoussoub, and H. Tehrani, Multiple solutions for a classical problem in the calculus of variations. *J. Diff. Eq.*, **131**, no. 2, 229–243 (1996).

[372] A.R. El Amrouss and M. Moussaoui, Minimax principles for critical-point theory in applications to quasilinear boundary-value problems. *Electron. J. Diff. Eq.*, no. 18, 9 pp. (2000) (online).

[373] Equadiff 6. Proceedings of the sixth Czechoslovak international conference on differential equations and their applications (Brno, August 26–30, 1985), J. Vosmanský and M. Zlámal (eds.). University J. E. Purkyně, Brno, 1986.

[374] J.F. Escobar, Positive solutions for some semilinear elliptic equations with critical Sobolev exponents. *Commun. Pure Appl. Math.*, **40**, no. 5, 623–657 (1987).

[375] P.C. Espinoza, Positive ordered solutions of a discrete analogue of a nonlinear elliptic eigenvalue problem. *SIAM J. Numer. Anal.*, **31**, no. 3, 760–767 (1994).

[376] L. Euler, Methodus inveniendi lineas curvas maximi minimive proprietate gaudentes sive solutio problematis isoperimetrici latissimo sensu accepti (Lausanne-Genève, 1744), *Opera, Lausanne-Genève, Ser. I*, **24** (ed. C. Carathéodory), Berne, 1952.

[377] F. Facchinei, Structural and stability properties of P_0 nonlinear complementarity problems. *Math. Oper. Res.*, **23**, no. 3, 735–745 (1998).

[378] E.R. Fadell, The relation between Ljusternik-Schnirelman category and the concept of genus. *Pacific J. Math.*, **89**, 33–42 (1980).

[379] _____, The equivariant Ljusternik-Schnirelman method for invariant functionals and relative cohomological index theories. *Méthodes topologiques en analyse non linéaire*, Grans (ed.), Sémin. Math. Sup., Montréal, no. **95**, 41–70, 1985.

[380] _____, Cohomological methods in non-free G-spaces with applications to general Borsuk-Ulam theorems and critical point theorems for invariant functionals. *Nonlinear*

functional analysis and its applications, S.P. Singh (ed.), Proc. Maratea 1985, NATO ASI Ser. **C 173**, Reidel, Dordrecht, 1–45, 1986.

[381] E.R. Fadell and S.Y. Husseini, Relative cohomological index theories. *Adv. in Math.*, **64**, 1–31 (1987).

[382] ———, An ideal cohomological index theoy with applications to Borsuk-Ulam and Bourgain-Yang theorems. *Ergod. Theor. Dyn. Syst.*, **8**, 73–85 (1988).

[383] ———, Category of loop spaces of open subsets in Euclidean spaces. *Nonlinear Anal.-Theor.*, **17**, no. 12, 1153–1161, (1991).

[384] E.R. Fadell and P.H. Rabinowitz, Bifurcation for odd potential operators and alternative topological index. *J. Funct. Anal.*, **26**, 48–67 (1977).

[385] ———, Generalized cohomological index theories for the Lie group actions with an applications to bifurcation questions for Hamiltonian systems. *Invent. Math.*, **45**, 139–174 (1978).

[386] E.R. Fadell, S.Y. Husseini, and P.H. Rabinowitz, Borsuk-Ulam theorems for arbitrary S^1 actions and applications. *Trans. Am. Math. Soc.*, **274**, 345–360 (1982).

[387] G. Fang, Topics on critical point theory. Ph. D. thesis, University of British Columbia, 1993.

[388] ———, The structure of the critical set in the general mountain pass principle. *Ann. Fac. Sci. Toulouse*, **3**, no. 6, 345–362 (1994).

[389] ———, Morse indices of critical manifolds generated by min-max methods with compact Lie group actions and applications. *Commun. Pure Appl. Math.*, **48**, no. 12, 1343–1368 (1995).

[390] ———, On the existence and the classification of critical points for nonsmooth functionals. *Can. J. Math.*, **47**, no. 4, 684–717 (1995).

[391] ———, Morse indices of degenerate critical orbits and applications – Perturbation methods in equivariant cases. *Nonlinear Anal.-Theor.*, **36 A**, no. 1, 101–118 (1999).

[392] G. Fang and N. Ghoussoub, Second order information on Palais-Smale sequences in the mountain pass theorem. *Manuscripta Math.*, **75**, no. 1, 81–95 (1992).

[393] ———, Morse type information on Palais-Smale sequences obtained by min-max principles. *Commun. Pure Appl. Math.*, **47**, 1595–1653 (1994).

[394] E. Feireisl, Time periodic solutions to a semilinear beam equation. *Nonlinear Anal.*, **12**, no. 3, 279–290 (1988).

[395] ———, Spatially localized free vibrations for a nonlinear string model. *Math. Method. Appl. Sci.*, **15**, no. 5, 331–343 (1992).

[396] P.L. Felmer, Variational methods in Hamiltonian systems. *Dynamical systems* (Temuco, 1991/1992), 151–178, Travaux en Cours, 52, Hermann, Paris, 1996.

[397] P.L. Felmer and E.A. De B. Silva, Homoclinic and periodic orbits for Hamiltonian systems. *Ann. Scuola Norm. Sup. Pisa*, **26**, no. 2, 285–301 (1998).

[398] Y. Feng, The study of nonlinear flexings in a floating beam by variational methods. Oscillations in nonlinear systems: Applications and numerical aspects. *J. Comput. Appl. Math.*, **52**, no. 1–3, 91–112 (1994).

[399] A. Floer, A refinement of the Conley index and an application to the stability of hyperbolic invariant sets. *Ergod. Theor. Dyn. Syst.*, **7**, 93–103 (1988).

[400] M. Flucher and J. Wei, Asymptotic shape and location of small cores in elliptic free-boundary problems. *Math. Z.*, **228**, no. 4, 683–703 (1998).

[401] A. Fonda, Periodic solutions for a conservative system of differential equations with a singularity of repulsive type. *Nonlinear Anal.*, **24**, no. 5, 667–676 (1995).

[402] I. Fonseca and W. Gangbo, *Degree theory in analysis and applications*. Oxford Lecture Series in Mathematics and its Applications, Oxford Science Publications, **2**, 1995.

[403] D. Fortunato and E. Jannelli, Infinitely many solutions for some nonlinear elliptic problems in symmetrical domains. *Proc. Roy. Soc. Edin. A*, **105**, 205–213 (1987).

[404] G. Fournier and M. Willem, Simple variational methods for unbounded potentials. *Topological fixed point theory and applications* (Tianjin, 1988), 75–82, Lecture Notes in Mathematics, **1411**, Springer, Berlin and New York, 1989.

[405] _____, Multiple solutions of the forced double pendulum equation. *Ann. Inst. H. Poincare–An.*, **6**, suppl., 259–281 (1989). (Contributions en l'honneur de J.-J. Moreau, eds. H. Attouch, J.-P. Aubin, F. Clarke, I. Ekeland, CRM Gauthier-Villars, Paris.)

[406] _____, The mountain circle theorem. Delay differential equations and dynamical systems (Claremont, CA, 1990), 147–160, Lecture Notes in Mathematics, Springer, Berlin, **1475**, 1991.

[407] G. Fournier, M. Timoumi, and M. Willem, The limiting case for strongly indefinite functionals. *Topol. Method. Nonlinear Anal.*, **1**, no. 2, 203–209 (1993).

[408] G. Fournier, D. Lupo, M. Ramos, and M. Willem, Limit relative category and critical point theory. *Dynamics reported*, C.K.R.T. Jones et al. (ed.), 1–24, Expositions in Dynamical Systems, New Series, **3**. Springer-Verlag, Berlin, 1994.

[409] M. Frigon, On a critical point theory for multivalued functionals and application to partial differential inclusions. *Nonlinear Anal.*, **31**, no. 5–6, 735–753 (1998).

[410] _____, On a new notion of linking and application to elliptic problems at resonance. *J. Diff. Eq.*, **153**, no. 1, 96–120 (1999).

[411] S. Fučik, *Solvability of nonlinear equations and boundary value problems*. Reidel, Dordrecht, 1980.

[412] S. Fučik, J. Nečas, J. Souček, and V. Souček, *Spectral analysis of nonlinear operators*. Lecture Notes in Mathematics, **346**, Springer-Verlag, Berlin, 1973.

[413] N. Fukagai and K. Narukawa, On a model equation of one-dimensional elasticity. *Adv. Math. Sci. Appl.*, **6**, no. 1, 31–65 (1996).

[414] S. Gaete and R.F. Manásevich, Existence of a pair of periodic solutions of an O.D.E. generalizing a problem in nonlinear elasticity, via variational methods. *J. Math. Anal. Appl.*, **134**, no. 2, 257–271 (1988).

[415] R.E. Gaines and J. Mawhin, *Coincidence degree and nonlinear differential equations*. Springer, Berlin, 1977.

[416] A.J. García and A.I. Peral, Multiplicity of solutions for elliptic problems with critical exponent or with a nonsymmetric term. *Trans. Am. Math. Soc.*, **323**, no. 2, 877–895 (1991).

[417] _____, On limits of solutions of elliptic problems with nearly critical exponent. *Commun. Part. Diff. Eq.*, **17**, no. 11–12, 2113–2126 (1992).

[418] L. Gasinski and N.S. Papageorgiou, An existence theorem for nonlinear hemivariational inequalities at resonance. *Bull. Aust. Math. Soc.*, **63**, no. 1, 1–14 (2001).

[419] _____, Nonlinear hemivariational inequalities at resonance. *J. Math. Anal. Appl.*, **244**, no. 1, 200–213 (2000).

[420] F. Gazzola and B. Ruf, Lower-order perturbations of critical growth nonlinearities in semilinear elliptic equations. *Adv. Diff. Eq.*, **2**, no. 4, 555–572 (1997).

[421] I. Gelfand and S. Fomin, *Calculus of variations*. Prentice-Hall, Englewood Cliffs, NJ, 1964.

[422] P.G. Georgiev, A short proof of a general mountain-pass theorem for locally Lipschitz functions. Preprint, IC 95/369, International Centre of Theoretical Physics., Trieste, 5 pp. (1995) (http://www.ictp.trieste.it/~pub_off).

[423] N. Ghoussoub, Location, multiplicity and Morse indices of min-max critical points. *J. Reine Angew. Math.*, **417**, 27–76 (1991).

[424] _____, A min-max principle with relaxed boundary condition. *Proc. Am. Math. Soc.*, **117**, no. 2, 439–447 (1992).

[425] _____, *Duality and perturbation methods in critical point theory*. Cambridge University Press, 1993.

[426] N. Ghoussoub and C. Chambers, Deformation from symmetry and multiplicity of solutions in non-homogenous problems. *J. Discrete Contin. Dyn. Syst.*, **8**, no. 1, 267–281 (2001).

[427] N. Ghoussoub and D. Preiss, A general mountain pass principle for locating and classifying critical points. *Ann. Inst. H. Poincaré*, **6**, no. 5, 321–330 (1989).

[428] J. Giacomoni and L. Jeanjean, A variational approach to bifurcation into spectral gaps. *Ann. Scuola Norm. Sup. Pisa*, **28**, no. 4, 651–674 (1999).

[429] F. Giannoni, Bounce trajectories with one bounce point. *Ann. Mat. Pur. Appl.*, **159**, no. 4, 101–115 (1991).

[430] D. Gilbarg and N. Trudinger, *Elliptic partial differential equations of second order*. Springer-Verlag, Berlin, 1977 (2nd ed. 1984).

[431] M. Girardi and M. Matzeu, Some results on solutions of minimal period to superquadratic Hamiltonian systems. *Nonlinear Anal.*, **7**, no. 5, 475–482 (1983).

[432] ———, Some results on periodic solutions of mountain pass type for Hamiltonian systems. Periodic solutions of Hamiltonian systems and related topics (Il Ciocco, 1986), 161–168, NATO Adv. Sci. Inst. Ser. C: Math. Phys. Sci., **209**, Reidel, Dordrecht and Boston, 1987.

[433] ———, Periodic solutions of convex autonomous Hamiltonian systems with a quadratic growth at the origin and superquadratic at infinity. *Ann. Mat. Pur. Appl.* **147**, no. 4, 21–72 (1987).

[434] ———, Some results on periodic solutions of superquadratic Hamiltonian systems. Differential equations and applications, Vol. I, II (Columbus, OH, 1988), 325–330, Ohio University Press, Athens, OH, 1989.

[435] ———, Dual Morse index estimates for periodic solutions of Hamiltonian systems in some nonconvex superquadratic cases. *Nonlinear Anal.*, **17**, no. 5, 481–497 (1991).

[436] ———, Periodic solutions of second order nonautonomous systems with the potentials changing sign. *Atti Accad. Naz. Lin.*, **4**, no. 4, 273–277 (1993).

[437] ———, Solutions of prescribed period for second order autonomous conservative systems. *Boll. Unione. Mat. Ital. A* (7), **8**, no. 2, 271–282 (1994).

[438] ———, On periodic solutions of the system $\ddot{x}(t) + b(t)(V_1'(x(t)) + V_2'(x(t))) = 0$ where $b(.)$ changes sign and V_1, V_2 have different superquadratic growths. Variational and local methods in the study of Hamiltonian systems (Trieste, 1994), 65–76, World Scientific Publishing, River Edge, NJ, 1995.

[439] S. Glasstone, K.J. Laidler, and H. Eyring, *The theory of rate processes*. McGraw-Hill, New York, 1941.

[440] D. Goeleven, A note on Palais-Smale condition in the sense of Szulkin. *Diff. Integral Eq.*, **6**, 1041–1043 (1993).

[441] D. Goeleven, D. Motreanu, and P.D. Panagiotopoulos, Semicoercive variational-hemivariational inequalities. *Appl. Anal.*, **65**, no. 1–2, 119–134 (1997).

[442] H. Goldstine, *A history of the calculus of variations. From the 17th century through the 19th century*. Springer-Verlag, New York, 1980.

[443] J.V. Gonçalves and C.O. Alves, Existence of positive solutions for m-Laplacian equations in $\mathbb{R}^N$ involving critical Sobolev exponents. *Nonlinear Anal.–Theor.*, **32**, no.1, 53–70 (1998).

[444] J.V. Gonçalves and S. Meira, On a class of semilinear elliptic problems near critical growth. *Int. J. Math. Math. Sci.*, **21**, no. 2, 321–330 (1998).

[445] J.V. Gonçalves and O.H. Miyagaki, Multiple nontrivial solutions of semilinear strongly resonant elliptic equations. *Nonlinear Anal.*, **19**, no. 1, 43–52 (1992).

[446] ———, Three solutions for a strongly resonant elliptic problem. *Nonlinear Anal.*, **24**, no. 2, 265–272 (1995).

[447] ———, Multiple positive solutions for semilinear elliptic equations in $\mathbb{R}^N$ involving subcritical exponents. *Nonlinear Anal.*, **32**, no. 1, 41–51 (1998).

[448] J.V. Gonçalves, J.C. de Pádua, and P.C. Carrião, Variational elliptic problems at double resonance. *Diff. Integral Eq.*, **9**, no. 2, 295–303 (1996).

[449] A. Göpfert and C. Tammer, A new maximal point theorem. *Z. Anal. Anwend.*, **14**, 379–390 (1995).

[450] ———, ε-approximate solutions and conical support points, a new maximal point theorem. *ZAMM*, **75**, 595–596 (1995).

[451] A. Göpfert, C. Tammer, and Zălinescu, New minimal point theorem in product spaces. *Z. Anal. Anwend.*, **18**, no. 3, 767–770 (1999).

[452] ———, On the vectorial Ekeland's variational principle and minimal points in product spaces. *Nonlinear Anal.–Theor.*, **39** A, no. 7, 909–922 (2000).

[453] C. Greco, Multiplicity of solutions for a class of not uniformly superlinear elliptic problems. (Italian) *Rend. Accad. Sci. Lett. Lombardo*, **118**, 181–190 (1987).

[454] ———, Infinitely many spacelike periodic trajectories on a class of Lorentz manifolds. *Rend. Sem. Mat. Univ. Padova*, **91**, 251–263 (1994).

[455] D. Gromoll and W. Meyer, On differentiable functions with isolated critical points. *Topology*, **8**, 361–369 (1969).

[456] M.R. Grossinho, Some existence and bifurcation results for nonlinear elliptic problems in strip-like domains. *Ric. Mat.*, **36**, no. 1, 127–138 (1987).

[457] M.R. Grossinho, F. Minhós, and S. Tersian, Positive homoclinic solutions for a class of second order differential equations. *J. Math. Anal. Appl.*, **240**, no. 1, 163–173 (1999).

[458] Y.G. Gu, Nontrivial solutions of semilinear elliptic equations of fourth order. Nonlinear functional analysis and its applications. Part 1 (Berkeley, CA, 1983), 463–471, Proc. Sympos. Pure Math., **45**, Part 1, American Mathematical Society, Providence, RI, 1986.

[459] ———, Nontrivial solutions of semilinear elliptic equations on unbounded domain. *Acta Math. Sci.* (English Ed.), **9**, no. 1, 83–92 (1989).

[460] D.J. Guo, J.X. Sun, and G.J. Qi, Some extensions of the mountain pass lemma. *Diff. Integral Eq.*, **1**, no. 3, 351–358 (1988).

[461] X.K. Guo, The multiplicity of positive solutions for the p-Laplace equation. (Chinese) *J. Guangxi Univ. Nat. Sci. Ed.*, **24**, no. 2, 102–105 (1999).

[462] Z. Guo, On the number of positive solutions for quasilinear elliptic eigenvalue problems. *Nonlinear Anal.*, **27**, no. 2, 229–247 (1996).

[463] P. Habets, R. Manásevich, and F. Zanolin, A nonlinear boundary value problem with potential oscillating around the first eigenvalue. *J. Diff. Eq.*, **117**, no. 2, 428–445 (1995).

[464] P. Habets, E. Serra, and M. Tarallo, Multiple periodic solutions for problems at resonance with arbitrary eigenvalues. *Topol. Method. Nonlinear Anal.*, **12**, no. 2, 293–307 (1998).

[465] S. Hadamard, *Leçons sur le calcul des variations*. Hermann, Paris, 1910.

[466] N. Halidias and N.S. Papageorgiou, Quasilinear elliptic problems with multivalued terms. *Czech. Math. J.*, **50**, no. 4, 803–823 (2000).

[467] A.R. Hall, *Philosophers at war: The quarrel between Newton and Leibniz*. Cambridge University Press, Cambridge, 1980.

[468] G. Hamel, Uber erzwungene Schwingungen bei endlichen amplituden. *Math. Ann.*, **86**, 1–13 (1922).

[469] D. Heidrich (ed.), *The reaction path in chemistry: Current approaches and perspectives*. Kluwer, Dodrecht, 1995.

[470] D. Hilbert, Über das Dirichletsche prinzip. *Jber. Deut. Math. Verein*, **8**, 184–188 (1900).

[471] S. Hilderbrandt, *The calculus of variations today, as reflected in the Oberwolfach meetings*. Perspect. Math., Anniversary of Oberwolfach 1984, Birkhäuser, 1985.

[472] S. Hill and L.D. Humphreys, Numerical mountain pass periodic solutions of a nonlinear spring equation. *Comput. Math. Appl.*, **35**, no. 12, 59–67 (1998).

[473] ———, Mountain pass solutions for a system of partial differential equations: an existence theorem with computational results. *Nonlinear Anal. Theor. Method.*, **39**, no. 6, 731–743 (2000).

[474] N. Hirano and N. Mizoguchi, Existence of solutions of minimal period of semilinear elliptic equations on strip-like domains. *Nonlinear Anal.*, **22**, no. 5, 567–571 (1994).

[475] ———, Nonradial solutions of semilinear elliptic equations on annuli. *J. Math. Soc. Jpn.*, **46**, no. 1, 111–117 (1994).

[476] J.B. Hiriart-Urruty, A short proof of the variational principle for approximate solutions of a minimization problem. *Am. Math. Mon.*, **90**, 206–207 (1983).

[477] R. Hochmuth, Nonlinear anisotropic boundary value problems – Regularity results and multiscale discretizations. *Nonlinear Anal.–Theor.*, **46A**, no.1, 1–18 (2001).

[478] H. Hofer, Variational and topological methods in partially ordered Hilbert spaces. *Math. Ann.*, **261**, no. 4, 493–514 (1982).

[479] _____ , Some theory of strongly indefinite functionals with applications. *Trans. Am. Math. Soc.*, **275**, 185–214 (1983).

[480] _____ , A note on the topological degree at a critical point of mountain pass-type. *Proc. Am. Math. Soc.*, **90**, no. 2, 309–315 (1984).

[481] _____ , A geometric description of the neighborhoods of a critical point given by the mountain-pass theorem. *J. London Math. Soc.*, **31**, no. 2, 566–570 (1985).

[482] _____ , The topological degree at a critical point of mountain-pass type. Nonlinear functional analysis and its applications, Part 1 (Berkeley, CA, 1983), 501–509, Sympos. Pure Math., **45**, Part 1, American Mathematical Society, Providence, RI, 1986.

[483] _____ , A strong form of the mountain pass theorem and applications. Nonlinear diffusion equations and their equilibrium states, I (Berkeley, CA, 1986), 341–350, *Math. Sci. Res. Inst. Publ.*, **12**, Springer, New York and Berlin, 1988.

[484] C.W. Hong, A critical point theorem and applications. (Chinese) *Acta Math. Sin.*, **27**, no. 2, 264–271 (1984).

[485] Y. Hong and Q. Yin, Periodic solutions of a class of second order Hamiltonian systems. *J. Math. Res. Expos.*, **16**, no. 1, 13–19 (1996).

[486] S. Hu, N.C. Kourogenis, and N.S. Papageorgiou, Nonlinear elliptic eigenvalue problems with discontinuities. *J. Math. Anal. Appl.*, **233**, no. 1, 406–424 (1999).

[487] L. D. Humphreys, Numerical mountain pass solutions of a suspension bridge equation. *Nonlinear Anal.*, **28**, no. 11, 1811–1826 (1997).

[488] C.U. Huy, P.J. McKenna, and W. Walter, Finite difference approximations to the Dirichlet problem for elliptic systems. *Numer. Math.*, **49**, 227–237 (1986).

[489] D.H. Hyers, G. Isac, and T.M. Rassias, *Topics in nonlinear analysis and applications.* World Scientific, Singapore, 1997.

[490] R. Ikehata, Palais-Smale condition for some semilinear parabolic equations. *RIMS Kokyuroku*, **1123**, 76–82 (2000).

[491] C. Imbusch, *Eine Anwendung des Mountain-Pass-Lemmas auf den Fragenkreis des Plateauschen Problems und eine Alternative zur Drei-Punkte-Bedingung.* (German) [An application of the mountain pass lemma to questions connected with the Plateau problem and an alternative to the three-point condition.] Bonner Mathematische Schriften [Bonn Mathematical Publications], **302**, Universität Bonn, Mathematisches Institut, Bonn, 1997.

[492] Instructional Workshop on Analysis and Geometry. Part III. Operator theory and nonlinear analysis. Australian National University, (Canberra, January 23–February 10, 1995), T. Cranny and J. Hutchinson (eds). Proceedings of the Centre for Mathematics and its Applications, Australian National University, **34**. Australian National University, Centre for Mathematics and its Applications, Canberra, 1996.

[493] A.D. Ioffe, Variational methods in local and global non-smooth analysis. Notes by Igor Zelenko. in *Nonlinear analysis, differential equations and control.* F.H. Clarke, et al. (ed.), Proceedings of the NATO Advanced Study Institute and seminaire de mathematiques superieures (Montreal, Canada, July 27–August 7, 1998), Kluwer Academic Publishers, Dordrecht, NATO ASI Ser., Ser. C, Math. Phys. Sci. **528**, 447–502 (1999).

[494] A.D. Ioffe and E. Schwartzman, Metric critical point theory. I, Morse regularity and homotopic stability of a minimum. *J. Math. Pure Appl.*, **75**, 125–153 (1996).

[495] _____ , Metric critical point theory. II, Deformation techniques. *New results in operator theory and its applications*, 132–144, *Oper. Theory Adv. Appl.*, **98**, Birkhäuser, Basel, 1997.

[496] Y. Jabri, A nonsmooth variational approach to differential problems. A case study of non-resonance under the first eigenvalue for a strongly nonlinear elliptic problem. *Nonlinear Anal.*, **52**, no. 2, 605–620 (2003).

344   Bibliography

[497] _____, The mountain pass theorem in applications. A survey. In preparation.

[498] Y. Jabri and M. Moussaoui, On the generalized linking principle. Preprint, University Mohamed I, Oujda, (1996) (http://xxx.lanl.gov/abs/Math.FA/9903189).

[499] _____, A Saddle point theorem without compactness and applications to semilinear problems, *Nonlinear Anal.–Theor.*, **32**, no. 3, 363–380 (1997).

[500] S. Jaffard, Analysis of the lack of compactness in the critical Sobolev embeddings. *J. Funct. Anal.*, **161**, no. 2, 384–396 (1999).

[501] N. Jakobowsky, A perturbation result concerning a second solution to the Dirichlet problem for the equation of prescribed mean curvature. *J. Reine Angew. Math.*, **457**, 1–21 (1994).

[502] _____, A result on large surfaces of prescribed mean curvature in a Riemannian manifold. *Calc. Var. Part. Diff. Eq.*, **5**, no. 1, 85–97 (1997).

[503] I.M. James, The Lusternik-Schnirelmann theorem reconsidered. *Topol. Appl.*, **44**, no. 1–3, 197–202 (1992).

[504] G. Jameson, *Ordered linear spaces.* Lecture Notes in Mathematics, **141**, Springer-Verlag, New York, 1970.

[505] J. Jang, On spike solutions of singularly perturbed semilinear Dirichlet problem. *J. Diff. Eq.*, **114**, no. 2, 370–395 (1994).

[506] L. Jeanjean, Existence of solutions with prescribed norm for semilinear elliptic equations. *Nonlinear Anal.*, **28**, no. 10, 1633–1659 (1997).

[507] _____, On the existence of bounded Palais-Smale sequences and application to a Landesman-Lazer-type problem set on $\mathbb{R}^N$. *Proc. Roy. Soc. Edin. A*, **129**, no. 4, 787–809 (1999).

[508] _____, Local conditions insuring bifurcation from the continuous spectrum. *Math. Z.*, **232**, no. 4, 651–664 (1999).

[509] L. Jeanjean and J.F. Toland, Bounded Palais-Smale mountain-pass sequences. *C.R. Acad. Sci. Paris*, **327**, no. 1, 23–28 (1998).

[510] M. Ji and G.Y. Wang, Minimal surfaces in Riemannian manifolds. *Mem. Am. Math. Soc.*, **104**, no. 495 (1993).

[511] J. Jost and X. Li-Jost, *Calculus of variations.* Cambridge Studies in Advanced Mathematics, **64**, Cambridge University Press, 1998.

[512] J. Jost and M. Struwe, Morse-Conley theory for minimal surfaces of varying topological type. *Invent. Math.*, **102**, no. 3, 465–499 (1990).

[513] Y. Kabeya and W.-M. Ni, Stationary Keller-Segel model with the linear sensitivity. Variational problems and related topics (Japanese) (Kyoto, 1997). *Sūrikaisekikenkyūsho Kōkyūroku*, **1025**, 44–65 (1998).

[514] S. Karlin, Positive operators. *J. Math. Mech.*, **8**, 907–937 (1962).

[515] T. Kato, *Perturbation theory for linear operators.* Grundlehren der Math. wissenschaften **132**, Springer, Berlin, 1993.

[516] G. Katriel, Mountain pass theorem and global homeomorphism theorems. *Ann. Inst. H. Poincaré–An.*, **11**, no. 2, 189–209 (1994).

[517] O. Kavian, *Introduction à la théorie des points critiques et applications aux problèmes elliptiques.* (French) [Introduction to critical point theory and applications to elliptic problems.] Mathématiques & Applications [Mathematics & Applications] Series, Springer-Verlag, Paris, 1993.

[518] J.L. Kazdan and F.W. Warner, Remarks on some quasilinear elliptic equations. *Commun. Pure Appl. Math*, **28**, no. 5, 567–597 (1975).

[519] H. Keller, *Differentiable calculus in locally convex spaces.* Lecture Notes in Mathematics, Springer-Verlag, **417**, 1974.

[520] S. Kesavan, *Topics in functional analysis and applications.* John Wiley & Sons, Inc., New York, 1989.

[521] J. Kim and K.W. Kim, Existence of positive solutions for pseudo-Laplacian equations involving critical Sobolev exponents. *Math. Jpn.*, **42**, no. 1, 75–86 (1995).

[522] K.W. Kim, Positive solutions for pseudo-Laplacian equations with critical Sobolev exponents. *Commun. Korean Math. Soc.*, **14**, no. 1, 81–97 (1999).

[523]   W. Kliesch, A mechanical string model of adiabatic chemical reactions. Max-Plank Institute, Preprint no. 28, 126 pp., Lecture Notes in Chemistry, Springer, 1997.

[524]   N. Koiso, Yang-Mills connections of homogeneous bundles. *Osaka J. Math.*, **27**, no. 1, 163–174 (1990).

[525]   _____, Yang-Mills connections of homogeneous bundles, II. *Einstein metrics and Yang-Mills connections* (Sanda, 1990), 79–84, Lecture Notes in Pure and Applied Mathematics, **145**, Dekker, New York, 1993.

[526]   P. Korman, An algorithm for computing unstable solutions of semilinear boundary value problems. *Computing*, **51**, no. 3–4, 327–334 (1993).

[527]   _____, On computation of solution curves for semilinear elliptic problems. *Numer. Funct. Anal. Optim.*, **16**, no. 1–2, 219–231 (1995).

[528]   _____, Monotone approximations of unstable solutions. *J. Comput. Appl. Math.*, **136**, no. 1–2, 309–315 (2001).

[529]   N.C. Kourogenis and N.S. Papageorgiou, Three nontrivial solutions for a quasilinear elliptic differential equation at resonance with discontinuous right hand side. *J. Math. Anal. Appl.*, **238**, no. 2, 477–490 (1999).

[530]   _____, Nonsmooth critical point theory and nonlinear elliptic equations at resonance. *Kodai Math. J.*, **23**, no. 1, 108–135 (2000).

[531]   _____, Nonsmooth critical point theory and nonlinear elliptic equations at resonance. *J. Aust. Math. Soc., Ser. A*, **69**, no. 2, 245–271 (2000).

[532]   _____, Multiple solutions for nonlinear discontinuous strongly resonant elliptic problems. *J. Math. Soc. Jpn.*, **53**, no. 1, 17–34 (2001).

[533]   M.A Krasnosel'skii, On the estimation of the number of critical points of functionals. *Usp. Math. Nauk*, **7**, no. 2 (48), 157–164 (1952).

[534]   _____, *Topological methods in the theory nonlinear integral equations.* Gostekhteoretizdat, Moscow, 1956 (English Translation, Pergamon Press, 1964.)

[535]   _____, *Positive solutions of operator equations.* P. Noordhoff, Groningen, The Netherlands, 1964.

[536]   M.A. Krasnosel'skii and P.P. Zabreiko, *Geometrical methods of nonlinear analysis.* Springer-Verlag, 1984.

[537]   W. Krawcewicz and W. Marzantowicz, Ljusternik-Schnirelman method for functionals invariant with respect to a finite group action. *J. Diff. Eq.*, **85**, 105–124 (1990).

[538]   _____, Some remarks on the Ljusternik-Schnirelman method for non-diffrentiable functionals invariant with respect to a finite group action. *Rocky Mt. J. Math.*, **20**, no. 4, 1041–1049 (1990).

[539]   M.G. Krein and M.A. Rutman, Linear operators leaving invariant a cone in a Banche space. *Rocky Mt. J. Math.*, **20**, no. 4, 1041–1049 (1990).

[540]   A. Kristaly and C. Varga, Cerami (C) condition and the mountain pass theorem for multivalued mappings. Preprint, 2001.

[541]   _____, Location and multiplicity results for multivalued functionals. Preprint, 2001.

[542]   W. Kryszewsky and A. Szulkin, On a semilinear Schrödinger equation with indefinite linear part. *Adv. Diff. Eq.*, **3**, no. 3, 441–472 (1998).

[543]   C. Kuratowski, *Topology.* Academic Press, **II**, New York and London, 1968.

[544]   I. Kuzin and S. Pohozaev, *Entire solutions of semilinear elliptic equations.* Progress in Nonlinear Differential Equations and Their Applications, **33**, Birkhäuser Verlag, Basel, 1997.

[545]   O. Ladyženskaya and N. Uralceva, *Linear and quasilinear equations of elliptic type.* Academic Press, New York, 1968.

[546]   V. Lakshmikantham and S. Leela, *Differential and integral inequalities: Theory and applications. Vol. I: Ordinary differential equations.* Mathematics in Science and Engineering, Vol. **55-I**. Academic Press, New York and London, 1969.

[547]   E. Lami Dozo and M.C. Mariani, Solutions to the Plateau problem for the prescribed mean curvature equation via the mountain pass lemma. *Stud. Appl. Math.*, **96**, no. 3, 351–358 (1996).

[548] E.M. Landesman, A.C. Lazer, and D.R. Meyers, On saddle point theorems in the calculus
       of variations. The Ritz algorithm and monotone convergence, *J. Math. Anal. Appl.*, **52**,
       591–614 (1975).

[549] E. Landesman, S. Robinson, and A. Rumbos, Multiple solutions of semilinear elliptic
       problems at resonance. *Nonlinear Anal.*, **24**, no. 7, 1049–1059 (1995).

[550] K. Lankers and G.F. Friesecke, Large-amplitude solitary waves in the 2D Euler equations
       for stratified fluids. *Nonlinear Anal.*, **29**, no. 9, 1061–1078 (1997).

[551] A.C. Lazer and P.J. McKenna, On the number of solutions of a nonlinear Dirichlet
       problem. *J. Math. Anal. Appl.*, **84**, 282–294 (1981).

[552] ———, Critical point theory and boundary value problems with nonlinearities crossing
       multiple eigenvalues. *Commun. Part. Diff. Eq.*, **10**, no. 2, 107–150 (1985).

[553] ———, Nonlinear flexings in a periodically forced floating beam. *Math. Method. Appl.
       Sci.*, **14**, no. 1, 1–33 (1991).

[554] A. Lazer and S. Solimini, Nontrivial solutions of operator equations and Morse indices
       of critical points of min-max type. *Nonlinear Anal.*, **10**, no. 4, 411–413 (1986).

[555] A. Lazer and S. Solimini, Nontrivial solutions of operator equations and Morse indices
       of critical points of min-max type. *Nonlinear Anal.*, **12**, no. 8, 761–775 (1988).

[556] H. Lebesgue, Sur le problème de Dirichlet. *Ren. Circ. Mat. Palermo*, **24**, 371–402
       (1907).

[557] C. Lefter, Critical point theorems for lower semicontinuous functionals. Preprint,
       IC/98/229, International Centre of Theoretical Physics (Trieste, December 1998) (`http:`
       `//www.ictp.trieste.it/~pub_off`).

[558] C. Lefter and D. Motreanu, Critical point methods in nonlinear eigenvalue problems
       with discontinuities. *Optimization, optimal control and partial differential equations*
       (Iaşi, 1992), 25–36, Internat. Ser. Numer. Math., **107**, Birkhäuser, Basel, 1992.

[559] L. Lefton and J. Santanilla, Positive solutions for a two-point nonlinear boundary value
       problem with applications to semilinear elliptic equations. *Diff. Integral Eq.*, **9**, no. 6,
       1293–1304 (1996).

[560] G. Leoni, Existence of solutions for holonomic dynamical systems with homogeneous
       boundary conditions. *Nonlinear Anal.*, **23**, no. 4, 427–445 (1994).

[561] P. Levy, Sur les fonctions de lignes implicites. *Bull. Soc. Math. France* **48** (1920).

[562] C.Y. Li, Periodic solutions of a class of superquadratic Lagrangian systems. (Chinese)
       J. Math. (Wuhan), **14**, no. 2, 217–222 (1994).

[563] G.B. Li, Nonzero critical points of the functional $I(u) = \int_\Omega F(x, u, Du, \cdots, D^m u)dx$
       in $W_0^m L_\phi(\Omega)$. (Chinese) *Acta Math. Sci. (Chinese)*, **7**, no. 2, 207–219 (1987).

[564] ———, The existence of infinitely many solutions of quasilinear partial differential
       equations in unbounded domains. *Acta Math. Sci. (English Ed.)*, **9**, no. 2, 175–188
       (1989).

[565] S. Li, An existence theorem on multiple critical points and its applications in nonlinear
       P.D.E. *Differential geometry and differential equations*, Proc. Symp., Changchun/China
       1982, 479–483 (1986).

[566] ———, A theorem on multiple critical points and its applications to nonlinear partial
       differential equations. *Acta Math. Sci.* (Chinese), **4**, no. 2, 135–140 (1984).

[567] S. Li and J.Q. Liu, An existence theorem for multiple critical points and its application.
       (Chinese) *Kexue Tongbao*, **29**, no. 17, 1025–1027 (1984).

[568] S. Li and Z.Q. Wang, Mountain pass theorem in order intervals and multiple solutions
       for the semilinear elliptic Dirichlet problems. *J. Anal. Math.*, **81**, 373–396 (2000).

[569] S. Li and M. Willem, Applications of local linking to critical point theory. *J. Math. Anal.
       Appl.*, **189**, 6–32 (1995).

[570] S. Li and Z. Zhang, Sign-changing and multiple solutions theorems for semilinear elliptic
       boundary value problems with jumping nonlinearities. *Acta Math. Sin.* (English Ser.),
       16, no. 1, 113–122 (2000).

[571] Y. Li and S. Shi, A generalization of Ekeland's ε-variational principle and of its Borwein-
       Preiss' smooth variant. *J. Math. Anal. Appl.*, **246**, no. 1, 308–319 (2000).

[572] Y. Li and Z.-Q. Wang, Gluing approximate solutions of minimum type on the Nehari manifold. *Electron. J. Diff. Eq.* Conf. 06, 215–223 (2001).

[573] Y. Li and J. Zhou, A minimax method for finding multiple critical points and its applications to emilinear PDEs. *SIAM J. Sci. Comput.*, **23**, 840–865 (2001).

[574] X.T. Liang, The existence of nontrivial solutions for a class of elliptic Euler equations. (Chinese) *Chinese Quart. J. Math.*, **2**, no. 4, 23–32 (1987).

[575] X. Li-Jost, Uniqueness of minimal surfaces in Euclidean and hyperbolic 3-space. *Math. Z.*, **217**, no. 2, 275–285 (1994).

[576] C.S. Lin, Uniqueness of least energy solutions to a semilinear elliptic equation in $\mathbb{R}^2$. *Manuscripta Math.*, **84**, no. 1, 13–19 (1994).

[577] F.-H. Lin, Solutions of Ginzburg-Landau equations and critical points of the renormalized energy. *Ann. Inst. H. Poincaré–An.*, **12**, no. 5, 599–622 (1995).

[578] J. Lions, *Quelques méthodes de résolution des problèmes aux limites non linéaires.* Dunod, Paris; Gauthiers-Villars, Paris, 1969.

[579] P.-L. Lions, The concentration-compactness principle in the calculus of variations. The locally compact case I. *Ann. Inst. H. Poincaré–An.*, **1**, no. 2, 109–145 (1984).

[580] _____, The concentration-compactness principle in the calculus of variations. The locally compact case II. *Ann. Inst. H. Poincaré–An.*, **1**, no. 4, 223–284 (1984).

[581] _____, The concentration-compactness principle in the calculus of variations. The limit case, part I. *Rev. Mat. Iberoam.*, **1**, no. 1, 145–201 (1985).

[582] _____, The concentration-compactness principle in the calculus of variations The limit case, part II. *Rev. Mat. Iberoam.*, **1**, no. 2, 45–121 (1985).

[583] _____, Solutions of Hartree-Fock equations for Coulomb systems. *Commun. Math. Phys.*, **109**, no. 1, 33–97 (1987).

[584] D. Liotard and J.-P. Penot, Critical paths and passes: Application to quantum chemistry. *Numerical methods in the study of critical phenomena*, Della Dora, (ed.) 213–221, Springer, Berlin, 1989.

[585] J.Q. Liu, The Morse index for a saddle point, *Syst. Sci. Math. Sci.*, **2**, no. 1, 32–39 (1989).

[586] _____, Nonlinear vibration of a beam. *Nonlinear Anal.*, **13**, no. 10, 1139–1148 (1989).

[587] _____, An existence theorem for multiple critical points and its application. (Chinese) *Kexue Tongbao*, **29**, no. 17, 1025–1027 (1984).

[588] Z. Liu and J. Sun, Invariant sets of descending flow in critical point theory with applications to nonlinear differential equations. *J. Diff. Eq.*, **172**, no. 2, 257–299 (2001).

[589] A.M. Ljapunov, Sur les figures d'équilibre peu différentes des ellipsoïdes d'une masse liquide homogène donnée d'un mouvement de rotation. *Zap. Akad. Nauk St. Petersburg*, 1–225 (1906). Reprinted in *Ann. Math. Stud.*, Princeton 1949.

[590] _____, Problème général de la stabilité du mouvement. *Ann. Fac. Sci. Toulouse*, **2**, 203–474 (1907).

[591] L. Ljusternick, *Topology of the calculus of variations in the large.* Translation Math. Monographs, **16**, American Mathematical Society Providence, RI, 1966.

[592] L. Ljusternick and L. Schnirelman, *Méthodes topologiques dans les problèmes variationnels.* Vol. 188, Actualités Sci. Industr., Paris, 1934.

[593] Y.M. Long, The minimal period problem of classical Hamiltonian systems with even potentials. *Ann. Inst. H. Poincaré–An.*, **10**, no. 6, 605–626 (1993).

[594] D. Lü, Positive solutions of second-order quasilinear elliptic equations. (Chinese) *Hunan Ann. Math.*, **8**, no. 1–2, 47–52 (1988).

[595] J.Z. Lu and X.K. Guo, Existence of positive solutions to quasilinear elliptic equations with critical growth in unbounded domains. (Chinese) *Hunan Ann. Math.*, **11**, no. 1–2, 149–168 (1991).

[596] W.D. Lu, The Dirichlet problem for a class of quasilinear elliptic equations of second order. (Chinese) *Sichuan Daxue Xuebao*, no. 1, 28–39 (1986).

[597] _____, An imbedding theorem for spaces of anisotropic integrable functions. (Chinese) *Acta Math. Sin.*, **27**, no. 3, 319–344 (1984).

[598] D.H. Luc, Critical values of functionals on Banach surfaces. Preprint IC/87/344, ICTP, (Trieste, November 1987) (http://www.ictp.trieste.it/~pub_off).

[599] C.D. Luning and W.L. Perry, Positive solutions of negative exponent generalized Edmen-Fowler boundary value problems. *SIAM J. Math. Anal.*, **12**, no. 6, 874–879 (1981).

[600] N.N. Lusin, *Integral and trigonometric series*. Moscow, 1951.

[601] L. Ma, A result on the Kazdan-Warner problem on S^n. *Bull. Sci. Math.*, **119**, no. 5, 409–418 (1995).

[602] _____ , Mountain pass on a closed convex set. *J. Math. Anal. Appl.*, **205**, no. 2, 531–536 (1997).

[603] T.F. Ma, Remarks on nontrivial solutions of a resonant elliptic problem. *Bol. Soc. Parana. Mat.*, **15**, no. 1–2, 51–54 (1995).

[604] Y. Ma, S. Peng, and Y. Deng, Existence of multiple positive solutions of critical semilinear elliptic problems in $\mathbb{R}^N$. *J. Central China Normal Univ. Natur. Sci.*, **32**, no. 1, 1–6 (1998).

[605] A.L. Maestripieri and M.C. Mariani, The prescribed mean curvature equation for a revolution surface with Dirichlet condition. *Bull. Belg. Math. Soc. Sim.*, **3**, no. 3, 257–265 (1996).

[606] P. Majer, Two variational methods on manifolds with boundary. *Topology*, **34**, no. 1, 1–12 (1995).

[607] G. Mancini and R. Musina, A free boundary problem involving limiting Sobolev exponents. *Manuscripta Math.*, **58**, no. 1–2, 77–93 (1987).

[608] E. Marinari, G. Parisi, D. Ruelle, and P. Windey, On the interpretation of $1/f$ noise. *Commun. Math. Phys.*, **89**, no. 1, 1–12 (1983).

[609] A. Marino and G. Prodi, Metodi perturbativi nella teoria di Morse. (Italian) *Boll. Unione Mat. Ital., IV. Ser.*, **11**, Suppl. Fasc. 3, 1–32 (1975).

[610] A. Marino and S. Scolozzi, Geodetichei con ostacolo. (Italian) *Boll. Unione Mat. Ital., B (6)*, **2**, 1–31 (1983).

[611] A. Marino, A.M. Micheletti, and A. Pistoia, Some variational results on semilinear problems with asymptotically nonsymmetric behaviour. *Nonlinear analysis*, 243–256, Quaderni, Scuola Norm. Sup., Pisa, (1991).

[612] W. Marzantowicz, A G-Ljusternik-Schnirelman category of spaces with action of a compact Lie group. *Topology*, **28**, 403–412 (1989).

[613] M. Marzocchi, Multiple solutions of quasilinear equations involving an area-type term. *J. Math. Anal. Appl.*, **196**, no. 3, 1093–1104 (1995).

[614] _____ , Nontrivial solutions of quasilinear equations in BVP. *Serdica Math. J.*, **22**, no. 3 (1996).

[615] A. Matkowski, Functional equations and Nemytskii operators. *Funkcial. Ekvac.*, **25**, 127–132 (1982).

[616] J. Matos and L. Sanchez, Two solutions for a nonlinear Dirichlet problem with positive forcing. *Math. Bohem.*, **121**, no. 1, 41–54 (1996).

[617] M. Matzeu, Mountain pass and linking type solutions for semilinear Dirichlet forms. *Recent trends in nonlinear analysis*, Progr. Nonlinear Equations Appl., **40**, Birkhäuser, Basel, 2000.

[618] M. Matzeu and M. Girardi, On periodic solutions of a class of second order nonautonomous systems with nonhomogeneous potentials indefinite in sign. *Rend. Sem. Mat. Univ. Padova*, **97**, 193–210 (1997).

[619] M. Matzeu and I. Scarascia, Convergence results for mountain pass periodic solutions of autonomous Hamiltonian systems. *Boll. Unione. Mat. Ital. A (7)* **10**, no. 2, 445–459 (1996).

[620] J. Mawhin, Compacité monotonie et convexité dans l'étude des problèmes aux limites semi-linéaires. *Séminaires d'Analyse Moderne, Univ. Sherbrooke*, no. 19, 1981.

[621] _____ , Remarks on: "Critical point theory and a theorem of Amaral and Pera" [Boll. Un. Mat. Ital. B (6) 3 (1984), no. 3, 583–598] by Ahmad and Lazer. *Boll. Unione. Mat. Ital. A (6)*, **3**, no. 2, 229–238 (1984).

[622] ———, Méthodes variationnelles et problèmes aux limites pour des systèmes lagrangiens d'équations différentielles. (French) [Variational methods and boundary value problems for Lagrangian systems of differential equations.] Proceedings of the fifth congress on differential equations and applications (Spanish) (Puerto de la Cruz, 1982), 79–90, Informes, 14, Univ. La Laguna, La Laguna, 1984.

[623] ———, *Problèmes de Dirichlet variationnels non linéaires.* (French) [Nonlinear variational Dirichlet problems.] Séminaires de Mathématiques supèrieures [Seminar on Higher Mathematics], Montréal, no. 104, 1986.

[624] ———, Points fixes, points critiques et problèmes aux limites. (French) [Fixed points, critical points and boundary value problems.] Séminaires de Mathématiques supèrieures [Seminar on Higher Mathematics], Montréal, no. 92, 1986.

[625] ———, Critical point theory and nonlinear differential equations. Differential equations and their applications, Equadiff 6, Proc. 6th Int. Conf., (Brno, Czechoslorakia, 1985), Lect. Notes Math. **1192**, 49–58 (1986).

[626] J. Mawhin and J. Ward, Nonresonance and existence for nonlinear elliptic boundary value problems. *Nonlinear Anal.–Theor.*, **6**, 677–684 (1981).

[627] J. Mawhin and M. Willem, Critical points of convex perturbations of some indefinite quadratic forms and semi-linear boundary value problems at resonance. *Ann. Inst. H. Poincaré*, **3**, 431–453 (1986).

[628] ———, *Critical points theory and Hamiltonian Systems.* Applied Mathematical Sciences, **74**, Springer-Verlag, New York and Berlin, 1989.

[629] J. Mawhin, J. Ward, and M. Willem, Variational methods and semilinear elliptic equations. *Arch. Rat. Mech. Anal.*, **95**, 269–277 (1986).

[630] V.G. Maz'ja, *Sobolev spaces.* Springer-Verlag, Berlin, 1985.

[631] L. McLinden, An application of Ekeland's theorem to minimax problems. *Nonlinear Anal.*, **6**, 189–196 (1982).

[632] R. Michalek, Existence of a positive solution of a general quasilinear elliptic equation with a nonlinear boundary condition of mixed type. *Nonlinear Anal.*, **15**, no. 9, 871–882 (1990).

[633] A.M. Micheletti and A. Pistoia, A multiplicity result for a class of superlinear elliptic problems. *Portugal. Math.*, **51**, no. 2, 219–229 (1994).

[634] ———, Nontrivial solutions for some fourth order semilinear elliptic problems. *Nonlinear Anal.*, **34**, no. 4, 509–523 (1998).

[635] J. Milnor, *Topology from the differentiable viewpoint.* University Press of Virginia, Charlottesville, VA, 1965.

[636] ———, *Morse theory.* Princeton University Press, Princeton, 1963.

[637] *Minimax results of Lusternik-Schnirelman type and applications.* Part 2 of the Proceedings of the NATO ASI "Variational Methods in Nonlinear Problems." (University of Montreal, Montreal, Quebec, July 7–25, 1986). Séminaire de Mathématiques Supérieures [Seminar on Higher Mathematics], **107**. Presses de l'Université de Montréal, Montreal, 1989.

[638] M. Mirea and D. Dăescu, Une forme plus général du théorème du col. (French) [A more general form of the mountain pass theorem.] *An. Univ. Craiova Ser. Mat. Inform.*, **21**, 59–64 (1995).

[639] E. Mirenghi and M. Tucci, Existence of T-periodic solutions for a class of Lagrangian systems. *Rend. Sem. Mat. Univ. Padova*, **83**, 19–32 (1990).

[640] P. Mironescu and V.D. Radulescu, A multiplicity theorem for locally Lipschitz periodic functionals. *J. Math. Anal. Appl.*, **195**, no. 3, 621–637 (1995).

[641] D. Mitrović and D. Žubrinić, *Fundamentals of applied functional analysis. Distributions – Sobolev spaces – nonlinear elliptic equations.* Pitman Monographs and Surveys in Pure and Applied Mathematics, **91**, Longman, Harlow, 1998.

[642] A.F. Monna, *Dirichlet's principle: A mathematical comedy of errors and its influence on the development of analysis.* Oosthoek, Scheltema & Holkema, Utrecht, 1975.

[643] P. Montecchiari, Multiplicity results for a class of semilinear elliptic equations on $\mathbb{R}^m$. *Rend. Sem. Mat. Univ. Padova*, **95**, 217–252 (1996).

[644] P. Montecchiari, M. Nolasco, and S. Terracini, A global condition for periodic Duffing-like equations. *Trans. Am. Math. Soc.*, **351**, no. 9, 3713–3724 (1999).

[645] S.A. Marano and D. Motreanu, Existence of two nontrivial solutions for a class of elliptic eigenvalue problems. *Arch. Math.*, **75**, no. 1, 53–58 (2000).

[646] C.B. Morrey Jr., Multiple integral problems in the calculus of variations and related topics. *Ann. Scuola Norm. Pisa*, (III) **14**, 1–61 (1960).

[647] _____, *Multiple integrals in the calculus of variations*. Springer-Verlag, 1966.

[648] D. Motreanu, Existence of critical points in a general setting. *Set-Valued Anal.*, **3**, no. 3, 295–305 (1995).

[649] _____, A multiple linking minimax principle. *Bull. Austral. Math. Soc.*, **53**, no. 1, 39–49 (1996).

[650] _____, A saddle point approach to nonlinear eigenvalue problems. *Math. Slovaca*, **47**, no. 4, 463–477 (1997).

[651] D. Motreanu and Z. Naniewicz, Discontinuous semilinear problems in vector-valued function spaces. *Diff. Integral Eq.*, **9**, no. 3, 581–598 (1996).

[652] D. Motreanu and P.D. Panagiotopoulos, Hysteresis: The eigenvalue problem for hemivariational inequalities. *Models of hysteresis* (Trento, 1991), 102–117, Pitman Res. Notes Math. Ser., **286**, Longman Sci. Tech., Harlow, 1993.

[653] _____, Double eigenvalue problems for hemivariational inequalities. *Arch. Rat. Mech. Anal.*, **140**, no. 3, 225–251 (1997).

[654] _____, *Minimax theorems and qualitative properties of the solutions of hemivariational inequalities*. Nonconvex Optimization and its Applications, **29**, Kluwer Academic Publishers, Dordrecht, 1999.

[655] D. Motreanu and V. Radulescu, Existence theorems for some classes of boundary value problems involving the p-Laplacian. *Panam. Math. J.*, **7**, no. 2, 53–66 (1997).

[656] _____, Existence results for inequality problems with lack of convexity. *Numer. Funct. Anal. Optim.*, **21**, no. 7–8, 869–884 (2000).

[657] V.B. Moroz and P.P. Zabrejko, A variant of the mountain pass theorem and its application to Hammerstein integral equations. *Z. Anal. Anwend.*, **15**, no. 4, 985–997 (1996).

[658] M. Morse, Relations between the critical points of a real function of n independent variables. *Trans. Am. Math. Soc.*, **27**, 345–396 (1925).

[659] _____, *The calculus of variations in the large*. Colloquium Publ. Amer. Math. Soc., **18** (1934).

[660] _____, Functional topology and abstract variational theory. *Proc. Natl. Acad. Sci. USA*, **22**, 313–319 (1936).

[661] _____, Functional topology and abstract variational theory. *Mémorial Sci. Math.*, **92**, Gauthiers-Villars, 1939.

[662] _____, *Variational analysis: Critical extremals and Sturmian extensions*. Wiley, New York, 1972.

[663] M. Morse and S. Cairns, *Critical point theory in global analysis*. Academic Press, New York, 1969.

[664] M. Morse and C. Tompkins, The existence of minimal surfaces of general critical types. *Ann. Math.*, *II. Ser.* **40**, 443–472 (1939).

[665] _____, On the existence of minimal surfaces of general critical types. *Proc. Natl. Acad. Sci. USA*, **25**, 153–158 (1939).

[666] _____, Unstable minimal surfaces of higher topological structure. *Duke Math. J.*, **8**, 350–375 (1941).

[667] _____, Corrections to our paper on the existence of minimal surfaces of general critical types. *Ann. Math.*, *II. Ser.*, **42**, 331 (1941).

[668] D. Motreanu and C. Varga, Some critical point results for locally Lipschitz functionals. *Commun. Appl. Nonlinear Anal.*, **4**, no. 3, 17–33 (1997).

[669]   V. Mustonen and J. Berkovitz, Nonlinear mappings of monotone type, I. Classification and degree theory. Report, University of Oulu, Finland, (1988).

[670]   K. Narukawa and T. Suzuki, Nonlinear eigenvalue problem for a modified capillary surface equation. *Funkc. Ekvac.*, **37**, no. 1, 81–100 (1994).

[671]   J. Nečas, *Introduction to the theory of nonlinear elliptic partial differential equations.* Teubner, Leipzig, 1983.

[672]   Z. Nehari, Characteristic values associated with a class of nonlinear second-order differential equations. *Acta Math.*, **105**, 141–175 (1961).

[673]   P.C. Nguyên and F. Schulz, Multiple solutions for a class of semilinear Dirichlet problems. *Houston J. Math.*, **17**, no. 1, 71–81 (1991).

[674]   W.-M. Ni, A nonlinear Dirichlet problem on the unit ball and its applications. *Indiana Univ. Math. J.*, **31**, no. 6, 801–807 (1982).

[675]   W.-M. Ni and J. Wei, On the location and profile of spike-layer solutions to singularly perturbed semilinear Dirichlet problems. *Commun. Pure Appl. Math.*, **48**, no. 7, 731–768 (1995).

[676]   W.-M. Ni, I. Takagi, and J. Wei, On the location and profile of spike-layer solutions to a singularly perturbed semilinear Dirichlet problem: Intermediate solutions. *Duke Math. J.*, **94**, no. 3, 597–618 (1998).

[677]   L.I. Nicolaescu, Existence and regularity for a singular semilinear Sturm-Liouville problem. *Diff. Integral Eq.*, **3**, no. 2, 305–322 (1990).

[678]   C.P. Niculescu and V. Rădulescu, A saddle point theorem for non-smooth functionals and problems at resonance. *Ann. Acad. Sci. Fenn. Math.*, **21**, no. 1, 117–131 (1996).

[679]   L. Nirenberg, *Topics in nonlinear functional analysis.* New York University Lecture Notes, 1974.

[680]   _____, The use of topological, functional analytic, and variational methods in nonlinear problems. A survey on the theoretical and numerical trends in nonlinear analysis. Proc. Third Sem. Funct. Anal. Appl. (Bari, 1978), II. Confer. Sem. Mat. Univ. Bari, No. 163–168, (1979), 391–398 (1980).

[681]   _____, Variational methods in nonlinear problems. Semin. Goulaouic-Meyer-Schwartz, Equations Deriv. Partielles 1980–1981, Expose no. 20, 5 pp. (1981).

[682]   _____, Variational and topological methods in nonlinear problems. *Bull. Am. Math. Soc. (N.S.)*, **4**, 267–302 (1981).

[683]   _____, *Variational methods in nonlinear problems.* Topics in calculus of variations (Montecatini Terme, 1987), 100–119, Lecture Notes in Mathematics, **1365**, Springer, Berlin-New York, 1989.

[684]   _____, Variational methods in nonlinear problems. *Proceedings of the international conference on the mathematical sciences after the year 2000* (Beirut, Lebanon, January 11–15, 1999), K. Bitar et al. (eds.) World Scientific, Singapore, 116–122, 2000.

[685]   Nonlinear diffusion equations and their equilibrium states. I. (Proceedings of the microprogram held in Berkeley, CA, August 25–September 12, 1986). W.-M. Ni, L. A. Peletier, and J. Serrin (eds.) Mathematical Sciences Research Institute Publications, **12**, Springer-Verlag, New York and Berlin, 1988.

[686]   Nonlinear functional analysis and its applications. (Proceedings of the summer research institute held at the University of California, Berkeley, CA, July 11–29, 1983). F. E. Browder (ed.). Proceedings of Symposia in Pure Mathematics, **45**, Part 2. American Mathematical Society, Providence, RI, 1986.

[687]   E.S. Noussair, C.A. Swanson, and J.F. Yang, Critical semilinear biharmonic equations in $\mathbb{R}^N$. *Proc. Roy. Soc. Edin. A*, **121**, no. 1–2, 139–148 (1992).

[688]   _____, Positive finite energy solutions of critical semilinear elliptic problems. *Can. J. Math.*, **44**, no. 5, 1014–1029 (1992).

[689]   _____, Quasilinear elliptic problems with critical exponents. *Nonlinear Anal.*, **20**, no. 3, 285–301 (1993).

[690]   W. Oettli and M. Théra, Equivalents of Ekeland's principle. *Bull. Aust. Math. Soc.*, **48**, no. 3, 385–392 (1993).

[691] D.C. Offin, Subharmonic oscillations for forced pendulum type equations. *Diff. Integral Eq.*, **3**, no. 5, 965–972 (1990).

[692] I. Ohnishi, A note of the existence of nonconstant critical points of free energy functionals in the gradient theory of phase transitions. *Adv. Math. Sci. Appl.*, **4**, no. 2, 517–528 (1994).

[693] R.S. Palais, Morse theory on Hilbert manifolds. *Topology*, **2**, 299–340 (1963).

[694] _____, Ljusternik-Schnirelman theory on Banach manifolds. *Topology*, **5**, 115–132 (1966).

[695] _____, *Foundations of global non-linear analysis*. Benjamin, New York, 1968.

[696] _____, Critical point theory and the minimax principle. *Nonlinear Funct. Anal. App.*, **15**, 185–212 (1970).

[697] _____, Nonlinear perturbations of linear elliptic boundary value problems at resonance. *J. Math. Mech.*, **19**, 609–623 (1970).

[698] _____, The principle of symmetric criticality, *Commun. Math. Phys.*, **69**, 19–30 (1979).

[699] R.S. Palais and S. Smale, A generalized Morse theory, *Bull. Am. Math. Soc.*, **70**, 165–171 (1964).

[700] R.S. Palais and C.L. Terng, *Critical point theory and submanifold geometry*. Lecture Notes in Mathematics, **1353**, Springer-Verlag, 1988.

[701] P.D. Panagiotopoulos, *Hemivariational inequalities*. Springer, Berlin, 1993.

[702] R. Panda, Solution of a semilinear elliptic equation with critical growth in $\mathbb{R}^2$. *Nonlinear Anal.*, **28**, no. 4, 721–728 (1997).

[703] A.A. Pankov, Invariant semilinear elliptic equations on a manifold of constant negative curvature. (Russian) *Funkt. Anal. Prilozhen.*, **26**, no. 3, 82–84 (1992); translation in *Funct. Anal. Appl.*, **26**, no. 3, 218–220 (1992).

[704] D. Pascali, Generalized subdifferentiable perturbations. *Differential equations and applications*, Proc. Int. Conf. (Columbus, OH, 1988), Vol. II, 277–281 (1989).

[705] D. Pascali, Some mountain climbing techniques in the theory of semilinear operator equations. *Libertas Math.*, **13**, 155–170 (1993).

[706] D. Passaseo, Multiplicity of critical points for some functionals related to the minimal surfaces problem. *Calc. Var. Part. Diff. Eq.*, **6**, no. 2, 105–121 (1998).

[707] E. Paturel, A new variational principle for a nonlinear Dirac equation on the Schwarzschild metric. *Commun. Math. Phys.*, **213**, no. 2, 249–266 (2000).

[708] B. Pellacci, Critical points for non differentiable functionals. *Boll. Unione Mat. Ital.*, *VII. Ser.*, **B 11**, no. 3, 733–749 (1997).

[709] S. Peng, The existence of positive solutions to a class of nonlinear elliptic equations with Neumann boundary. (Chinese) *J. Hebei Norm. Univ., Nat. Sci. Ed.*, **24**, no. 3, 286–288 (2000).

[710] Y.D. Peng and Y.B. Deng, The existence of nontrivial solutions of critical semilinear biharmonic equations. (Chinese) *Acta Math. Sci.*, **17**, no. 4, 452–465 (1997).

[711] J.P. Penot, Méthode de descente: point de vue topologique et géométrique. Notes de cours 3ème cycle, Univ. Pau, 1975.

[712] _____, The drop theorem, the petal theorem and Ekeland's variational principle. *J. Nonlinear Anal.*, **10**, 813–822 (1986).

[713] K. Perera, Applications of local linking to asymptotically linear elliptic problems at resonance. *Nonlinear Diff. Eq. Appl.*, **6**, no. 1, 55–62 (1999).

[714] K. Perera and M. Schechter, Morse index estimates in saddle point theorems without a finite-dimensional closed loop. *Indiana Univ. Math. J.*, **47**, no. 3, 1083–1095 (1998).

[715] _____, Nontrivial solutions of elliptic semilinear equations at resonance. *Manuscripta. Math.*, **101**, no. 3, 301–311 (2000).

[716] *Periodic solutions of Hamiltonian systems and related topics.* (Proceedings of the NATO advanced research workshop held in Il Ciocco, October 13–17, 1986). P.H. Rabinowitz, A. Ambrosetti, I. Ekeland, and E.J. Zehnder (eds.). NATO Advanced Science Institute Series C: Mathematical and Physical Sciences, **209**. D. Reidel Publishing Co., Dordrecht and Boston, 1987.

[717] S. S. Petrova, P.A. Nekrasov and the mountain pass method. (Russian) *Voprosy Istor. Estestvoznan. Tekhn.*, no. 2, 107–109 (1994).

[718] K.M. Pflüger, Semilinear elliptic problems with nonlinear boundary conditions in unbounded domains. *Z. Anal. Anwend.*, **14**, no. 4, 829–851 (1995).

[719] _____, Periodic solutions of non-linear anisotropic partial differential equations. *Math. Method. Appl. Sci.*, **19**, no. 5, 363–374 (1996).

[720] R. Phelps, Subreflexive normed linear spaces. *Arch. Math.*, **8**, 444–450 (1957).

[721] _____, Support cones and their generalizations. *Proc. Sym. Pure Math.*, AMS, **VII**, 393–401 (1963).

[722] _____, Support cones in Banach spaces and their applications. *Adv. Math.*, **13**, 1–19 (1974).

[723] _____, *Convex functions, monotone operators and differentiability*. Lecture Notes in Mathematics, **1364**, 2nd ed., Springer-Verlag, Berlin, 1993.

[724] M. del Pino and P.L. Felmer, Local mountain passes for semilinear elliptic problems in unbounded domains. *Calc. Var. Part. Diff. Eq.*, **4**, no. 2, 121–137 (1996).

[725] H. Prado and P. Ubilla, Existence of nonnegative solutions for generalized *p*-Laplacians. Reaction diffusion systems (Trieste, 1995), 289–298, Lecture Notes in Pure and Applied Mathematics, **194**, Dekker, New York, 1998.

[726] P. Pucci and J. Serrin, Extensions of the mountain pass theorem. *J. Funct. Anal.*, **59**, 185–210 (1984).

[727] _____, A mountain pass theorem. *J. Diff. Eq.*, **60**, no. 1, 142–149 (1985).

[728] _____, The structure of the critical set in the mountain pass theorem. *Trans. Am. Math. Soc.*, **91**, no. 1, 115–132 (1987).

[729] G.J. Qi, A generalization of the mountain pass lemma. (Chinese) *Kexue Tongbao*, **31**, no. 10, 724–727 (1986).

[730] _____, Extension of the mountain pass lemma. *Kexue Tongbao (English Ed.)*, **32**, no. 12, 798–801 (1987).

[731] P. Quittner, On positive solutions of semilinear elliptic problems. *Comment. Math. Univ. Carolin.*, **30**, no. 3, 579–585 (1989).

[732] P.J. Rabier, Ehresmann fibrations and Palais-Smale conditions for morphisms of Finsler manifolds. *Ann. Math.*, **146**, no. 3, 647–691 (1997).

[733] P.H. Rabinowitz, Some aspects of nonlinear eigenvalue problems, Rocky Mountain Consortium Symposium on Nonlinear Eigenvalue Problems (Santa Fe, NM, 1971), *Rocky Mt. J. Math.*, **3**, 161–202 (1973).

[734] _____, Variational methods for nonlinear eigenvalue problems. *Eigenvalues of nonlinear problems*, Prodi ed., CIME Ceremonese, Rome, 140–195, 1974.

[735] _____, A note on the topological degree for potential operators, *J. Math. Anal. Appl.*, **51**, 483–492 (1975).

[736] _____, Some critical point theorems and applications to semilinear elliptic partial differential equations. *Ann. Scuola. Norm. Sup. Pisa*, **5**, 412–424 (1977).

[737] _____, Some critical point theorems and applications to semilinear elliptic partial differential equations, *Ann. Scuola. Norm. Sup. Pisa*, **5**, 215–223 (1978).

[738] _____, Some minimax theorems and applications to nonlinear partial differential equations. *Nonlinear analysis*, Academic Press, New York, 161–177, 1978.

[739] _____, A minimax principle and applications to elliptic partial differential equations, *Nonlinear P.D.E.s and applications*, 97–115, Lecture Notes in Mathematics, Springer, Berlin, **648** (1978).

[740] _____, Critical points of indefinite functionals and periodic solutions of differential equations, *Proceeding of the international congress of mathematics*, Helsinki, 791–796, 1978.

[741] _____, Periodic solutions of Hamiltonian systems. *Commun. Pure Appl. Math.*, **31**, no. 2, 157–184 (1978).

[742] _____, A variational method for finding periodic solutions of differential equations. *Nonlinear evolution equations*, M.G. Crandall (ed.), 225–251, Academic Press, New York, 1978.

[743] _____, On subharmonic solutions of Hamiltonian systems. *Commun. Pure Appl. Math.*, **33**, 609–633 (1980).

[744] _____, The mountain pass theorem: Theme and variations. Differential Eqns. (Sao Paulo, 1981), H. de Figueiredo (ed.), Lecture Notes in Mathematics, Springer, Berlin and New York, **957**, 237–271, 1982.

[745] _____, Multiple critical points of perturbed symmetric functionals. *Trans. Am. Math. Soc.*, **272**, 753–770 (1982).

[746] _____, Some aspects of critical point theory, *M.R.C. Technical report* no. 2465, 1983.

[747] _____, Global aspects of bifurcation. Topological methods in bifurcation theory, *Sem. Math, Sup. Sem. Sci. OTAN (NATO advanced study Inst.)*, **91**, 63–112 (1985).

[748] _____, Minimax methods in critical point theory with applications to differential equations. C.B.M.S. Regional conference ser. math, American Mathematical Society, **65**, 1986.

[749] _____, Minimax methods for indefinite functionals. Nonlinear functional analysis and its applications, Part 2 (Berkeley, CA, 1983), 287–306, Proc. Sympos. Pure Math., **45**, Part 2, American Mathematical Society, Providence, RI, 1986.

[750] _____, *On a class of functionals invariant under a $\mathbb{Z}^n$ action*, CMS Technical Summary Report, Univ. of Wisconsin, 1987.

[751] _____, Homoclinic orbits for a class of Hamiltonian systems. *Proc. Roy. Soc. Edin. Sect. A*, **114**, no. 1–2, 33–38 (1990).

[752] _____, A note on a semilinear elliptic equation on $\mathbb{R}^n$. *Nonlinear analysis*, 307–317, Quaderni, Scuola Norm. Sup., Pisa, 1991.

[753] _____, Critical point theory and applications to differential equations: A survey. *Topological nonlinear analysis*, 464–513, Progr. Nonlinear Differential Equations Appl., *15*, Birkhäuser Boston, Boston, 1995.

[754] _____, Multibump solutions of differential equations: An overview. *Chinese J. Math.*, **24**, no. 1, 1–36 (1996).

[755] P.H. Rabinowitz and K. Tanaka, Some results on connecting orbits for a class of Hamiltonian systems. *Math. Z.*, **206**, no. 3, 473–499 (1991).

[756] V.D. Radulescu, Mountain pass theorems for non-differentiable functions and applications. *Proc. Jpn. Acad. Ser. A Math. Sci.*, **69**, no. 6, 193–198 (1993).

[757] _____, Mountain-pass type theorems for nondifferentiable convex functions. *Rev. Roumaine Math. Pure. Appl.* **39**, no. 1, 53–62 (1994).

[758] _____, A Lyusternik-Schnirelman type theorem for locally Lipschitz functionals with applications to multivalued periodic problems. *Proc. Jpn. Acad. Ser. A*, **71**, no. 7, 164–167 (1995)

[759] M. Ramaswamy and P.N. Srikanth, Multiplicity result for an ODE via Morse index. *Houston J. Math.*, **15**, no. 4, 595–599 (1989).

[760] M. Ramos, A critical point theorem suggested by an elliptic problem with asymmetric nonlinearities. *J. Math. Anal. Appl.*, **196**, no. 3, 938–946 (1995).

[761] M. Ramos and C. Rebelo, A unified approach to min-max critical point theorems. *Portugal. Math.*, **51**, no. 4, 489–516 (1994).

[762] Q.K. Ran, Multiplicity of positive solutions of a class of degenerate elliptic equations. *J. Shanghai Jiaotong Univ.* (Chin. Ed.) **33**, no. 6, 657–660 (1999).

[763] M. Reeken, Stability of critical points under small perturbations I and II, *Manuscripta Math.*, **7**, 387–411 (1972); **8**, 69–92 (1973).

[764] _____, Stability of critical values and isolated critical continua, *Manuscripta Math.*, **12**, 163–193 (1974).

[765] N. Ribarska, T. Tsachev, and M. Krastanov, On the general mountain pass principle of Ghoussoub-Preiss, *Math. Balkanica*, (N.S.) **5**, no. 4, 350–358 (1991).

[766] _____, Deformation lemma on C^1-Finsler manifolds and applications. *Comp. Rend. Bulg. Acad. Sci.*, **47**, no. 1, 13–16 (1994).

[767] _____, Deformation lemma, Ljusternik-Schnirellmann theory and mountain pass theorem on C^1-Finsler manifolds. *Serdica Math. J.*, **21**, no. 3, 239–266 (1995).

[768] _____, Speculating about mountains. *Serdica Math. J.*, **22**, 341–358 (1996).

[769] _____, A saddle point theorem without a finite-dimensional closed loop. *C.R. Acad. Bulg. Sci.*, **51**, no. 11–12, 13–16 (1998).

[770] _____, The intrinsic mountain pass principle. *Topol. Method. Nonlinear Anal.*, **12**, no. 2, 309–322 (1998).

[771] _____, The intrinsic mountain pass principle. *C.R. Acad. Sci. Paris. I Math.*, **329**, no. 5, 399–404 (1999).

[772] _____, A note on: "On a critical point theory for multivalued functionals and application to partial differential inclusions" [*Nonlinear Anal.*, **31**, no. 5–6, 735–753 (1998)] by M. Frigon. *Nonlinear Anal.–Theor.*, **43**, no. 2, 153–158 (2001).

[773] B. Ricceri, On a classical existence theorem for nonlinear elliptic equations. *Constructive, experimental, and nonlinear analysis.* (Selected papers of a workshop, Limoges, France, September 22–23, 1999), M. Thera (ed.), American Mathematical Society, Providence, RI, publ. for the Canadian Mathematical Society. CMS Conf. Proc., **27**, 275–278 (2000).

[774] R.T. Rockafellar, *Convex analysis*. Princeton University Press, 1969.

[775] _____, *La théorie des sous-gradients et ses applications à l'optimisation. Fonctions convexes et nonconvexes.* Collection de la chaire Aisenstadt, Les presses de l'université de Montréal, Montréal, 1979.

[776] _____, *The theory of subgradients and its applications to problems of optimization. Convex and nonconvex problems.* Heldermann, Berlin, 1981.

[777] R.T. Rockafellar and R.J.-B. Wets, *Variational analysis*. Grundlehren der Mathematischen Wissenschaften, **317**, Springer, 1998.

[778] E.E. Rothe, Gradient mappings, *Bull. Am. Math. Soc.*, **59**, 5–19 (1953).

[779] _____, Some remarks on critical point theory in Hilbert space. *Proc. Symposia Nonlinear Problems*, Univ. of Wisconsin Press, 233–256, 1963.

[780] _____, Some remarks on critical point theory in Hilbert space (continuation), *J. Math. Anal. Appl.*, **20**, 515–520 (1967).

[781] _____, Weak topology and calculus of variations, *Calculus of variations, classical and modern*, CIME 1966, Ceremonese Rome, 207–237, 1968.

[782] _____, Morse theory in Hilbert spaces, *Rocky Mt. J. Math.*, **3**, 251–274 (1973).

[783] W. Rother, The existence of infinitely many solutions all bifurcating from $\lambda = 0$. *Proc. Roy. Soc. Edin. A*, **118**, no. 3–4, 295–303 (1991).

[784] W. Rudin, *Functional analysis*, McGraw-Hill, 1973.

[785] _____, *Real and complex analysis*, 2nd ed., McGraw-Hill, 1974.

[786] B. Ruf, Perturbation of symmetry and generalized mountain pass theorems. in *Variational methods in nonlinear analysis* (Erice, 1992), 137–147, Gordon and Breach, Basel, 1995.

[787] B. Ruf and S. Solimini, On a class of superlinear Sturm-Liouville problems with arbitrarily many solutions. *SIAM J. Math. Anal.*, **17**, no. 4, 761–771 (1986).

[788] B. Ruf and P.N. Srikanth, Multiplicity results for superlinear elliptic problems with partial interference with the spectrum. *J. Math. Anal. Appl.*, **118**, no. 1, 15–23 (1986).

[789] A.J. Rumbos, A multiplicity result for strongly nonlinear perturbations of elliptic boundary value problems. *J. Math. Anal. Appl.*, **199**, no. 3, 859–872 (1996).

[790] J. Santanilla, Existence and nonexistence of positive radial solutions of an elliptic Dirichlet problem in an exterior domain. *Nonlinear Anal.*, **25**, no. 12, 1391–1399 (1995).

[791] D.H. Sattinger, Monotone methods in nonlinear elliptic and parabolic boundary value problems. *Indiana Univ. Math. J.*, **21**, 979–1000 (1972).

[792] H.H. Schaefer, *Topological vector spaces*. Springer-Verlag, New York, 1971.

[793] M. Schechter, The Hampwile theorem for nonlinear eigenvalues. *Duke Math. J.*, **59**, no. 2 (1989).

[794] _____, Nonlinear elliptic boundary value problems at strong resonance. *Am. J. Math.*, **112**, no. 3, 439–460 (1990).

[795] _____, The mountain pass alternative. *Adv. Appl. Math.*, **12**, no. 1, 91–105 (1991).

[796] _____, A saddle point theorem applied to semilinear boundary value problems. *Panam. Math. J.*, **1**, no. 2, 1–25 (1991).

[797] _____, A variation of the mountain pass lemma and applications. *J. London Math. Soc.* **44**, no. 3, 491–502 (1991).

[798] _____, New saddle point theorems. *Proceedings of an international symposium on generalized functions and their applications*, Varanasi, India, December 23–26, 1991.

[799] _____, The mountain cliff theorem. Differential equations and mathematical physics (Birmingham, AL, 1990), 263–279, Math. Sci. Engrg., **186**, Academic Press, Boston, 1992.

[800] _____, A bounded mountain pass lemma without the (PS) condition and applications. *Trans. Am. Math. Soc.*, **331**, no. 2, 681–703 (1992).

[801] _____, A generalization of the saddle point method and applications. *Ann. Polnici Math.*, **57**, 269–281 (1992).

[802] _____, Splitting subspaces and critical points. *Appl. Anal.*, **49**, 33–48 (1993).

[803] _____, Critical points over splitting subspaces. *Nonlinearity*, **6**, 417–427 (1993).

[804] _____, Splitting subspaces and critical points. *Appl. Anal.*, **49**, 33–48 (1993).

[805] _____, Strong resonance problems for elliptic semilinear boundary value problems. *J. Oper. Theor.*, **30**, no. 2, 301–314 (1993).

[806] _____, The Hampwile alternative. *Commun. Appl. Nonlinear Anal.*, **1**, no. 4, 13–46 (1994).

[807] _____, Superlinear elliptic boundary value problems. *Manuscripta Math.*, **86**, no. 3, 253–265 (1995).

[808] _____, The intrinsic mountain pass. *Pacific J. Math.*, **171**, no. 2, 529–544 (1995).

[809] _____, Critical points when there is no saddle point geometry. *Topol. Method Nonlinear Anal.*, **6**, 295–308 (1995).

[810] _____, The saddle point alternative. *Am. J. Math.*, **117**, 1603–1626 (1995).

[811] _____, Weak linking. Preprint, **970202**, Univerity of California Irvine (1996).

[812] _____, New techniques in critical point theory. *Partial differential equations and applications* (Collected papers in honor of Carlo Pucci on the occasion of his 70th birthday), P. Marcellini et al. (ed.), Dekker, New York, *Lect. Notes Pure Appl. Math.*, **177**, 289–294 (1996).

[813] _____, New linking theorems. *Rend. Sem. Mat. Univ. Padova*, **99**, 255–269 (1998).

[814] _____, Infinite-dimensional linking. *Duke Math. J.*, **94**, no. 3, 573–595 (1998).

[815] _____, Critical point theory with weak-to-weak linking. *Commun. Pure Appl. Math.*, **51**, no. 11–12, 1247–1254 (1998).

[816] _____, *Linking methods in critical points*. Birkhäuser, 1999.

[817] M. Schechter and K. Tintarev, Nonlinear eigenvalues and mountain pass methods. *Topol. Method. Nonlinear Anal.*, **1**, no. 2, 183–201 (1993).

[818] _____, Pairs of critical points produced by linking subsets with applications to semilinear elliptic problems. *Bull. Soc. Math. Belg. Sér. B*, **44**, no. 3, 249–261 (1992).

[819] I. Schindler, Quasilinear elliptic boundary value problems on unbounded cylinders and a related mountain-pass lemma. *Arch. Rat. Mech. Anal.*, **120**, no. 4, 363–374 (1992).

[820] _____, A critical value function and applications to semilinear elliptic equations on unbounded domains. *Nonlinear Anal.–Theor.*, **24**, no. 6, 947–959 (1995).

[821] I. Schindler and K. Tintarev, Abstract concentration compactness and elliptic equations on unbounded domains. *Nonlinear analysis and its applications to differential equations* (Lisbon, 1998), 369–380, Progr. Nonlinear Differential Equations Appl., **43**, Birkhäuser Boston, Boston, 2001.

[822] E. Schmidt, Zur Theorie der linearen und nichlinearen Integralgleichungen, III. *Math. Ann.*, **65**, 370–399 (1908).

[823] F. Schuricht, Bifurcation from minimax solutions by variational inequalities. *Math. Nachr.*, **154**, 67–88 (1991).

[824] J.T. Schwartz, Generalizing the Ljusternik-Schnirelman theory of critical points. *Commun. Pure Appl. Math.*, **17**, 307–315 (1964).

[825] _____, *Nonlinear functional analysis*. Gordon and Breach, New York, 1969.

[826] E. Séré, Existence of infinitely many homoclinic orbits in Hamiltonian systems. *Math. Z.*, **209**, no. 1, 27–42 (1992).

[827] ———, Looking for the Bernoulli shift. *Ann. Inst. H. Poincaré–An.*, **10**, no. 5, 561–590 (1993).

[828] I. Shafrir, A deformation lemma. *C.R. Acad. Sci. Paris*, t. **313**, no. 9, Série I, 599–602, (1991).

[829] V.L. Shapiro, Superlinear quasilinearity and the second eigenvalue. *Nonlinear Anal–Theor.*, **44A**, no. 1, 81–96 (2001).

[830] ———, Quasilinearity below the 1st eigenvalue. *Proc. Am. Math. Soc.*, **129**, no. 7, 1955–1962 (2001).

[831] Y.T. Shen, The nontrivial solution of quasilinear elliptic equation in $W_0^{1,p}$. *Sci. Sin. Ser. A*, **27**, no. 7, 720–730 (1984).

[832] ———, A nontrivial solution of the quasilinear elliptic Euler equation. (Chinese) *Acta Math. Sin.*, **28**, no. 3, 375–381 (1985).

[833] ———, Some remarks on nontrivial solutions of second-order quasilinear elliptic equations. (Chinese) *Hunan Ann. Math.*, **8**, no. 1–2, 20–24 (1988).

[834] Y.T. Shen and X.K. Guo, The nontrivial critical points of the functional $\int_\Omega F(x, u, Du)\, dx$. (Chinese) *Acta Math. Sci.*, **10**, no. 3, 249–258 (1990).

[835] S. Shi, Ekeland's variational principle and the mountain pass lemma. *Acta. Math. Sin.*, *(N.S.)*, **1**, no. 4, 348–355 (1985).

[836] M. Shiffman, The Plateau problem for minimal surfaces which are not relative minima. *Bull. Am. Math. Soc.*, **44**, 637, (1938).

[837] ———, The Plateau problem for minimal surfaces of arbitrary topological structure. *Am. J. Math.*, **61**, 853–882 (1939).

[838] ———, The Plateau problem for non-relative minima. *Ann. Math. II*, **40**, 834–854 (1939).

[839] ———, The plateau problem for non-relative minima. *Proc. Natl. Acad. Sci. USA*, **25**, 215–220 (1939).

[840] ———, Unstable minimal surfaces with several boundaries. *Ann. Math. II*, **43**, 197–222 (1942).

[841] ———, Unstable minimal surfaces with any rectifiable boundary. *Proc. Natl. Acad. Sci. USA*, **28**, 103–108 (1942).

[842] E.A. De Be Silva, Critical point theorems and applications to differential equations. Ph.D. thesis, Wisconsin University, 1988.

[843] ———, Linking theorems and applications to semilinear elliptic problems at resonance. *Nonlinear Anal.*, **16**, no. 5, 455–477 (1991).

[844] ———, Critical point theorems and applications to a semilinear elliptic problem. *NoDEA–Nonlinear Diff.*, **3**, no. 2, 245–261 (1996).

[845] E.A. De Be Silva and S.H.M. Soares, Liouville-Gelfand type problems for the N-Laplacian on bounded domains of $\mathbb{R}^N$. *Ann. Scuola Norm. Sup. Pisa*, 4 **28**, no. 1, 1–30 (1999).

[846] E.A. De Be Silva and M.A. Teixeira, A version of Rolle's theorem and application. *Bol. Soc. Bras. Mat.*, **29**, no. 2, 301–327 (1998)

[847] M. Sion, On general minimax theorems. *Pacific J. Math.*, **8**, 171–176 (1958).

[848] S. Smale, Morse theory and nonlinear generalization of the Dirichlet problem. *Ann. Math.*, **17**, 307–315 (1964).

[849] ———, Global variational analysis. *Bull. Am. Math. Soc.*, **83**, 683–693 (1964).

[850] D. Smets and M. Willem, Solitary waves with prescribed speed on infinite lattices. *J. Funct. Anal.*, **149**, no. 1, 266–275 (1997).

[851] S.L. Sobolev, On a theorem of functional analysis. *Mat. Sb. (N.S.)*, **4**, 471–497 (1938).

[852] ———, *Applications of functional analysis in mathematical physics*. Transl. Math. Mon., **7**, American Mathematical Society, Providence, RI, 1963.

[853] G.S. Spradlin, A Hamiltonian system with an even term. *Topol. Method. Nonlinear Anal.*, **10**, no. 1, 93–106 (1997).

[854] ———, An elliptic partial differential equation with a symmetrical almost periodic term. *Calc. Var. Part. Diff. Eq.*, **9**, no. 3, 233–247 (1999).

[855] ———, An elliptic equation with spike solutions concentrating at local minima of the Laplacian of the potential. *Electron. J. Diff. Eq.*, no. 32, 14 pp. (2000) (online).

[856] M. Squassina, Existence and multiplicity results for quasilinear elliptic differential systems. *Electron. J. Diff. Eq.*, no. 14, 12 pp. (1999) (online).

[857] ———, Weak solutions to general Euler's equation via nonsmooth critical point theory. *Ann. Fac. Sci. Toulouse Math.*, **9**, no. 1, 113–131 (2000).

[858] ———, Multiplicity results for perturbed symmetric quasilinear elliptic systems. *Quad. Se. Mat. Brescia*, **4/2000**, Brescia, 2000.

[859] ———, Perturbed S^1-symmetric Hamiltonian systems. *Appl. Math. Lett.*, in press (2000); *Quad. Se. Mat. Brescia*, **11/2000**, Brescia, 2000.

[860] ———, An eigenvalue problem for elliptic systems. *New York J. Math.*, **6**, 95–106 (2000).

[861] ———, On the multiplicity of solutions for a fully nonlinear Emden-Fowler equation. *Electron. J. Diff. Eq.*, no. 63, 10 pp. (2001) (online).

[862] G. Stampacchia, Le problème de Dirichlet pour les équations elliptiques du second ordre à coefficients discontinus. *Ann. Inst. Fourier*, **15**, 189–259 (1965).

[863] F. C. Şt. Cîrstea, Existence of nontrivial weak solutions for a class of quasilinear problems. *An. Univ. Craiova Ser. Mat. Inform.*, **24**, 73–87 (1998).

[864] G. Strang and G.J. Fix, *An analysis of the finite element method.* Prentice-Hall, Englewood Cliffs, NJ, 1973.

[865] G. Ströhmer, Instabile Minimalflächen mit halbfreiem Rand. (German) [Unstable minimal surfaces with partially free boundary.] *Analysis*, **2**, no. 1–4, 315–335 (1982).

[866] ———, Instabile Lösungen der Eulerschen Gleichungen gewisser Variationsprobleme. (German) [Unstable solutions of the Euler equations of certain variational problems.] *Arch. Rat. Mech. Anal.*, **79**, no. 3, 219–239 (1982).

[867] ———, Unstable solutions of the Euler equations of certain variational problems. *Math. Z.*, **186**, no. 2, 179–199 (1984).

[868] M. Struwe, Infinitely many critical points for functionals which are not even and applications to superlinear boundary value problems. *Manuscripta Math.*, **32**, 335–364 (1980).

[869] ———, Multiple solutions of anticoercive boundary value problems for a class of ordinary differential equations of second order. *J. Diff. Eq.*, **37**, 285–295 (1980).

[870] ———, Infinitely many solutions of superlinear boundary value problems with rotational symmetry. *Arch. Math.*, **36**, 360–369 (1981).

[871] ———, Superlinear boundary value problems with rotational symmetry. *Arch. Math.*, **39**, 233–240 (1982).

[872] ———, A note on a result of Ambrosetti and Mancini. *Ann. Mat. Pur. Appl.*, **131**, 107–115 (1982).

[873] ———, Multiple solutions of differential equations without the Palais-Smale conditions. *Math. Ann.*, **261**, 399–412 (1982).

[874] ———, Generalized Palais-Smale conditions and applications. *Vorlesungsr. SFB* **72**, Approximation Optimierung, Univ. Bonn 17, 45 pp. (1983).

[875] ———, A global compactness result for elliptic boundary value problems involving limiting nonlinearities. *Math. Z.*, **187**, no. 4, 511–517 (1984).

[876] ———, On a critical point theory for minimal surfaces spanning a wire in $\mathbb{R}^n$. *J. Reine Angew. Math.*, **349**, 1–23 (1984).

[877] ———, Large H-surfaces via the mountain-pass-lemma. *Math. Ann.*, **187**, no. 3, 441–459 (1985).

[878] ———, Functional analytic aspects of the Plateau problem. *Mathematics and mathematical engineering* (Delft, 1985). Delft Progr. Rep. **10**, no. 4, 271–281 (1985).

[879] ———, A generalized Palais-Smale condition and applications. *Nonlinear functional analysis and its applications*, Part 2 (Berkeley, CA, 1983), 401–411, Proc. Sympos. Pure Math., **45**, Part 2, American Mathematical Society, Providence, RI, 1986.

[880]   _____, A Morse theory for annulus-type minimal surfaces. *J. Reine Angew. Math.*, **308**, 1–27 (1986).

[881]   _____, *Plateau's problem and the calculus of variations*. Math. Notes, **35**, Princeton University Press, Princeton, 1989.

[882]   _____, *Variational methods, applications to nonlinear partial differential equations and Hamiltonian systems*, Springer-Verlag, Berlin, 1990.

[883]   _____, The parametric Plateau problem and related topics. The problem of Plateau, 258–284, World Scientific Publishing, River Edge, NJ, 1992.

[884]   C.A. Stuart, Bifurcation into spectral gaps. *Bull. Belg. Math. Soc. Sim.*, 1995, suppl., 59 pp.

[885]   C.A. Stuart and H.S. Zhou, A variational problem related to self-trapping of an electromagnetic field. *Math. Method. Appl. Sci.*, **19**, no. 17, 1397–1407 (1996).

[886]   _____, Applying the mountain pass theorem to an asymptotically linear elliptic equation on $\mathbb{R}^N$. *Commun. Part. Diff. Eq.*, **24**, no. 9–10, 1731–1758 (1999).

[887]   J.X. Sun, The Schauder condition in the critical point theory. *Kexue Tongbao*, **31**, no. 17, 1157–1162 (1986).

[888]   T. Suzuki, *Semilinear elliptic equations*. GAKUTO International Series. Mathematical Sciences and Applications, **3**, Gakkōtosho Co., Ltd., Tokyo, 1994.

[889]   _____, Positive solutions for semilinear elliptic equations on expanding annuli: Mountain pass approach. *Funkcial. Ekvac.*, **39**, no. 1, 143–164 (1996).

[890]   C.A. Swanson and L.S. Yu, Critical p-Laplacian problems in $\mathbb{R}^N$. *Ann. Mat. Pur. Appl., IV. Ser.*, **169**, 233–250 (1995).

[891]   A. Szulkin, Minimax principles for lower semicontinuous functions and applications to nonlinear boundary value problems. *Ann. Inst. H. Poincaré*, **3**, no. 2, 77–109 (1986).

[892]   _____, Ljusternik-Schnirelman theory on C^1-manifolds. *Ann. Inst. H. Poincaré*, **5**, 119–139 (1988).

[893]   _____, A Morse theory and existence of periodic solutions to partial differential equations. *Bull. Soc. Math. France*, 1988.

[894]   _____, Critical point theory of Ljusternik-Schnirelmann type and applications to partial differential equations. *Minimax results of Lusternik-Schnirelman type and applications* (Montreal, PQ, 1986), 35–96, Sém. Math. Sup., **107**, Presses Univ. Montréal, Montreal, PQ, 1989.

[895]   _____, *A relative category and applications to critical point theory for strongly indefinite functionals*. Report no. 1, Dep. Math., Univ. Stockholm, 1989.

[896]   W. Takahashi, Existence theorems generalizing fixed point theorems for multivalued mappings. *Fixed point theory and applications* (Proc. Int. Conf., Marseille-Luminy, France, 1989), Pitman Res. Notes Math. Ser., **252**, 397–406 (1991).

[897]   K. Tananka, Infinitely many solutions for the equation $u_{tt} - u_{xx} \pm |u|^{p-1}u = f(x, t)$ II. *Trans. Am. Math. Soc.*, **307**, 615–645 (1988).

[898]   _____, Morse indices at critical points related to the symmetric mountain pass theorem and applications. *Commun. Part. Diff. Eq.*, **14**, no. 1, 99–128 (1989).

[899]   _____, Multiple positive solutions for some nonlinear elliptic systems. *Topol. Method Nonlinear Anal.*, **10**, no. 1, 15–45 (1997).

[900]   C.-L. Tang, An existence theorem of nontrivial solutions of semilinear equations in reflexive Banach spaces and its applications. *Acad. Roy. Belg. Bull. Cl. Sci.*, **7**, no. 1–6, 87–100 (1997).

[901]   C.-L. Tang and Q.-J. Gao, Elliptic resonant problems at higher eigenvalues with an unbounded nonlinear term. *J. Diff. Eq.*, **146**, no. 1, 56–66 (1998).

[902]   N. Tarfulea, One result of Cauchy-Lagrange type. *An. Univ. Timişoara Ser. Mat.-Inform.*, **32**, no. 2, 123–127 (1994).

[903]   S. Terracini, Non-degeneracy and chaotic motions for a class of almost-periodic Lagrangean systems. *Nonlinear Anal.–Theor.*, **37**, no. 3, 337–361 (1999).

[904]   S.A. Tersian, A minimax theorem and applications to nonresonance problems for semilinear equations. *Nonlinear Anal.*, **10**, 651–668 (1986).

[905] ———, Some existence results for nontrivial solutions of Hammerstein integral equations. *Differential equations* (Plovdiv, 1991), 205–219, World Scientific Publishing, River Edge, NJ, 1992.

[906] ———, A note on Palais-Smale condition and mountain-pass principle for locally Lipschitz functionals. *Variational methods in nonlinear analysis* (Erice, 1992), 193–203, Gordon and Breach, Basel, 1995.

[907] ———, Nontrivial solutions of semilinear Schrödinger equations on $\mathbb{R}^n$ and strip-like domains. *Appl. Anal.*, **56**, no. 3–4, 335–350 (1995).

[908] ———, On nontrivial solutions of semilinear Schrödinger equations on $\mathbb{R}^n$. *Rend. Accad. Sci. Lett. Lombardo*, **129**, no. 1–2, 97–109 (1996).

[909] S. Tersian and J. Chaparova, Periodic and homoclinic solutions of extended Fisher-Kolmogorov equations. *J. Math. Anal. Appl.*, **260**, no. 2, 490–506 (2001).

[910] S.A. Tersian and P.P. Zabrejko, Hammerstein integral equations with nontrivial solutions. *Res. Math.*, **19**, no. 1–2, 179–188 (1991).

[911] F. de Thélin, Résultats d'existence et de non-existence pour la solution positive et bornée d'une e.d.p. elliptique non linéaire. (French) [Existence and nonexistence results for a positive bounded solution of a nonlinear elliptic partial differential equation.] *Ann. Fac. Sci. Toulouse Math.*, **8**, no. 3, 375–389 (1986/87).

[912] K. Thews, A reduction method for some nonlinear Dirichlet problems. *Nonlinear Anal.*, **3**, 795–813 (1979).

[913] G. Tian, On the mountain pass lemma. *Kexue Tongbao*, **28**, no. 14, 833–835 (1983).

[914] ———, On the mountain-pass lemma. *Kexue Tongbao*, **29**, no. 9, 1150–1154 (1984).

[915] M. Timoumi, Oscillations de systèmes hamiltoniens surquadratiques non coercitifs. (French) [Oscillations of noncoercive superquadratic Hamiltonian systems.] *Demonstratio Math.*, **27**, no. 2, 293–300 (1994).

[916] K. Tintarev, Level set maxima and quasilinear elliptic problems. *Pacific J. Math.*, **153**, no. 1, 185–200 (1992).

[917] ———, A theorem of the "mountain impasse" type and semilinear elliptic problems on manifolds. *J. Diff. Eq.*, **113**, no. 1, 234–245 (1994).

[918] ———, Mountain Impasse theorem and spectrum of semilinear elliptic equations. *Trans. Am. Math. Soc.*, **336**, no. 2, 621–629 (1994).

[919] ———, Second eigenfunctions of nonlinear eigenvalue problems. *J. Math. Anal. Appl.*, **196**, no. 1, 361–383 (1995).

[920] ———, Mountain pass and impasse for nonsmooth functionals with constraints. *Nonlinear Anal. Theor. Method. Appl.*, **27**, no. 9, 1049–1054 (1996).

[921] ———, Isotopic linking and critical points of functionals. *Nonlinear Anal. Theor. Method. Appl.*, **30**, no. 7, 4145–4149 (1997).

[922] I. Todhunter, *A history of the calculus of variations during the nineteenth century.* Chelsea, 1962; reprint of the 1861 edition.

[923] L. Tonelli, *Fondamenti di calcolo delle variazioni*, **1–3**, Bologna, 1921–1923.

[924] A.J. Tromba, On the number of simply connected minimal surfaces spanning a curve. *Mem. Am. Math. Soc.*, **12**, no. 194 (1977).

[925] ———, Degree theory on oriented infinite-dimensional varieties and the Morse number of minimal surfaces spanning a curve in $\mathbb{R}^n$. *Manuscripta Math.*, **48**, no. 1–3, 139–161 (1984).

[926] ———, Degree theory on oriented infinite-dimensional varieties and the Morse number of minimal surfaces spanning a curve in $\mathbb{R}^n$. I. $n \geq 4$. *Trans. Am. Math. Soc.*, **290**, no. 1, 385–413 (1985).

[927] ———, On the Morse number of embedded and non-embedded minimal immersions spanning wires on the boundary of special bodies in $\mathbb{R}^3$. *Math. Z.*, **188**, 149–170 (1985).

[928] A.J. Tromba and K.D. Elworthy, Differential structures and Fredholm maps on Banach manifolds. *Global analysis* (Berkeley, CA, 1968), 45–94, Proc. Sympos. Pure Math., XV, American Mathematical Society, Providence, RI, 1970.

[929] K. Tso, On the existence of convex hypersurfaces with prescribed mean curvature. *Ann. Scuola Norm. Sup. Pisa*, **16**, no 4, no. 2, 225–243 (1989).

[930] R. Turner, Superlinear Sturm-Liouville problems. *J. Diff. Eq.*, **13**, 157–171 (1973).

[931] P. Ubilla, Homoclinic orbits for a quasi-linear Hamiltonian system. *J. Math. Anal. Appl.*, **193**, no. 2, 573–587 (1995).

[932] Vacantiecursus 1985 Variatierekening. (Dutch) [1985 holiday course: calculus of variations.] CWI Syllabi, 7. Mathematisch Centrum, Amsterdam, 1985.

[933] M. Vainberg, On the structure of an operator. *Dokl. Akad. Nauk SSSR*, **92**, no. 3, 457–460 (1953)

[934] _____, On a form of S-property for functions, *Moskov. Oblast. Pedagog. Inst. Uc. Zap.* **21**, 65–72 (1954)

[935] _____, *Variational methods for the the study of nonlinear operators*. Holden Day, San Francisco, 1964.

[936] _____, *Variational methods and methods of monotone operators in the theory of nonlinear equations*, Wiley, 1973.

[937] E.W.C. van Groesen, On mountain passes to a proof of God? (Dutch) 1985 holiday course: Calculus of variations, 237–245, CWI Syllabi, **7**, Math. Centrum, Amsterdam, 1985.

[938] _____, Applications of natural constraints in critical point theory to boundary value problems on domains with rotation symmetry. *Arch. Math.* (Basel), **44**, no. 2, 171–179 (1985).

[939] J. Vélin and F. de Thélin, Existence and nonexistence of nontrivial solutions for some nonlinear elliptic systems. *Rev. Mat. Univ. Complut. Madrid*, **6**, no. 1, 153–194 (1993).

[940] S. Villegas, A Neumann problem with asymmetric nonlinearity and a related minimizing problem. *J. Diff. Eq.*, **145**, no. 1, 145–155 (1998).

[941] E. Vitillaro, Some new results on global nonexistence and blow-up for evolution problems with positive energy. *Rend. Ist. Mat. Univ. Trieste*, **31**, Suppl. 2, 245–275 (2000).

[942] T. Wang, A minimax principle without differentiability. *J. Nanjing Univ., Math. Biq.*, **6**, no. 2, 46–52 (1989).

[943] _____, Lusternik-Schnirelman category theory on closed subsets of Banach manifolds. *J. Math. Anal. Appl.*, **149**, no. 2, 412–423 (1990).

[944] _____, A minimax principle without differentiability. *J. Math. Res. Expo.*, **11**, no. 1, 111–116 (1991).

[945] X.D. Wang and X.T. Liang, Existence of nontrivial solutions in nonisotropic Sobolev spaces for a class of quasilinear elliptic equations. (Chinese) *Math. Appl.*, **8**, no. 3, 289–296 (1995).

[946] X.J. Wang, Neumann problems of semilinear elliptic equations involving critical Sobolev exponents. *Diff. Eq.*, **93**, no. 2, 283–310, (1991).

[947] Z.Q. Wang, A note on the deformation theorem. *Acta Math. Sin.*, **30**, 106–110 (1987).

[948] _____, Multiple solutions for indefinite functionals and applications to asymptotically linear problems. *Math. Sin.*, New series, **5**, 101–113 (1989).

[949] _____, On a superlinear elliptic equation. *Ann. Inst. H. Poincare–An.*, **8**, 43–57 (1991).

[950] J.R. Ward Jr., A boundary value problem with a periodic nonlinearity. *Nonlinear Anal.*, **10**, no. 2, 207–213 (1986).

[951] G. Warnecke, Uber das homogene Dirichlet-Problem bei nichtlinearen partiellen Differentialgleichungen vom Typ der Boussinesq-Gleichung. (German) [On the homogeneous Dirichlet problem for nonlinear partial differential equations of the Boussinesq type.] *Math. Method. Appl. Sci.*, **9**, no. 4, 493–519 (1987).

[952] M. Willem, *Lectures on critical point theory*. Trabalho de mathematica, **199**, Univ. Brasilia, 1983.

[953] _____, Periodic oscillations of odd second order Hamiltonian systems. *Boll. Unione Mat. Ital. B*, **3**, no. 2, 293–304 (1984).

[954] _____, *An introduction to critical point theory, minimax methods and periodic solutions of Hamiltonian systems*. (Lecture SMR/5 given at College on variational problems in

analysis held at the International Centre of Theoretical Physics on January 11–February 5, 1988.)

[955] _____, Bifurcation, symmetry and Morse theory. *Boll. Unione Mat. Ital.*, **7**, 3–B, 17–27 (1989).

[956] _____, *Un lemme de déformation quantitatif en calcul des variations.* (French) [A quantitative deformation lemma in the calculus of variations.] Institut de mathématiques pures et appliquées [Applied and Pure Mathematics Institute], Recherche de mathématiques [Mathematics Research] no. 19, Catholic University of Louvain, May 1992.

[957] _____, *Minimax theorems.* Progress in Nonlinear Differential Equations and their Applications, **24**, Birkhäuser Boston, Boston, 1996.

[958] N.B. Willms and G.M.L. Gladwell, Saddle points and overdetermined problems for the Helmholtz equation. *Z. Angew. Math. Phys.*, **45**, no. 1, 1–26 (1994).

[959] M.W. Wong, Existence of weak solutions of a nonlinear pseudo-differential equation. *Commun. Appl. Nonlinear Anal.*, **1**, no. 1, 69–86 (1994).

[960] R. Woodhouse, *A history of the calculus of variations in the eighteenth century.* Chelsea, New York, 1964; reprint of the 1910 edition.

[961] A. Wu, Existence of multiple non-trivial solutions for nonlinear p-Laplacian problems on $\mathbb{R}^N$. *Proc. Roy. Soc. Edin. A*, **129**, no. 4, 855–883 (1999).

[962] S.P. Wu, A homoclinic orbit for Lagrangian systems. *Syst. Sci. Math. Sci.*, **8**, no. 1, 75–81 (1995).

[963] S. Wu and J. Liu, Homoclinic orbits for second order Hamiltonian system with quadratic growth. A Chinese summary appears in *Gaoxiao Yingyong Shuxue Xuebao A*, **10**, no. 4, 473 (1995); *Appl. Math. J. Chinese Univ. B*, **10**, no. 4, 399–410 (1995).

[964] S. Wu and H. Yang, A note on homoclinic orbits for second order Hamiltonian system. A Chinese summary appears in *Gaoxiao Yingyong Shuxue Xuebao A*, **13**, no. 3, 363 (1998); *Appl. Math. J. Chinese Univ. B*, **13**, no. 3, 251–262 (1998).

[965] J.B. de M. Xavier and O.H. Miyagaki, Remarks on a resonant problem with an unbounded nonlinearity. *J. Math. Anal. Appl.* **209**, no. 1, 255–273 (1997).

[966] D.L. Xu and S.P. Wu, A note on radial solutions of the p-Laplacian equation. (Chinese) *Gaoxiao Yingyong Shuxue Xuebao A*, **10**, no. 3, 245–248 (1995).

[967] H.X. Xu, On multiple solutions of quasilinear elliptic equations with parameters. (Chinese) *Hunan Daxue Xuebao*, **15**, no. 1, Special Issue of Math., 248–252 (1988).

[968] _____, The problem of multiple solutions for systems of second-order quasilinear elliptic Euler equations in $W_0^{1,p}(\Omega, \mathbb{R}^n)$. (Chinese) *Chinese Ann. Math. A*, **10**, no. 5, 538–544 (1989).

[969] J. Xu, Positive solutions of elliptic equations in unbounded domain. (Chinese) *J. Eng. Math., Xi'an*, **17**, no. 2, 87–91 (2000).

[970] R.Y. Xue, Positive solutions of nonlinear elliptic equations with mixed boundary conditions. *Acta Math. Sci. (English Ed.)*, **12**, no. 3, 292–303 (1992).

[971] N. Yamashita, J. Imai, and M. Fukushima, The proximal point algorithm for the P_0 complementarity problem. *Complementarity: Applications, algorithms and extensions.* (Papers from the international conference on complementarity (ICCP 99), Madison, WI, June 9–12, 1999). M.C. Ferris (ed.), et al., Kluwer Academic Publishers, Dordrecht; *Appl. Optim.*, **50**, 361–379 (2001).

[972] J.F. Yang, Existence of nontrivial solutions to quasilinear elliptic equations in unbounded domains. (Chinese) *Acta Math. Sci. (Chinese)*, **9**, no. 4, 429–438 (1989).

[973] _____, Remarks on the equilibrium shape of a tokamak plasma. *Proc. Roy. Soc. Edin. A*, **123**, no. 6, 1059–1070 (1993).

[974] J.F. Yang and X.P. Zhu, On the existence of nontrivial solution of a quasilinear elliptic boundary value problem for unbounded domains. I. Positive mass case. *Acta Math. Sci. (English Ed.)*, **7**, no. 3, 341–359 (1987).

[975] L. Yang and Z.H. Yang, Existence of solutions to second-order semilinear elliptic equations with critical Sobolev exponents. (Chinese) *J. Chengdu Univ. Sci. Tech.*, no. 4, 39–49 (1992).

[976] X.P. Yang, Existence of multiple solutions to quasilinear elliptic obstruction problems. (Chinese) *Acta Math. Sci. (Chinese)*, **11**, no. 1, 54–60 (1991).

[977] K. Yosida, *Functional analysis and its application*, Springer-Verlag, 1965 (6th ed. 1980).

[978] P.P. Zabreiko and M.A. Krasnosel'skii, Sovability of nonlinear operator equations. *Funct. Anal. Appl.*, **5**, 206–208 (1971).

[979] P.P. Zabrejko and S.A. Tersian, On the variational method for solvability of nonlinear integral equations of Hammerstein type. *C.R. Acad. Bulg. Sci.*, **43**, no. 6, 9–11 (1990).

[980] E. Zeidler, *Lectures on Ljusternik-Schnirelman theory for indefinite nonlinear eigenvalue problems and its applications.* Fučik and Kufner (eds.), 176–219 (1979).

[981] ———, Ljusternik-Schnirelman theory for indefinite and non necessarily odd nonlinear operators and its applications. *Nonlinear Anal.*, **4**, 451–489 (1980).

[982] ———, *Nonlinear functional analysis and its applications. III. Variational methods and optimization.* Translated from the German by Leo F. Boron. Springer-Verlag, New York and Berlin, 1985.

[983] ———, *Nonlinear functional analysis and its applications. I. Fixed-point theorems.* Translated from the German by P.R. Wadsack. Springer-Verlag, New York and Berlin, 1984. Second printing 1992.

[984] ———, *Nonlinear functional analysis and its applications. II/A. Linear monotone operators.* Translated from the German by the author and L.F. Boron. Springer-Verlag, New York and Berlin, 1990.

[985] ———, Ljusternik-Schnirelman theory on general level sets. *Math. Nachr.*, **129**, 235–259 (1986).

[986] ———, *Applied functional analysis. Main principles and their applications.* Applied Mathematical Sciences, **109**, Springer-Verlag, New York, 1995.

[987] W.P. Zeimer, *Weakly differentiable functions, Sobolev spaces and functions of bounded variations.* Springer-Verlag, New York, Berlin, and Heidelberg, 1989.

[988] G.Q. Zhang, A variant mountain pass lemma. *Sci. Sin. A*, **26**, 1241–1255 (1983).

[989] G.Y. Zhang, Neumann boundary value problems for Laplace equations with critical growth. (Chinese) *Acta Sci. Nat. Univ. Sunyatseni*, **36**, no. 2, 32–37 (1997).

[990] K. Zhang, A two-well structure and intrinsic mountain pass points. *Calc. Var. Part. Diff. Eq.*, **13**, no. 2, 231–264 (2001).

[991] P.H. Zhao and C.K. Zhong, Multiplicity of positive solutions of a class of elliptic equations. (Chinese) *J. Lanzhou Univ. Nat. Sci.*, **34**, no. 1, 10–14 (1998).

[992] P.H. Zhao and C.K. Zhong, Multiplicity of positive solutions to nonlinear elliptic equations. (Chinese) *J. Lanzhou Univ. Nat. Sci.*, **32**, no. 3, 17–21 (1996).

[993] ———, Multiple positive solutions of semilinear elliptic equation involving critical Sobolev exponents. *J. Math. Stud.*, **29**, no. 4, 21–28 (1996).

[994] C.K. Zhong, Remarks on the mountain pass lemma. (Chinese) *J. Lanzhou Univ. Nat. Sci.*, **28**, no. 2, 25–29 (1992).

[995] ———, The mountain pass lemma in variational inequalities. (Chinese) *J. Lanzhou Univ. Nat. Sci.*, **31**, no. 2, 1–5 (1995).

[996] ———, A generalization of Ekeland's variational principle and application to the study of the relation between the weak P.S. condition and coercivity. *Nonlinear Anal.–Theor.*, **29**, no. 12, 1421–1431 (1997).

[997] ———, On Ekeland's variational principle and a minimax theorem. *J. Math. Anal. Appl.*, **205**, no. 1, 239–250 (1997).

[998] ———, A generalization of Ekeland's variational principle and applications. (Chinese) *Acta Math. Sin.*, **40**, no. 2, 185–190 (1997).

[999] C.K. Zhong and P.H. Zhao, Locally Ekeland's variational principle and some surjective mapping theorems. *Chin. Ann. Math. B*, **19**, no. 3, 273–280 (1998).

[1000] C.Q. Zhou, Existence of nontrivial solution for Dirichlet problem of quasilinear elliptic equation with critical growth condition. (Chinese) *Hunan Daxue Xuebao*, **21**, no. 4, 12–20 (1994).

[1001] H. Zhou and Z. Zhang, A note on applying mountain pass theorem. *J. Central China Norm. Univ. Nat. Sci.*, **32**, no. 4, 383–388 (1998).

[1002] H.-S. Zhou, Positive solution for a semilinear elliptic equation which is almost linear at infinity. *Z. Angew. Math. Phys.*, **49**, no. 6, 896–906 (1998).

[1003] H.-S. Zhou, The existence of positive solutions to nonautonomous field equations on $\mathbb{R}^N$. (Chinese) *Acta Math. Sci.*, **9**, no. 1, 113–119 (1989).

[1004] _____, Positive solutions for a Dirichlet problem. *Acta Math. Appl. Sin., Engl. Ser.*, **17**, no. 3, 340–349 (2001).

[1005] X.P. Zhu, and J.F. Yang, The quasilinear elliptic equation on unbounded domain involving critical Sobolev exponent. *J. Part. Diff. Eq.*, **2**, no. 2, 53–64 (1989).

[1006] X.P. Zhu, and H.S. Zhou, Existence of multiple positive solutions of inhomogeneous semilinear elliptic problems in unbounded domains. *Proc. Roy. Soc. Edin. A*, **115**, no. 3–4, 301–318 (1990).

Index